T0137390

METHODS IN MOLECULAR BIOLOGY

Series Editor
John M. Walker
School of Life and Medical Sciences
University of Hertfordshire
Hatfield, Hertfordshire, AL10 9AB, UK

For further volumes:
http://www.springer.com/series/7651

Plant Senescence

Methods and Protocols

Edited by

Yongfeng Guo

Tobacco Research Institute, Chinese Academy of Agricultural Sciences,
Qingdao, Shandong, China

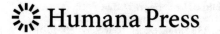

Editor
Yongfeng Guo
Tobacco Research Institute
Chinese Academy of Agricultural Sciences
Qingdao, Shandong, China

ISSN 1064-3745   ISSN 1940-6029   (electronic)
Methods in Molecular Biology
ISBN 978-1-4939-8540-1   ISBN 978-1-4939-7672-0   (eBook)
https://doi.org/10.1007/978-1-4939-7672-0

This Humana Press imprint is published by Springer Nature
The registered company is Springer Science+Business Media, LLC
The registered company address is: 233 Spring Street, New York, NY 10013, U.S.A.

Preface

As the final phase of plant development, senescence is of great significance for the life cycle of plants as well as for agriculture. Plant senescence is a type of senescence distinct from senescence processes in animals and other systems (Chapter 1), mainly because of the nutrient remobilization process which, as a result of evolutionary fitness, recycles nutrients in senescing tissues to sink organs, a critical step for yield formation in agriculture. Nutrients in senescing organs are released through massive degradation of macromolecules such as proteins, polysaccharides, lipids, and nucleotides. The macromolecule degradation and nutrient remobilization processes are driven by active expression of thousands of senescence-associated genes (SAGs) (Chapter 2), which makes plant senescence distinct from many other plant developmental processes.

Besides age, senescence of plants can be induced by a great number of factors including reproductive growth, phytohormones, and environmental cues. There are also many factors which function to inhibit plant senescence (Chapter 2). Thus, a complex regulatory network involving transduction of various signals, gene expression, and metabolism is responsible for the initiation of plant senescence. Protocols related to plant senescence induced by ethylene (Chapter 8), abscisic acid (Chapters 9 and 11), jasmonic acid, salicylic acid, strigolactones, peptide hormones (Chapter 11), shade/darkness (Chapter 12), salt (Chapter 13), pathogen (Chapter 15), oxidative stress (Chapter 16), nutritional status (Chapter 14), and postharvest conditions (Chapter 17) are summarized in this volume.

Once senescence is initiated, plants undergo a similar senescence execution process, the so-called senescence syndrome. Protocols commonly used in characterizing molecular, cellular, and biochemical aspects of the senescence syndrome are included in this volume. These are methods in studying chlorophyll catabolism (Chapter 18), programmed cell death (Chapter 21), cell wall polysaccharides (Chapter 25), vacuolar proteases (Chapter 22), chloroplast protein degradation (Chapter 24), and autophagy (Chapter 23). Methods for studying different types of senescence, including leaf senescence (Chapter 3), flower senescence (Chapter 4), fruit ripening (Chapter 7), nodule senescence (Chapter 5), and floral organ abscission (Chapters 6 and 26), are described. Also included are techniques for studying plant senescence in general, such as gene identification and functional analysis (Chapter 19), MicroRNAs (Chapter 20), map-based cloning (Chapter 17), virus-induced gene silencing (Chapter 4), electron microscopy (Chapter 26), transcriptome (Chapter 27), and metabolome (Chapter 28).

The techniques covered in this volume are either developed specially for plant senescence studies or optimized for studying senescing plants. Every method includes not only a detailed description of the procedures but also thoughtful notes for correct interpretation, special cautions, or alternative options. The aim is to provide a useful handbook of standard protocols for plant molecular biologists working on senescence.

Qingdao, Shandong, China *Yongfeng Guo*

Contents

Contributors

AKHTAR ALI • *Tobacco Research Institute, Chinese Academy of Agricultural Sciences, Qingdao, Shandong, China; Nuclear Institute for Food and Agriculture, Peshawar, Pakistan*

SHOKI AOYAMA • *Faculty of Science and Graduate School of Life Science, Hokkaido University, Sapporo, Japan*

SALMA BALAZADEH • *Max Planck Institute of Molecular Plant Physiology, Potsdam-Golm, Germany; Institute of Biochemistry and Biology, University of Potsdam, Potsdam-Golm, Germany*

STEFAN BIEKER • *ZMBP, Department of General Genetics, University Tuebingen, Tuebingen, Germany*

JUDY A. BRUSSLAN • *Department of Biological Sciences, California State University, Long Beach, CA, USA*

SNEŽANA BUDIMIR • *Institute for Biological Research "Siniša Stanković", University of Belgrade, Belgrade, Serbia*

MELINKA A. BUTENKO • *Department of Molecular Biosciences, University of Oslo, Oslo, Norway*

DAVID CHAGNÉ • *The New Zealand Institute for Plant & Food Research Limited, Food Industry Science Centre, Palmerston North, New Zealand*

CHENXIA CHENG • *Department of Ornamental Horticulture, China Agricultural University, Beijing, China; Beijing Key Laboratory of Development and Quality Control of Ornamental Crops, China Agricultural University, Beijing, China*

LORENZA COSTA • *Instituto de Fisiología Vegetal (INFIVE)—Consejo Nacional de Investigaciones Científicas y Técnicas (CONICET), Universidad Nacional de La Plata, Buenos Aires, Argentina*

ROSS CROWHURST • *The New Zealand Institute for Plant & Food Research Limited, Mt. Albert Research Centre, Auckland, New Zealand*

XUEFEI CUI • *Tianjin Key Laboratory of Protein Science, College of Life Sciences, Nankai University, Tianjin, China; Department of Plant Biology and Ecology, College of Life Sciences, Nankai University, Tianjin, China*

ADITI DAS • *Institute of Plant and Microbial Biology, University of Zurich, Zurich, Switzerland*

PAUL DIJKWEL • *Institute of Fundamental Sciences, Massey University, Palmerston North, New Zealand*

ALEXANDER ERBAN • *Max Planck Institute of Molecular Plant Physiology, Potsdam-Golm, Germany*

ALISDAIR R. FERNIE • *Max Planck Institute of Molecular Plant Physiology, Potsdam-Golm, Germany*

SUSHENG GAN • *Plant Biology Section, School of Integrative Plant Science, College of Agriculture and Life Sciences, Cornell University, Ithaca, NY, USA*

XIAOMING GAO • *Tobacco Research Institute, Chinese Academy of Agricultural Sciences, Qingdao, Shandong, China*

JUNPING GAO • *Department of Ornamental Horticulture, China Agricultural University, Beijing, China; Beijing Key Laboratory of Development and Quality Control of Ornamental Crops, China Agricultural University, Beijing, China*

PATRICK GIAVALISCO • *Max Planck Institute of Molecular Plant Physiology, Potsdam-Golm, Germany*

QINGQIU GONG • *Tianjin Key Laboratory of Protein Science, College of Life Sciences, Nankai University, Tianjin, China; Department of Plant Biology and Ecology, College of Life Sciences, Nankai University, Tianjin, China*

JUAN JOSÉ GUIAMÉT • *Instituto de Fisiología Vegetal (INFIVE)—Consejo Nacional de Investigaciones Científicas y Técnicas (CONICET), Universidad Nacional de La Plata, Buenos Aires, Argentina*

YONGFENG GUO • *Tobacco Research Institute, Chinese Academy of Agricultural Sciences, Qingdao, Shandong, China*

HONGWEI GUO • *The State Key Laboratory of Protein and Plant Gene Research, College of Life Sciences, Peking-Tsinghua Center of Life Sciences, Peking University, Beijing, China; Department of Biology, South University of Science and Technology of China, Shenzhen, Guangdong, China*

LUZIA GUYER • *Institute of Plant and Microbial Biology, University of Zurich, Zurich, Switzerland*

RAINER HOEFGEN • *Max Planck Institute of Molecular Plant Physiology, Potsdam-Golm, Germany*

STEFAN HÖRTENSTEINER • *Institute of Plant and Microbial Biology, University of Zurich, Zurich, Switzerland*

DONALD A. HUNTER • *The New Zealand Institute for Plant & Food Research Limited, Food Industry Science Centre, Palmerston North, New Zealand*

DUŠICA JANOŠEVIĆ • *Institute of Botany and Botanical Garden "Jevremovac", Faculty of Biology, University of Belgrade, Belgrade, Serbia*

RUBINA JIBRAN • *The New Zealand Institute for Plant & Food Research Limited, Food Industry Science Centre, Palmerston North, New Zealand; Institute of Fundamental Sciences, Massey University, Palmerston North, New Zealand*

HAI-CHUN JING • *Key Laboratory of Plant Resources, Institute of Botany, Chinese Academy of Sciences, Beijing, China*

AAKANSHA KANOJIA • *Institute of Fundamental Sciences, Massey University, Palmerston North, New Zealand*

DAMI KIM • *Department of Plant Science, Plant Genomics and Breeding Institute, Research Institute of Agriculture and Life Sciences, Seoul National University, Seoul, Republic of Korea*

YINGZHEN KONG • *Tobacco Research Institute, Chinese Academy of Agricultural Sciences, Qingdao, Shandong, China*

JOACHIM KOPKA • *Max Planck Institute of Molecular Plant Physiology, Potsdam-Golm, Germany*

XIAOHONG KOU • *School of Chemical Engineering and Technology, Tianjin University, Tianjin, China*

LIN LI • *Institute of Plant Biology, State Key Laboratory of Genetic Engineering, School of Life Sciences, Fudan University, Shanghai, China*

KE LI • *College of Life Sciences, Hebei Normal University, Shijiazhuang, Hebei, China*

WEI LI • *Tobacco Research Institute, Chinese Academy of Agricultural Sciences, Qingdao, Shandong, China*

ZHONGHAI LI • *The State Key Laboratory of Protein and Plant Gene Research, College of Life Sciences, Peking-Tsinghua Center of Life Sciences, Peking University, Beijing, China*

WEI LIU • *State Key Laboratory for Biology of Plant Diseases and Insect Pests, Institute of Plant Protection, Chinese Academy of Agricultural Sciences, Beijing, China*

NAN MA • *Department of Ornamental Horticulture, China Agricultural University, Beijing, China; Beijing Key Laboratory of Development and Quality Control of Ornamental Crops, China Agricultural University, Beijing, China*

DANA E. MARTÍNEZ • *Instituto de Fisiología Vegetal (INFIVE)—Consejo Nacional de Investigaciones Científicas y Técnicas (CONICET), Universidad Nacional de La Plata, Buenos Aires, Argentina*

YING MIAO • *Center for Molecular Cell and Systems Biology, College of Life Sciences, Fujian Agriculture and Forestry University, Fuzhou, Fujian, China*

BERND MUELLER-ROEBER • *Max Planck Institute of Molecular Plant Physiology, Potsdam-Golm, Germany; Institute of Biochemistry and Biology, University of Potsdam, Potsdam-Golm, Germany*

MALCOLM O'NEILL • *Complex Carbohydrate Research Center, University of Georgia, Athens, GA, USA*

NAM-CHON PAEK • *Department of Plant Science, Plant Genomics and Breeding Institute, Research Institute of Agriculture and Life Sciences, Seoul National University, Seoul, Republic of Korea*

MAREN POTSCHIN • *ZMBP, Department of General Genetics, University Tuebingen, Tuebingen, Germany*

FANGJUN QI • *State Key Laboratory for Biology of Plant Diseases and Insect Pests, Institute of Plant Protection, Chinese Academy of Agricultural Sciences, Beijing, China*

YUJUN REN • *Center for Molecular Cell and Systems Biology, College of Life Sciences, Fujian Agriculture and Forestry University, Fuzhou, Fujian, China*

YASUHITO SAKURABA • *Department of Plant Science, Plant Genomics and Breeding Institute, Research Institute of Agriculture and Life Sciences, Seoul National University, Seoul, Republic of Korea*

TAKEO SATO • *Faculty of Science and Graduate School of Life Science, Hokkaido University, Sapporo, Japan*

JOS H.M. SCHIPPERS • *Key Laboratory of Plant Resources, Institute of Botany, Chinese Academy of Sciences, Beijing, China; Molecular Ecology, Institute for Biology I, RWTH Aachen University, Aachen, Germany*

CHUN-LIN SHI • *Department of Molecular Biosciences, University of Oslo, Oslo, Norway*

YI SONG • *Institute of Plant Biology, State Key Laboratory of Genetic Engineering, School of Life Sciences, Fudan University, Shanghai, China*

KERRY SULLIVAN • *The New Zealand Institute for Plant & Food Research Limited, Food Industry Science Centre, Palmerston North, New Zealand*

AIZHEN SUN • *The National Key Laboratory of Plant Molecular Genetics and National Center for Plant Gene Research (Shanghai), CAS Center for Excellence in Molecular Plant Sciences, Institute of Plant Physiology and Ecology, Shanghai Institutes for Biological Sciences, Chinese Academy of Sciences, Shanghai, China*

TAKAYUKI TOHGE • *Max Planck Institute of Molecular Plant Physiology, Potsdam-Golm, Germany*

BRANKA UZELAC • *Institute for Biological Research "Siniša Stanković", University of Belgrade, Belgrade, Serbia*

SCOTT W. VANDE WETERING • *Department of Biological Sciences, California State University, Long Beach, CA, USA*

GENG WANG • *College of Life Sciences, Hebei Normal University, Shijiazhuang, Hebei, China*

MUTSUMI WATANABE • *Max Planck Institute of Molecular Plant Physiology, Potsdam-Golm, Germany*

XIAO-YUAN WU • *Key Laboratory of Plant Resources, Institute of Botany, Chinese Academy of Sciences, Beijing, China*

MENGSHI WU • *School of Chemical Engineering and Technology, Tianjin University, Tianjin, China*

YAN XIA • *Key Laboratory of Plant Resources, Institute of Botany, Chinese Academy of Sciences, Beijing, China*

XIUYING XIA • *School of Life Science and Biotechnology, Dalian University of Technology, Dalian, Liaoning, China*

JUNJI YAMAGUCHI • *Faculty of Science and Graduate School of Life Science, Hokkaido University, Sapporo, Japan*

ULRIKE ZENTGRAF • *ZMBP, Department of General Genetics, University Tuebingen, Tuebingen, Germany*

WEIBO ZHAI • *State Key Laboratory for Biology of Plant Diseases and Insect Pests, Institute of Plant Protection, Chinese Academy of Agricultural Sciences, Beijing, China*

ZENGLIN ZHANG • *Tobacco Research Institute, Chinese Academy of Agricultural Sciences, Qingdao, Shandong, China*

WENWEI ZHANG • *State Key Laboratory for Biology of Plant Diseases and Insect Pests, Institute of Plant Protection, Chinese Academy of Agricultural Sciences, Beijing, China*

YANYAN ZHANG • *College of Tourism and Food Science, Shanghai Business School, Shanghai, China*

KEWEI ZHANG • *College of Chemistry and Life Sciences, Zhejiang Normal University, Jinhua, Zhejiang, China*

LIMING ZHAO • *Key Laboratory of Plant Resources, Institute of Botany, Chinese Academy of Sciences, Beijing, China; Max Planck Institute of Molecular Plant Physiology, Potsdam University, Potsdam-Golm, Germany*

NA ZHENG • *State Key Laboratory for Biology of Plant Diseases and Insect Pests, Institute of Plant Protection, Chinese Academy of Agricultural Sciences, Beijing, China*

JING ZHENG • *Tianjin Key Laboratory of Protein Science, College of Life Sciences, Nankai University, Tianjin, China; Department of Plant Biology and Ecology, College of Life Sciences, Nankai University, Tianjin, China*

JINXIN ZHENG • *Tianjin Key Laboratory of Protein Science, College of Life Sciences, Nankai University, Tianjin, China; Department of Plant Biology and Ecology, College of Life Sciences, Nankai University, Tianjin, China*

CHUNJIANG ZHOU • *College of Life Sciences, Hebei Normal University, Shijiazhuang, Hebei, China*

GONGKE ZHOU • *Qingdao Institute of Bioenergy and Bioprocess Technology, Chinese Academy of Sciences, Qingdao, Shandong, China*

Part I

Introduction of Plant Senescence

Chapter 1

Concepts and Types of Senescence in Plants

Susheng Gan

Abstract

Concepts, classification, and the relationship between different types of senescence are discussed in this chapter. Senescence-related terminology frequently used in yeast, animal, and plant systems and senescence processes at cellular, organ, and organismal levels are clarified.

Key words Senescence, Aging, Mitotic senescence, Postmitotic senescence, Organismal senescence, Organ senescence, Monocarpic senescence, Polycarbic senescence, Natural senescence, Artificial senescence

1 Introduction

Senescence is a universal phenomenon in living organisms, and the word senescence has been used by scientists working on variety of systems such as yeast, fruit fly, worm, human being, and plants. However, the meaning of the word to scientists working on different organisms can be different, and the difference can be subtle in some cases and very obvious in some other cases. Here I try to discuss and clarify the term and its related classification at cellular and organismal levels to avoid possible confusion, with an emphasis on plants.

2 Senescence vs. Aging

The word senescence derives from two Latin words: "senex" and "senescere." "Senex" means "old"; this Latin root is shared by "senile," "senior," and even "senate." In ancient Rome the "Senatus" was a "council of elders" that is composed of the heads of patrician families. "Senescere" means "to grow old." The Merriam-Webster online dictionary defines senescence as "the state of being old or the process of becoming old."

Yongfeng Guo (ed.), *Plant Senescence: Methods and Protocols*, Methods in Molecular Biology, vol. 1744,
https://doi.org/10.1007/978-1-4939-7672-0_1, © Springer Science+Business Media, LLC 2018

Aging is also the process of getting older, often in a sense of "declining" in strength, vigor, viability, etc. Because of this, aging has been regarded as a synonym of senescence, and the two words have often been used interchangeably. That being said, it should be pointed out that aging tends to refer to those declining changes as function of time in both living organisms and nonliving organisms. For example, a piece of plastic sheet can become fragile or fragmentary after many years of exposure to air; this process of "becoming fragile" is aging. To certain degree, aging is a "passive" process. In contrast, senescence tends to refer to the declining process that may be caused by, in addition to age/time, many environmental cues (such as drought stress, extreme temperatures) and internal factors (such as reproductive growth, various plant hormones).

At the cellular level, senescence can be classified into two categories: mitotic senescence and postmitotic senescence as discussed below.

3 Mitotic Senescence vs. Postmitotic Senescence

As shown in the cartoon in Fig. 1, a cell's life history consists of mitotic and postmitotic processes [1]. A cell may undergo a certain number of mitotic divisions to produce daughter cells. After a limited number of divisions (e.g., about 40 divisions in human fibroblasts), the cell can no longer divide mitotically. Once a cell ceases mitotic division permanently, it is called *mitotic senescence*. That is, the outcome of mitotic senescence is not cell death but the cell's inability to divide mitotically. In literature concerning yeast, germline cells, and mammalian cells in culture, this type of senescence is often referred to as *cellular senescence* (sometimes called *cellular aging*) or *replicative senescence* (sometimes called *replicative aging*) [2–5]. If a cell keeps dividing and fails to undergo mitotic senescence (e.g., cancer cells), it is said "immortalized." Therefore, mitotic senescence is a mechanism to suppress cancer development.

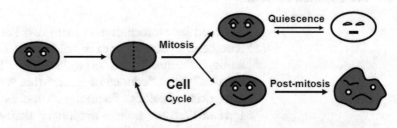

Fig. 1 Illustration of a cell's life history. The cell undergoes two types of senescence: mitotic and postmitotic senescence. When the cell stops dividing, it is called mitotic senescence or replicative senescence or proliferative senescence. Postmitotic senescence is the active degenerative and attrition process of the cell that can no longer undergo cell division. Modified from (Gan, 2003) [1]

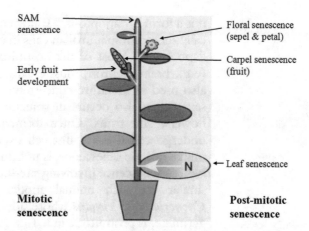

Fig. 2 Plants exhibit both mitotic senescence and postmitotic senescence

An example of mitotic senescence in plants is the arrest of shoot apical meristem (SAM); SAM consists of non-differentiated, germline-like cells that can divide for finite times to produce cells that will be then differentiated to form new organs such as leaves and flowers (Fig. 2). The arrest of SAM is also called *proliferative senescence* in plant literature [6]. This is similar to replicative senescence in yeast and animal cells in culture as discussed above. Another example of mitotic senescence may be the arrest of mitotic cell division at early stage of fruit development (Fig. 2). Fruit size is a function of cell number, cell size, and intercellular space, and cell number is the major factor. The number of cells is determined at the very early stage of fruit development and remained unchanged [7].

If a cell stops mitosis *temporarily* due to unfavorable conditions (environmentally and/or physiologically) but retains its mitotic capacity and can reenter into mitotic cycles to produce more daughter cells, the temporarily un-dividing or resting status or process is called *cell quiescence* [8] (Fig. 1). The SAM cells will stop dividing under unfavorable conditions. For example, the SAM of many trees will stop proliferative process when they perceive short-day photoperiod signal; short day often means the winter season is coming. These meristem cells retain their division capability during winter time and will resume division activity when spring is coming. Therefore, the short-day-induced cell quiescence is an evolutionary fitness strategy. Ethylene and ABA may play a role in regulating the temporary "arrest" of the meristem [9]. The meristem in an axillary bud suppressed by auxin is also in the quiescent status.

A mitotically senescent non-dividing cell may undergo active degenerative process leading to the ultimate death; this process is called *postmitotic senescence* (Fig. 1). If the degeneration is solely a function of age, "aging" is the right word to describe it (see above "Senescence vs. Aging"). If the degenerative process is very quick, it is a form of "apoptosis." It however should be noted that mitotically senescent mammalian cells in culture appear to be resistant to "apoptosis." Most of the postmitotic cells are somatic in nature (e.g., brain, neuron, muscle cells); thus "somatic senescence" is also used in literature concerning animals. In plants, postmitotic senescence also occurs in somatic tissues/organs such as leaves, flowers, and fruits. Once formed, cells in these organs rarely undergo cell division but cell expansion; thus their senescence, unlike mitotic senescence, is not due to an inability to divide. This type of senescence involving predominantly somatic tissues is very similar to such animal model systems as *Drosophila* and *Caenorhabditis elegans* whose adult body, with the exception of germline, is postmitotic in nature [1].

It should be reiterated that mitotically senescent cells (and quiescent cells) are not dead and mitotic senescence has nothing to do with cell death, but the outcome of postmitotic senescence is cell death.

4   Organismal Senescence vs. Organ Senescence

At the organismal level, when an organism's ability to respond to stress declines, its homeostasis becomes increasingly imbalanced, and its risk of disease increases with age, leading to the ultimate death of the whole organism, this is the aging of the whole organism and is often referred to as organismal senescence. Although cellular senescence may contribute to organismal senescence [2], the latter is much more inclusive, for example, many age-related diseases such as Alzheimer's disease are parts of organismal aging. In literature concerning plants, organismal senescence is *whole plant senescence*. Among the most studied whole plant senescence is *monocarpic senescence*. Annuals (e.g., *Arabidopsis*), biennials (e.g., wheat), and some perennials (e.g., bamboo) possess a monocarpic life pattern, which is characterized by only a single reproductive event in the life cycle. After flowering (and setting seeds or fruits), the whole plant will senesce and die. Monocarpic senescence includes three coordinated processes: (1) senescence of the SAM, a form of mitotic senescence or proliferative senescence as discussed above; (2) senescence of somatic organs and tissues, a form of postmitotic senescence (see below); and (3) senescence of axillary buds to prevent formation of new shoots/branches. This third aspect of whole plant senescence has received little attention in the senescence research community [10, 11].

Organ senescence refers to the senescence of individual organs in plants, including *leaf senescence, flower senescence,* and *fruit senescence* (Fig. 2). In most papers concerning flower senescence, petals are the major foci (thus better called *petal senescence*) although sepals, the outer whirl of a flower and often green (very similar to leaves), are also undergoing senescence process. Fruit senescence is often called *fruit ripening*. Root is an important organ and is believed to undergo senescence as well (*root senescence*). Another important type of organ senescence is root *nodule senescence*. Root nodule is a special nitrogen-fixing symbiotic organ consisting of roots of Fabaceae plants and their associated bacteria called rhizobia.

5   Monocarpic Senescence vs. Polycarbic Senescence

Monocarpic senescence is indeed the whole plant senescence as discussed above (cf. organismal senescence vs. organ senescence) [10, 11]. In contrast, *polycarpic senescence* refers to senescence in polycarpic plants, mostly perennials, that flower and set seeds/fruits many times in their whole lifetime. It could be simply the senescence processes in organs such as leaves and flowers in, for example, apple tree plants, which is distinct from monocarpic senescence. In fact, the term polycarpic senescence has rarely been used.

6   Natural Senescence vs. Artificial Senescence

Senescence can be influenced by many environmental cues and endogenous factors. For example, stresses such as drought, flooding, extreme temperature, pathogen infection, and nutrient deficiency can readily induce leaf senescence. In the absence of the environmental stresses, senescence will still occur; this type of senescence is called *natural senescence*. Plants grown under ideal conditions in a growth chamber will generally undergo natural senescence, which is primarily controlled developmentally, and thus this type of senescence is also called *developmental senescence*. Strictly speaking, senescence of plants in wild sets cannot be named natural senescence because the plants often suffer from one or more environmental stresses.

In contrast, *artificial senescence* refers to senescence of a plant, often its detached organs such as leaves, that are subject to stress (es) (e.g., darkness) and/or treatment with senescence-inducing plant growth regulators (e.g., abscisic acid, ethylene, and jasmonic acid). Artificial senescence is operational and mainly used in laboratory settings.

Acknowledgments

Our research on plant senescence has been supported by grants from the National Science Foundation, US Department of Energy Basic Bioscience Program, USDA NRI, US-Israel BARD, and Cornell Genomics Initiative. I thank past and current members of the Gan Laboratory for useful discussions. This chapter is modified from Gan (2003).

References

1. Gan S (2003) Mitotic and postmitotic senescence in plants. Sci Aging Knowledge Environ 2003:RE7

2. Ben-Porath I, Weinberg RA (2005) The signals and pathways activating cellular senescence. Int J Biochem Cell Biol 37:961–976

3. Patil CK, Mian IS, Campisi J (2005) The thorny path linking cellular senescence to organismal aging. Mech Ageing Dev 126:1040–1045

4. Sedivy JM (1998) Can ends justify the means? Telomeres and the mechanisms of replicative senescence and immortalization in mammalian cells. Proc Natl Acad Sci U S A 95:9078–9081

5. Takahashi Y, Kuro-O M, Ishikawa F (2000) Aging mechanisms. Proc Natl Acad Sci U S A 97:12407–12408

6. Hensel LL, Nelson MA, Richmond TA et al (1994) The fate of inflorescence meristems is controlled by developing fruits in *Arabidopsis*. Plant Physiol 106:863–876

7. Tanksley SD (2004) The genetic, developmental, and molecular bases of fruit size and shape variation in tomato. Plant Cell 16(Suppl):S181–S189

8. Stuart JA, Brown MF (2006) Energy, quiescence and the cellular basis of animal life spans. Comp Biochem Physiol A Mol Integr Physiol 143:12–23

9. Ruonala R, Rinne PL, Baghour M et al (2006) Transitions in the functioning of the shoot apical meristem in birch (Betula pendula) involve ethylene. Plant J 46:628–640

10. Davies PJ, Gan S (2012) Towards an integrated view of monocarpic plant senescence. Russ J Plant Physiol 59:467–478

11. Guo Y, Gan S (2011) AtMYB2 regulates whole plant senescence by inhibiting cytokinin-mediated branching at late stages of development in Arabidopsis. Plant Physiol 156:1612–1619

Chapter 2

Initiation, Progression, and Genetic Manipulation of Leaf Senescence

Akhtar Ali, Xiaoming Gao, and Yongfeng Guo

Abstract

As a representative form of plant senescence, leaf senescence has received the most attention during the last two decades. In this chapter we summarize the initiation of leaf senescence by various internal and external signals, the progression of senescence including switches in gene expression, as well as changes at the biochemical and cellular levels during leaf senescence. Impacts of leaf senescence in agriculture and genetic approaches that have been used in manipulating leaf senescence of crop plants are discussed.

Key words Leaf senescence, Signals, Hormones, Transcription factors, Chlorophyll, Protein degradation, Nutrient recycling, IPT, AtNAP

1 Introduction

Leaf senescence is the terminal phase of the developmental process of a leaf that involves differential gene expression, active degradation of cellular structures and nutrients remobilization. The onset and progression of senescence are age-dependent and can be influenced by various internal and external signals [1, 2]. During leaf senescence macromolecules such as proteins, lipids, and nucleic acids are degraded, and nutrients released from the catabolism of these macromolecules as well as other nutrients are recycled to actively growing new buds, young leaves, and developing fruits and seeds. Therefore, leaf senescence is also termed as a nutrient mining and recycling process that is distinct from many other developmental programs [1, 3].

Plant hormones affect the timing of leaf senescence by regulating plant development and stress responses. Different hormones regulate the onset and progression of leaf senescence either directly or indirectly by altering the developmental program or by altering plants' response to stresses [4]. Execution of the leaf senescence process has a direct impact on agriculture via modifying agronomic traits of crop plants, including biomass, seed yield, seed protein

Yongfeng Guo (ed.), *Plant Senescence: Methods and Protocols*, Methods in Molecular Biology, vol. 1744,
https://doi.org/10.1007/978-1-4939-7672-0_2, © Springer Science+Business Media, LLC 2018

composition, and tolerance to abiotic stresses [5]. Premature senescence is a major cause of losses in grain filling and biomass yield due to leaf yellowing and deteriorated photosynthesis. Senescence is also responsible for the losses resulted from short shelf life of many vegetables and fruits. The substantial changes occurring during leaf senescence are regulated genetically under highly complex molecular networks. The genetically identified regulatory factors include transcription regulators, receptors, and signaling components for hormones and stress responses, as well as regulators of metabolism [1, 2].

Based on current understanding of the molecular regulatory mechanisms underlying leaf senescence, several approaches have been developed for genetic manipulation of leaf senescence. Application of these strategies on various crop species suggests great potential of senescence manipulation in agricultural improvement [5].

2   Initiation of Leaf Senescence

Leaf senescence can be regulated by various internal signals and environmental cues. Among the internal factors are age and phytohormones. Aging is a passive process with chances of death increasing over time due to lower level of resistance against stresses [6]. Leaf senescence is believed to be initiated by developmental age. However, the onset of senescence occurs under a complex mechanism executed by various signaling pathways (Fig. 1).

2.1 Age- and Development- Dependent Senescence

Leaf development goes through expansion, maturity, and senescence phases [7]. The initiation of leaf senescence mainly depends on developmental changes. Senescence induction occurs only upon a certain developmental stage [4], after which senescence can be induced and regulated by various hormonal and environmental cues. It has been observed that even in the absence of biotic and abiotic stresses, leaf senescence ultimately occurs and progresses in an age-dependent manner [8]. For monocarpic plants senescence overlaps with the reproductive phase [9], the regulation of senescence is under correlative control, and the onset of whole-plant senescence is initiated with the development of reproductive sink that remobilizes nutrients from vegetative tissues [3]. For instance, mechanical removal of sink tissues delayed senescence in oilseed rape, soybean, and wheat and decreased nitrogen remobilization in oilseed rape and soybean [10–12]. A delay in onset of leaf senescence is associated with slow rate of flower development in pea [13]. Furthermore, senescence could be delayed by inhibiting kernel set in maize [14]. These results support the notion that plant genomes are optimized for reproduction which determines the onset of leaf and whole-plant senescence in monocarpic plants. Age and developmental signals

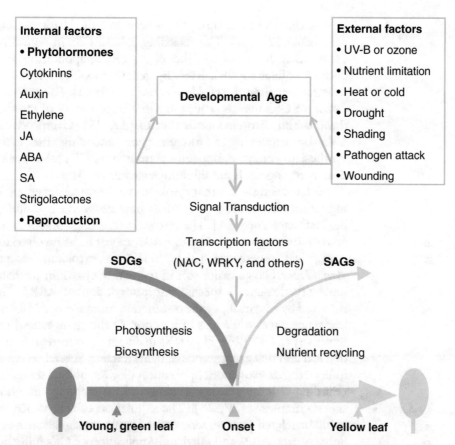

Fig. 1 The leaf senescence regulatory network [16]

may influence the initiation and execution of senescence via changes in reproduction growth, cell differentiation, and the levels of phytohormones [15].

2.2 Phytohormones Regulating Leaf Senescence

Plant hormones play a key role in developmental processes as well as in the integration of environmental signals. A number of phytohormones have been shown to play important roles in the regulation of leaf senescence [17]. These hormones may interact with each other and with a whole range of developmental, environmental, and metabolic signals to regulate the senescence process [18]. Advances made through transcriptomic studies have revealed gene expression data for hormone biosynthesis, signaling, and responses during senescence. Combination of physiological and genetic information will help in creating a model for hormonal and developmental control of leaf senescence.

2.2.1 Cytokinins

Cytokinins (CKs) are adenine derivatives with isoprenoid or aromatic side chains, known for their strong effects on delay of leaf senescence. The most abundant CKs in higher plants are isopentenyladenine and zeatin. An inverse association exits between cytokinin levels and the progression of senescence [19]. The decline in cytokinin levels is thought to be a key signal for the onset of senescence [20, 21]. The retarding role of CKs in senescence has been widely documented either by exogenous application or endogenously enhancing their levels by genetic modification and even in mutants impaired in cytokinin signaling [5, 22]. Exogenous application of CKs retards senescence in a large variety of monocotyledonous and dicotyledonous species [22, 23]. Overproduction of CKs by expressing a foreign gene encoding the cytokinin-synthesizing enzyme, isopentenyl transferase (IPT), has been a predominant approach of inhibiting senescence [5].

It has been shown that cytokinin biosynthesis genes and signaling genes such as the type A ARRs have reduced transcription during leaf senescence [24]. The cytokinin signaling components CKI1 (cytokinin independent 1) and ARR2 appear to be involved in regulating leaf senescence [25]. AHK3, a cytokinin receptor in *Arabidopsis*, plays a major role in regulating cytokinin-mediated leaf longevity through a specific phosphorylation of ARR2 [26]. As reported by Kim et al., a gain-of-function mutation of AHK3 delayed leaf senescence, while loss of function of this gene caused reduced sensitivity to CKs [26]. How CKs maintain the photosynthetic phase of a leaf and delay senescence remains unclear. Several recent studies indicated that exogenous application of CKs inhibits senescence via retarding chlorophyll degradation and maintaining the photosynthetic complexes [27, 28]. The regulation of senescence by CKs could be related to other senescence-regulating signals such as auxin, light, sugars, ABA, and ethylene. Application of CKs influences the expression of genes involved in these signaling pathways [29, 30].

2.2.2 Auxin

Exploring the role of auxin in the regulation of senescence is complicated because of its importance in various aspects of plant growth and development [2]. Earlier studies showed that exogenous application of auxin delays leaf senescence [31, 32]. Expression of *SAG12*, a marker gene for developmental senescence, was markedly decreased by auxin treatment in detached senescing leaves [33]. Although the change of auxin concentration during senescence varies between different studies [34, 35], overexpression of *YUCCA6*, a gene encoding the rate-limiting enzyme in de novo auxin biosynthesis, delayed leaf senescence by increasing auxin levels [36]. Mutation of ARF1 and ARF2, repressors of auxin-responsive transcription, can delay senescence [37, 38]. RNA-Seq analysis performed in cotton showed that most auxin pathway-related genes were downregulated, suggesting that auxin may act as a negative regulator of leaf senescence [39].

2.2.3 Gibberellic Acid

Gibberellic acid (GA) is a pentacyclic diterpene well known for its role in cell elongation, seed germination, reproductive growth, senescence, and tolerance against various environmental stresses [40]. It was first reported that GA retarded senescence of excised leaf tissues from *Taraxacum officinale* by maintaining chlorophyll levels and RNA synthesis [41]. GA concentrations in leaves of romaine lettuce declined with the progression of senescence [42]. A recent study showed that leaf senescence was accelerated in an *Arabidopsis* mutant deficient in the GA signaling repressor DELLA, while overexpression of the DELLA-encoding gene *RGL1* caused prolonged leaf longevity [43]. This study also showed that the GA-deficient mutant *ga1* and the triple GA receptor mutant *gid1a gid1b gid1c* displayed dramatically delayed leaf senescence [43], supporting a negative effect of GA on leaf senescence.

2.2.4   Ethylene

Ethylene is a key senescence-promoting hormone. Ethylene application accelerates leaf and flower senescence, while inhibitors of ethylene perception or biosynthesis delay leaf senescence [44]. The biosynthesis of ethylene increases during the early stages of senescence initiation, and the expression of genes encoding ethylene biosynthetic enzymes is upregulated in senescing tissues [45]. Jabran et al. suggested that ethylene-dependent senescence occurs only in leaves that reached certain age [4], because senescence cannot be induced until a defined developmental age is reached in *Arabidopsis* [46, 47]. This view is genetically supported by the altered responses of a series of *old* (onset of leaf death) mutants exposed to various ethylene treatments [46].

Mutations in components of the ethylene signaling pathway exhibit differential timing of the onset of senescence. The *etr1* (ethylene resistant 1) mutation in the ethylene receptor ETR1 delays leaf senescence in *Arabidopsis* and petunia [47, 48]. Ethylene signaling components EIN2 (ethylene insensitive 2) and EIN3 positively regulate leaf senescence [49, 50]. Transcription factor EIN3 directly promotes chlorophyll degradation by enhancing the promoter activities of chlorophyll catabolic genes (NYE1, NYC1, and PAO) to activate their expressions [51]. Moreover, ORE1, a direct target of EIN3 and a positive regulator of leaf senescence, in turn activates the expression of the ethylene biosynthetic gene *ACS2* during senescence, suggesting a positive feedback of senescence to ethylene biosynthesis [51].

Ethylene also exerts its role in regulating leaf senescence by interacting with other phytohormones [44]. EIN2 is also important for senescence controlled by abscisic acid and jasmonates, illustrating the cross talk that occurs between these hormones [52]. A recent study showed that the strigolactone biosynthesis genes *MAX3* (more axillary growth 3) and *MAX4* were induced upon treatments with ethylene [53]. In addition, strigolactone application strongly accelerated leaf senescence in the presence of ethylene [53].

2.2.5   Abscisic Acid

In addition to its role in stress responses and stomata aperture, abscisic acid (ABA) functions in promoting leaf senescence [54]. The level of endogenous ABA increases during leaf senescence in a number of plant species [55, 56], and exogenously applied ABA can induce the expression of senescence-associated genes (*SAGs*) and accelerate leaf yellowing [57, 58]. Microarray studies revealed that genes involved in the key steps of ABA biosynthesis and signaling are upregulated during leaf senescence in *Arabidopsis* [24]. ABA-insensitive mutants, *abi1-1* and *snrk2.2/2.3/2.6*, were shown to exhibit stay-green phenotypes after ABA treatments [57]. Three ABA-responsive element (ABRE)-binding transcription factors, ABF2, ABF3, and ABF4, have been shown to promote leaf senescence through direct activation of the chlorophyll degrading NYE1 [57]. An ABA-inducible receptor-like kinase, RPK1, was shown to promote both ABA-induced leaf senescence and developmental leaf senescence in *Arabidopsis* [59]. In a recent study, a pathway of senescence induction by ABA was elucidated. It was shown that ABA promotes leaf senescence by activating SnRK2s, which subsequently phosphorylates ABFs and RAV1 transcription factors. The phosphorylated ABFs and RAV1 then upregulate the expression of *SAGs* [60].

2.2.6 Brassinosteroids

Brassinosteroids (BRs) are plant steroids essential for regulating various physiological processes including plant senescence [61]. BR application has been shown to accelerate senescence in a number of plant species [62–64]. Mutants in BR biosynthesis and BR signaling also support a role of BRs in senescence. For example, the *det2* mutant that is defective in BR biosynthesis shows a delayed senescence phenotype [65]. Due to the strong developmental defects of BR-related mutants, molecular genetic analysis of the role of BRs in senescence is somewhat challenging.

2.2.7 Jasmonates

Jasmonates are oxylipins, major component of fragrant oils, including jasmonic acid (JA) and methyl jasmonate (MeJA). Jasmonates are considered important senescence inducers because JA induces leaf senescence in a variety of plant species [66]. In *Arabidopsis*, endogenous JA level increases with the aging of leaves, and a number of genes associated with the JA biosynthesis pathway are upregulated during leaf senescence [67]. A subset of TCP transcription factors that are targets of miR319 have been shown to be positive regulators of leaf senescence through activating the JA biosynthesis pathway [68].

Components of the JA signaling pathway have been shown to be important in senescence regulation. JA promotes senescence in *Arabidopsis* leaves but unable to induce precocious leaf senescence in the JA-insensitive mutant *coi1* which is deficient in JA perception [67]. The bHLH subgroup IIIe factors MYC2, MYC3, and MYC4 are key signaling components in the JA signaling pathway and function in JA-induced leaf senescence [69]. JAZ proteins are key transcriptional repressors of the JA signaling pathway and play negative roles in leaf senescence. Mutations in JAZ4, JAZ8, and

ZAZ7 all cause early senescence, while overexpression of these genes leads to delayed senescence phenotypes [70, 71].

COI1-dependent JA repression of Rubisco activase has been suggested to be an important mechanism underlying JA-induced leaf senescence [72]. In addition, extensive cross talks exist between JA and other senescence-regulating hormones including salicylic acid, ethylene, and auxin [66].

2.2.8 Salicylic Acid

Salicylic acid (SA) is involved in disease resistance and programmed cell death [73]. It has been shown that the SA level increased at early stages and was maintained during the late stages of leaf senescence in common sage (*Salvia officinalis*) [74]. Exogenous applications of methyl SA promoted drought-induced leaf senescence [74]. The SA 3-hydroxylase (S3H) involved in SA catabolism plays a role in leaf senescence regulation. The *s3h* mutant accumulates very high levels of SA and exhibits a precocious senescence phenotype. Conversely, the gain-of-function plants contain extremely low levels of SA and display a significantly extended leaf longevity [75]. In addition, several SA signaling components, including NPR1 and PAT14, have been found to play a role in leaf senescence regulation [76, 77].

2.2.9 Strigolactones

Strigolactones (SLs) are terpenoid lactones that regulate shoot branching and also involved in leaf senescence [78]. One of the leaf senescence regulator ORE9, which was identified through mutant screening, also regulates shoot branching and was later shown to be identical to the SL signaling component MAX2 [79, 80]. Mutation of the MAX2/ORE9 ortholog in rice, D3, leads to delayed leaf senescence [81]. SL biosynthesis mutants in petunia and *Lotus japonicus* also exhibit delayed senescence phenotypes [82, 83]. Exogenously applied GR24 (SL analogue) accelerate leaf senescence in SL-deficient mutants in rice [84]. SL biosynthesis genes MAX3 and MAX4 are drastically induced by ethylene treatment, and leaf senescence can only be promoted by the application of SLs in the presence of ethylene, indicating cross talk between SLs and ethylene in senescence regulation [53].

2.3 Stress Induction of Leaf Senescence

Under natural conditions, plants are continuously exposed to multiple stresses. As a mechanism of survival under harsh conditions, plant senescence can be induced and affected by a variety of environmental stresses including nutrient deprivation, extended darkness, drought, cold, high temperature, salt stress, and wounding [1, 2]. The senescence-inducing stresses usually share molecular, biochemical, cellular, and physiological events between each other and with senescence. The production, signaling, and scavenging of reactive oxygen species (ROS), for example, are associated with various stress responses and senescence [85]. Cross talk between senescence-inducing stresses is a common phenomenon.

2.3.1 Salinity Stress

Salinity stress is a major abiotic stress limiting plant growth and development. Plants' responses to salinity stress include growth reduction, changes in biomass allocation, leaf senescence, and eventually death of plants [86]. High concentration NaCl treatments of detached mature leaves in sweet potato accelerated leaf senescence in a dose-dependent manner [87]. The leaf senescence induced by salt was shown to be accompanied by a decrease in chlorophyll content, reduction of photosynthetic efficiency, an elevation of H_2O_2 level, and expression of *SAGs* [87]. Extensive overlaps between the transcriptomes of salt-induced senescence and developmental senescence were observed in *Arabidopsis* leaves [86]. A gene encoding the chlorophyll catabolism enzyme pheophorbide a oxygenase (PAO) was isolated in pepper to be involved in salt-induced leaf senescence [88], providing a possible mechanism of salt induction of senescence. Expression of *PAO* can be induced by salt, ABA, MeJA, and SA [88], suggesting multi-pathway cross talk.

2.3.2 Drought Stress

Drought stress may cause leaf wilting, cell membrane damage, and premature senescence, which helps to reduce water loss at the whole-plant level and to remobilize nutrients from the senescing leaves to younger leaves or sink organs [89]. Under drought conditions, significant declines of photosynthesis rate and chlorophyll content were observed when leaf water potential in sorghum plants started to drop, indicating the initiation of leaf senescence, which was confirmed by the expression changes of senescence marker genes [90]. A plastid-nucleus located protein WHIRLY1 has been shown to be a key regulator of drought-induced senescence. Under drought conditions, the HvWHIRLY1 RNAi barley plants showed a stay-green phenotype when significant leaf senescence was induced in wild type plants [91].

As the key hormone involved in drought response via regulating stomata closure, ABA is also important in the regulation of drought-induced senescence [54]. A transcription factor ABIG1 (ABA insensitive growth 1) has been isolated to function downstream of ABA to restrict growth and promote leaf senescence in *Arabidopsis* [92]. Drought-induced expression of the ABA receptor gene *PYL9* dramatically increased drought resistance and promoted drought-induced leaf senescence in both *Arabidopsis* and rice [60]. It is suggested that leaf senescence may benefit drought resistance by helping in generating an osmotic potential gradient and causing water to preferentially flow to developing tissues [60].

2.3.3 Heat Stress

Under heat stress, plants exhibit many physiological effects, including protein denaturation, increase in membrane fluidity and level of reactive oxygen species, inactivation of chloroplast and mitochondrial enzyme activities, as well as decline of photosystem

II-mediated electron transport and photosynthetic activities. These changes lead to an acceleration of leaf senescence [93]. Creeping bentgrass (*Agrostis stolonifera*) plants exposed to heat stress (38/33 °C, day/night) for 28 days showed significant decline in leaf chlorophyll content mainly due to accelerated chlorophyll degradation [94]. Exogenous application of ethylene inhibitors, cytokinins, or nitrogen suppressed heat-induced leaf senescence in creeping bentgrass [95], suggesting cross talk between heat stress and other senescence-regulating factors.

2.4 Networking Senescence-Inducing Signals

Extensive cross talk between senescence-inducing signals has been reported. This is also evident by the substantial overlaps of gene expression between developmental senescence and senescence induced by various signals [24, 96, 97].

In order to study the interactions between different senescence-promoting signals on their regulation of *SAG* expression, He et al. [98] analyzed the effects of ethylene, JA, ABA, BRs, dehydration, and darkness on GUS expression of 147 *Arabidopsis* enhancer trap lines in which the GUS reporter was expressed in senescing leaves but not in non-senescing ones. Their results indicated that each senescence-promoting factor upregulates a subset of *SAGs* and that the GUS expression in certain lines is induced by one or more senescence-promoting factors. Based on this data, the authors proposed a leaf senescence regulatory network in which *SAGs* induced by only one stimulus are placed in the upstream portion of the regulatory network. If a gene is regulated by multiple stimuli, it is believed that this gene functions in the downstream portion of the regulatory network [98] (Fig. 2).

Guo and Gan [97] analyzed microarray expression data from 27 different treatments known to promote senescence and compared them with the transcriptome of developmental leaf senescence. At the early stages of treatments, limited similarity in gene expression patterns was observed between developmental senescence and the senescence-promoting treatments. Once the senescence program is initiated and yellowing is visible, usually after a prolonged period of treatments, a great proportion of *SAGs* induced by developmental leaf senescence are shared by gene expression profiles in response to hormone/stress treatments [97]. Gene expression profiles of developmental and artificially induced senescence differ significantly at early senescence stages, but share considerable similarities during later stages of senescence, indicating that although different signals that lead to the initiation of senescence may do so through distinct signal transduction pathways, senescence processes induced either developmentally or by different senescence-promoting treatments may share common execution events [97] (Fig. 2).

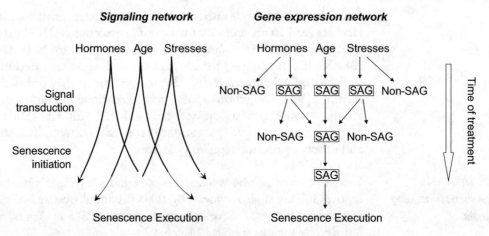

Fig. 2 Signal transduction and gene expression networks in leaf senescence [99]

3   Genetic Regulation of Leaf Senescence

Leaf senescence is a genetically regulated process. Upon the onset of senescence, genes implicated in photosynthesis are downregulated, whereas a subset of genes termed as senescence-associated genes (*SAGs*) are transcriptionally upregulated. Identification and functional characterization of *SAGs* are of vital importance in understanding the molecular, genetic, biochemical, and physiological processes underlying leaf senescence. Research efforts have made by scientists in this field culminated in the isolation and functional characterization of hundreds of *SAGs* from various plant species. Genetic and genomic accomplishments made in the past two decades have led to major advancements in understanding the molecular mechanisms of leaf senescence, and thousands of putative *SAGs* have been identified [1, 2, 99].

Advanced genome sequencing and whole transcriptome profiling tools enabled investigators to study the global transcriptional reprogramming during leaf senescence. Changes in the expression of about 20% of all the genes have been shown during leaf senescence in *Arabidopsis* [100]. Buchanan-Wollaston et al. identified 827 *SAGs* with a threefold upregulated expression upon onset of leaf senescence through microarray analysis using the Affymetrix *Arabidopsis* whole-genome GeneChip (ATH1) [24]. A high-resolution time course profile of gene expression of *Arabidopsis* leaves before and after the onset of senescence categorized 6323 differentially expressed genes [101]. More recently, a comprehensive analysis of leaf senescence transcriptomes indicated that the chloroplast transcriptome, but not the mitochondrial transcriptome, shows major changes during leaf aging, with a shared expression pattern with nuclear transcripts encoding chloroplast-targeted proteins. Tight temporal and distinct interorganellar coordination of various transcriptomes might be critical for proper execution of leaf senescence [102].

The drastic switch of gene expression during leaf senescence is driven by transcription factors. For clarification of the massive regulatory network of gene expression, identification and functional characterization of transcription factors are essential. Intensive transcriptomic analyses by various groups have identified NAC, WRKY, AP2/EREBP, MYB, C2H2 zinc-finger, bZIP, and GRAS being the largest transcription factor families that are upregulated during leaf senescence [101, 103]. Large numbers of transcription factors in these families have been characterized to play an important role in regulating leaf senescence.

The NAC transcription factor family has been widely investigated for its involvement in the signaling pathways that regulate senescence [16, 104]. AtNAP, one of the NAC family transcription factor genes, has been shown to be transcriptionally upregulated during senescence of *Arabidopsis* leaves. AtNAP knockout mutants display a significant delay in leaf senescence, while inducible overexpression of this gene causes precocious senescence [105]. Downstream targets of AtNAP include SAG113, a protein phosphatase 2C-encoding gene which functions to regulate ABA-mediated stomatal movement and water loss during leaf senescence [106, 107]. The AtNAP ortholog in rice, OsNAP, was shown to have the same regulatory role in rice leaf senescence [108]. It is suggested that OsNAP serves as an important link between ABA and leaf senescence since expression of *OsNAP* is specifically induced by ABA and repressed in ABA-deficient mutants. OsNAP was shown to positively regulate leaf senescence by directly targeting genes related to chlorophyll degradation and nutrient transport [108]. The NAC family member AtORE1/ANAC092 has also been identified as a positive regulator of leaf senescence [109]. Both ORE1 and AtNAP function downstream of the ethylene signaling proteins EIN2 and EIN3. As a transcription factor downstream to EIN2, EIN3 positively regulates leaf senescence by activating ORE1 and AtNAP [110]. Genetic analyses suggested that ORE1 and AtNAP act in distinct and overlapping signaling pathways downstream of EIN3 [110].

WRKY family transcription factors also play a central role in regulating leaf senescence in plants [16, 111]. The most well-studied WRKY53 is a positive senescence regulator. WRKY53 knockout plants show a delayed senescence phenotype [112]. Expression of WRKY53 is upregulated at an early stage and then downregulated again at later stages of leaf senescence. WRKY53 is thus believed to play a regulatory role in the early events of senescence [113]. Sixty-three genes, including at least six other members of the WRKY gene family, have been isolated as putative target genes of WRKY53 via genomic pull-down assays, suggesting that this WRKY transcription factor acts as an upstream control element in a transcription factor signaling cascade leading to leaf senescence [112].

4 Progression of Leaf Senescence

Progression of leaf senescence involves both degenerative and nutrient recycling processes. The most distinct feature of leaf senescence is the drastic metabolic shift from anabolism to catabolism with raise in the number of catabolic genes expressed in senescing leaves reached to almost twofold that of anabolic genes [103]. It is characterized by loss of chlorophyll and degradation of proteins, nucleic acids, lipids, and nutrient remobilization. Leaf senescence occurs in an orderly manner, beginning with the degeneration of the chloroplast and hydrolysis and remobilization of macromolecules, followed by the degeneration of the mitochondrion and nucleus [99, 114].

4.1 Loss of Chlorophyll

As the result of chlorophyll loss, leaf yellowing is the first visible symptom of leaf senescence. The rapid chlorophyll loss associated with chloroplast degeneration is frequently used as a biomarker for the start of senescence. Breakdown of chloroplastic proteins provides a major source of nitrogen, and chlorophylls need to be removed from the chlorophyll-binding proteins to make them available for degradation [115]. Unbound chlorophyll is also toxic to the cell and needs to be degraded properly to avoid free radical production [116]. Chlorophyll degradation begins in the chloroplast and involves a complex enzymatic pathway crossing several subcellular compartments [115, 117]. The chlorophyll degradation pathway consists of several enzymatic steps that produce different short-lived intermediates and phyllobilins that ultimately accumulate in the vacuole of senescing cells [118]. More recently, regulatory mechanism of chlorophyll breakdown during leaf senescence is widely investigated in various crop plants. In *Arabidopsis*, NYE1 and its close homologue NYE2 have been isolated as key regulators of chlorophyll degradation [119, 120]. In rice, the SGRL (stay green rice like) gene was reported to regulate chlorophyll degradation since overexpression of *SGRL* accelerates chlorophyll degradation in dark-induced senescence leaves [121]. A novel protein RLS1 (rapid leaf senescence 1) with NB-ARM domains is also implicated in rapid loss of chlorophyll content during senescence in rice [122].

4.2 Lipid Degradation

Lipid degradation is activated during leaf senescence [123]. A study on the metabolic profile of *Arabidopsis* leaf development indicated that chloroplast lipids (galactosyldiacylglycerols) are depleted during senescence [124]. The degradation products of lipids are metabolized and converted to phloem-mobile sucrose for translocation out of the senescing leaves [123, 125]. Lipid degradation may also have a regulatory role in senescence. Phospholipase Dα is known to be involved in the regulation of phytohormone-induced leaf senescence [126]. The *Arabidopsis* gene *SAG101* encoding an acyl hydrolase, which catalyzes the release of oleic acids from triolein in an in vitro assay, plays a regulatory role in leaf senescence [127].

4.3 Protein Degradation

Protein degradation is the most important process and plays a critical role in nitrogen recycling during leaf senescence. Transcriptomic studies have shown substantial upregulation of genes involved in proteolytic activities [101, 103]. Serine proteases and CysProt are the most abundant enzymes associated with leaf senescence, while aspartic, threonine, and metalloproteases are also detected [128, 129]. Plant protease activities have been detected in different cellular compartments including nuclei, chloroplasts, cytosol, ER, vacuoles, mitochondria, apoplast, cell wall, and special vesicles [128].

The largest source of mobilizable proteins within the cell is in the chloroplast, where Rubisco and the chlorophyll-binding light-harvesting proteins of PSII (LHCIIs) are the major proteins [117]. Chloroplast proteins contain up to 80% of total leaf nitrogen and are the major source of nitrogen for mobilization from senescing leaves [130]. During senescence chloroplast proteins seem to be degraded via coordinated actions of chloroplast proteases, senescence-associated vacuoles (SAVs), and the ubiquitin/26S proteasome pathway [128]. Genes encoding plastidial proteases such as DegP, Clp, and FtsH proteases are upregulated in senescing leaves [129]. The metalloprotease FtsH6, a member of the FtsH VAR1/VAR2 group, has been shown to degrade Lhcb3 during dark-induced senescence and high-light acclimation in barley and *Arabidopsis* [131]. SAVs are specific lytic compartments for degradation of chloroplastic proteins during senescence. The detection of a number of chloroplast proteins in SAV-enriched fractions supports the roles of SAVs in protein degradation [132, 133]. In the cytoplasm, proteins are targeted and degraded by the ubiquitin/proteasome system. The increased expression of genes encoding proteins involved in ubiquitin-dependent proteolysis [134] and identification of delayed senescence mutants in this category [79] support an important role of this system during senescence.

In addition to SAVs, after the initial degradation step within the chloroplast, photosynthetic proteins are believed to be transported to the central vacuole via the autophagic pathway [135, 136]. Senescence-induced autophagy has been studied in senescent leaves and cotyledons, where carbon and nitrogen are efficiently remobilized [137, 138]. The decrease in number and size of chloroplasts during senescence is affected in *Arabidopsis* mutant *atg4a4b-1* [138]. It has been shown that Rubisco and stroma-targeted fluorescent proteins can be mobilized to the central vacuole through an ATG gene-dependent autophagic process without prior chloroplast destruction [139].

4.4 Nitrogen Remobilization

Progression of leaf senescence is characterized by the transition from nutrient assimilation to nutrient remobilization [140]. Nitrogen availability often limits plant growth and reproduction. For survival, plants adopt an important strategy by relocating nitrogen from source (older senescing leaves) into younger, growing organs (new leaves, buds). The efficient use of nitrogen is essential for the plant life cycle. In cereal crops like wheat, rice, and barley, approximately 90% of the nitrogen is remobilized from the vegetative plant parts to the grain [141].

Autophagy, which is involved in chloroplast protein degradation, has been shown to play a critical role in nitrogen remobilization to seeds during leaf senescence. Nitrogen remobilization efficiency is largely reduced in autophagy mutants (*atg*) compared with the wild type [142]. It was evaluated that autophagy is involved in 40% of the ^{15}N remobilization flux from rosette leaves to the seeds of *Arabidopsis* plants grown under low-nitrate conditions [135].

As a result of protein degradation, in senescing leaves, a large amount of ammonium is produced, which is re-assimilated into the amide group of glutamate, which is then transferred to the GS/GOGAT cycle by the concerted reaction of glutamine synthetase (GS1) and glutamate synthase (NADH-GOGAT) to produce glutamine, which is believed to be the major mobile amino acid to be transported to sink tissues through long-distance transportation [52]. Genes encoding enzymes related to the GS/GOGAT cycle have been shown to be upregulated during leaf senescence [103, 140]. In consistent with this, two recent studies on metabolomes of leaf senescence in *Arabidopsis* [124] and sunflower [143], respectively, both showed increased accumulation of glutamine during senescence.

5 Genetic Manipulation of Leaf Senescence for Agriculture Improvement

In agricultural aspects, leaf senescence has great impact on crop production either by altering photosynthesis duration or by modifying the nutrient remobilization efficiency and also causes postharvest spoilage such as leaf yellowing and nutrient loss in vegetable crops. The timing of leaf senescence affects crop production by controlling agronomic traits like biomass and grain number/weight. In annual crop plants, early induction of senescence reduces crop production, while delaying leaf senescence and extending the duration of active photosynthesis could substantially increase the instant photoassimilate source and hence increase the grain yield [9]. Being a crucial trait, significant advances have been made in understanding the regulatory mechanisms of leaf senescence. The basic findings achieved have been made practically applicable for manipulating leaf senescence in various agronomic and horticultural crops by devising various strategies [5], such as (1) plant hormone biology-based technology, (2) senescence-specific transcription factor biology-based technology, and (3) translation initiation factor biology-based technology.

5.1 Plant Hormone Biology-Based Technology

Isopentenyl transferase (IPT) is known for catalyzing the rate-limiting step in the biosynthesis of cytokinins, which are inhibitors of leaf senescence. Gan and Amasino have designed an autoregulated cytokinin production system in which the promoter of the senescence-specific *SAG12* is fused to *IPT* to form a senescence-inhibition system [20]. The onset of senescence activates the *SAG12* promoter to direct *IPT* expression, resulting in the biosynthesis of cytokinins. In turn, cytokinins inhibit senescence and in this way attenuating the *SAG12* promoter activity, thus prevent overproduction of cytokinins [20]. Transgenic tobacco harboring this system displays a significant delay in leaf senescence without any abnormalities in growth and development, a marked increase in grain yield and biomass, and enhanced tolerance to drought and other stresses [20, 144]. Delay in leaf senescence of PSAG12:IPT tobacco was shown to be associated with a longer period of high photosynthetic activities [20]. The successful delay in senescence and the improved yield of the tobacco plants motivated scientists to transfer the IPT-based technology to agronomically and horticulturally important crop plants. A suppression of senescence was observed in all plants transformed with the PSAG12:IPT construct [5]. For example, senescence-regulated expression of the IPT gene delayed leaf senescence in rice, both with regard to chlorophyll degradation and photosynthetic capacity [145, 146]. This system has been widely explored and successfully used in various plant species for manipulating senescence and has been successful in getting, for example, increased fruit weight in tomato [147], increased biomass in tobacco [148], higher grain yield in rice [149], higher yield in peanut [150], higher lint yield and biomass in cotton [151], and improved turf quality in creeping bentgrass [152].

5.2 Senescence-Specific Transcription Factor Biology-Based Technology

The NAC family transcription factor AtNAP plays a key role in regulating leaf senescence by transcriptionally activating/repressing genes involved in the execution of senescence in *Arabidopsis* [105]. Knockout mutation of AtNAP led to nearly 50% increase in leaf's photosynthetic longevity [153]. AtNAP homologous genes have so far been reported in a variety of plant species including rice, maize, wheat, soybean, kidney bean, peach, tomato, petunia, potato, poplar, *Festuca arundinacea*, and bamboo [5]. Like AtNAP, the homologues of rice, kidney bean, bamboo, and cotton have been shown to be expressed in senescing leaves but not in non-senescing ones and, more importantly, the homologues function as AtNAP orthologs because they complemented the *Arabidopsis nap* null mutants [105, 154, 155], suggesting a conserved role of theses NAP homologues in different species and a potential approach of delaying leaf senescence by targeting these genes.

RNAi knocking down of the AtNAP ortholog in rice, OsNAP, caused a 5–7-day delay in senescence [108]. When grown in the field, the OsNAP-knockdown transgenic plants showed both significantly slower decrease in functional photosynthetic capacity and an extended grain-filling period compared with wild type. As a result, the best RNAi line showed a 23.0% increase in seed-setting ratio and a 3.1% increase in 1000 grain weight, resulting in a 10.3% increase in grain yield [108]. Similarly, RNA silencing of the maize homologue, ZmNAP, caused a stay-green phenotype and a 15–30% increase in 1000 grain weight in maize [156]. Knocking down of the NAP ortholog in cotton, GhNAP, also caused a significant delay in leaf senescence. Compared with wild type, lint yield and percentage of the GhNAPi plants were increased by >15% and 9%, respectively [155].

5.3 Translation Initiation Factor Biology-Based Technology

Senescence can also be regulated at the translational and posttranslational levels. One of the eukaryotic translation initiation factors, eIF-5A, has been shown to be involved in programmed cell death associated with plant senescence [157, 158]. The mechanism of the regulation of plant senescence by eIF-5A is not clear, but it has been suggested that this translation initiation factor acts as a nucleocytoplasmic transporter, translocating newly synthesized senescence-associated mRNAs from the nucleus to the cytoplasm for subsequent translation [159]. The activity of eIF-5A needs the posttranslational modification of a specific lysine residue to form hypusine, the first step of which is catalyzed by deoxyhypusine synthase (DHS). Both eIF-5A and DHS are upregulated during senescence [159]. Antisense suppression of DHS delays senescence in *Arabidopsis* leaves [157, 158].

DHS and different isoforms of eIF-5A have been isolated from a variety of plant species [160]. Transgenic tomato overexpressing the 3′-untranslated region of tomato DHS produced fruits that ripened normally but exhibited delayed postharvest softening and senescence [161]. Moreover, these transgenic plants showed higher activity of PSII as compare to wild type [161]. Transgenic poplar trees expressing *TaeIF5A1* showed high protein levels coupled with improved abiotic stress tolerance. Moreover, *TaeIF5A1*-transformed plants exhibited lower electrolyte leakage and higher chlorophyll content under salt stress conditions [162].

6   Conclusion

Significant progress has been made in leaf senescence research. It has been demonstrated that initiation and progression of leaf senescence can be affected by age, phytohormones, and environmental cues. Perception of signals often leads to changes in gene expression, and the upregulation of thousands of *SAGs* causes the senescence syndrome: loss of chlorophyll, lipid degradation, protein degradation, mobilization and reutilization of nutrients, and cell death. The research on the signal transduction pathway of leaf senescence is more and more thorough, and the signal regulation network is gradually emerging by using integrated approaches involving many molecular genetics tools, especially systems biology approaches. A number of strategies for leaf senescence manipulation have been developed based on current knowledge. Some of the strategies have been widely used for agricultural improvement.

Acknowledgments

Research in the Guo Lab has been supported by the Fundamental Research Funds for Central Non-profit Scientific Institution (2013ZL024, Y2017JC27), the National Natural Science Foundation of China (31571494), and the Agricultural Science and Technology Innovation Program (ASTIP-TRIC02).

References

1. Guo YF, Gan SS (2005) Leaf senescence: signals, execution, and regulation. Curr Top Dev Biol 71:83–112
2. Lim PO, Kim HJ, Gil Nam H (2007) Leaf senescence. Annu Rev Plant Biol 58:115–136
3. Noodén LD (1988) The phenomenon of senescence and aging. In: Noodén LD, Leopold AC (eds) Senescence and aging in plants. Academic Press, San Diego, pp 1–50
4. Jibran R, Hunter D, Dijkwel P (2013) Hormonal regulation of leaf senescence through integration of developmental and stress signals. Plant Mol Biol 82:547–561
5. Guo Y, Gan SS (2014) Translational researches on leaf senescence for enhancing plant productivity and quality. J Exp Bot 65:3901–3913
6. Leopold AC (1975) Aging, senescence, and turnover in plants. Bioscience 25:659–662
7. Guiboileau A, Sormani R, Meyer C et al (2010) Senescence and death of plant organs: nutrient recycling and developmental regulation. C R Biol 333:382–391
8. Hensel LL, Grbic V, Baumgarten DA et al (1993) Developmental and age-related processes that influence the longevity and senescence of photosynthetic tissues in Arabidopsis. Plant Cell 5:553–564
9. Gregersen PL, Culetic A, Boschian L et al (2013) Plant senescence and crop productivity. Plant Mol Biol 82:603–622
10. Patterson TG, Brun WA (1980) Influence of sink removal in the senescence pattern of wheat. Crop Sci 20:19–23
11. Crafts-Brandner SJ, Egli DB (1987) Sink removal and leaf senescence in soybean : cultivar effects. Plant Physiol 85:662–666
12. Htwe NMPS, Yuasa T, Ishibashi Y et al (2011) Leaf senescence of soybean at reproductive stage is associated with induction of autophagy-related genes, GmATG8c, GmATG8i and GmATG4. Plant Prod Sci 14:141–147

13. Kelly MO, Davies PJ (1986) Genetic and photoperiodic control of the relative rates of reproductive and vegetative development in peas. Ann Bot 58:13–21

14. Borras L, Maddonni GA, Otegui ME (2003) Leaf senescence in maize hybrids: plant population, row spacing and kernel set effects. Field Crop Res 82:13–26

15. Thomas H, Stoddart JL (1980) Leaf senescence. Annu Rev Plant Physiol 31:83–111

16. Li W, Guo YF (2014) Transcriptome, transcription factors and transcriptional regulation of leaf senescence. J Bioinf Comp Genomics 1:1

17. Schippers JHM, Jing HC, Hille J et al. (2007) Developmental and hormonal control of leaf senescence. Senescence processes in plants, 145–170

18. Beaudoin N, Serizet C, Gosti F et al (2000) Interactions between abscisic acid and ethylene signaling cascades. Plant Cell 12:1103–1115

19. Van Styaden J, Cook E, Nooden LD (1988) Cytokinins and senescence. In: Noodén LD, Leopold AC (eds) Senescence and aging in plants. Academic Press, San Diego, pp 281–328

20. Gan S, Amasino RM (1995) Inhibition of leaf senescence by autoregulated production of cytokinin. Science 270:1986–1988

21. Nooden LD, Singh S, Letham DS (1990) Correlation of xylem sap cytokinin levels with monocarpic senescence in soybean. Plant Physiol 93:33–39

22. Gan SS, Amasino RM (1996) Cytokinins in plant senescence: from spray and pray to clone and play. Bioessays 18:557–565

23. Hwang I, Sheen J, Muller B (2012) Cytokinin signaling networks. Annu Rev Plant Biol 63:353–380

24. Buchanan-Wollaston V, Page T, Harrison E et al (2005) Comparative transcriptome analysis reveals significant differences in gene expression and signalling pathways between developmental and dark/starvation-induced senescence in Arabidopsis. Plant J 42:567–585

25. Hwang I, Sheen J (2001) Two-component circuitry in Arabidopsis cytokinin signal transduction. Nature 413:383–389

26. Kim HJ, Ryu H, Hong SH et al (2006) Cytokinin-mediated control of leaf longevity by AHK3 through phosphorylation of ARR2 in Arabidopsis. Proc Natl Acad Sci U S A 103:814–819

27. Li Z, Su D, Lei B et al (2015) Transcriptional profile of genes involved in ascorbate glutathione cycle in senescing leaves for an early senescence leaf (esl) rice mutant. J Plant Physiol 176:1–15

28. Talla SK, Panigrahy M, Kappara S et al (2016) Cytokinin delays dark-induced senescence in rice by maintaining the chlorophyll cycle and photosynthetic complexes. J Exp Bot 67:1839–1851

29. Trivellini A, Cocetta G, Vernieri P et al (2015) Effect of cytokinins on delaying petunia flower senescence: a transcriptome study approach. Plant Mol Biol 87:169–180

30. Brenner WG, Romanov GA, Köllmer I et al (2005) Immediate-early and delayed cytokinin response genes of Arabidopsis thaliana identified by genome-wide expression profiling reveal novel cytokinin-sensitive processes and suggest cytokinin action through transcriptional cascades. Plant J 44:314–333

31. Sacher JA (1957) Relationship between auxin and membrane-integrity in tissue senescence and abscission. Science 125:1199–1200

32. Nooden LD, Kahanak GM, Okatan Y (1979) Prevention of monocarpic senescence in soybeans with auxin and cytokinin: an antidote for self-destruction. Science 206:841–843

33. Noh YS, Amasino RM (1999) Identification of a promoter region responsible for the senescence-specific expression of SAG12. Plant Mol Biol 41:181–194

34. Quirino BF, Normanly J, Amasino RM (1999) Diverse range of gene activity during Arabidopsis thaliana leaf senescence includes pathogen-independent induction of defense-related genes. Plant Mol Biol 40:267–278

35. Dela Fuente RK, Leopold AC (1968) Lateral movement of auxin in phototropism. Plant Physiol 43:1031–1036

36. Kim JI, Murphy AS, Baek D et al (2011) YUCCA6 over-expression demonstrates auxin function in delaying leaf senescence in Arabidopsis thaliana. J Exp Bot 62:3981–3992

37. Lim PO, Lee IC, Kim J et al (2010) Auxin response factor 2 (ARF2) plays a major role in regulating auxin-mediated leaf longevity. J Exp Bot 61:1419–1430

38. Ellis CM, Nagpal P, Young JC et al (2005) AUXIN RESPONSE FACTOR1 and AUXIN RESPONSE FACTOR2 regulate senescence and floral organ abscission in Arabidopsis thaliana. Development 132:4563–4574

39. Lin M, Pang C, Fan S et al (2015) Global analysis of the Gossypium hirsutum L. Transcriptome during leaf senescence by RNA-Seq. BMC Plant Biol 15:43

40. Rodrigues C, Vandenberghe LPD, De Oliveira J et al (2012) New perspectives of gibberellic acid production: a review. Crit Rev Biotechnol 32:263–273

41. Fletcher RA, Osborne DJ (1965) Regulation of protein and nucleic acid synthesis by gibberellin during leaf senescence. Nature 207:1176–1177

42. Aharoni N, Richmond AE (1978) Endogenous gibberellin and abscisic acid content as related to senescence of detached lettuce leaves. Plant Physiol 62:224–228

43. Chen L, Xiang S, Chen Y et al (2017) Arabidopsis WRKY45 interacts with the DELLA protein RGL1 to positively regulate age-triggered leaf senescence. Mol Plant 10:1174–1189. https://doi.org/10.1016/j.molp.2017.07.008

44. Iqbal N, Khan NA, Ferrante A et al (2017) Ethylene role in plant growth, development and senescence: interaction with other phytohormones. Front Plant Sci 8:475

45. Hunter DA, Yoo SD, Butcher SM et al (1999) Expression of 1-Aminocyclopropane-1-carboxylate oxidase during leaf ontogeny in white clover. Plant Physiol 120:131–142

46. Jing HC, Schippers JH, Hille J et al (2005) Ethylene-induced leaf senescence depends on age-related changes and OLD genes in Arabidopsis. J Exp Bot 56:2915–2923

47. Grbic V, Bleecker AB (1995) Ethylene regulates the timing of leaf senescence in Arabidopsis. Plant J 8:595–602

48. Wang H, Stier G, Lin J et al (2013) Transcriptome changes associated with delayed flower senescence on transgenic petunia by inducing expression of etr1-1, a mutant ethylene receptor. PLoS One 8:e65800

49. Oh SA, Park JH, Lee GI et al (1997) Identification of three genetic loci controlling leaf senescence in Arabidopsis thaliana. Plant J 12:527–535

50. Li ZH, Peng JY, Wen X et al (2013) ETHYLENE-INSENSITIVE3 is a senescence-associated gene that accelerates age-dependent leaf senescence by directly repressing miR164 transcription in Arabidopsis. Plant Cell 25:3311–3328

51. Qiu K, Li ZP, Yang Z et al (2015) EIN3 and ORE1 accelerate Degreening during ethylene-mediated leaf senescence by directly activating chlorophyll catabolic genes in Arabidopsis. PLoS Genet 11:e1005399

52. Kim JH, Chung KM, Woo HR (2011) Three positive regulators of leaf senescence in Arabidopsis, ORE1, ORE3 and ORE9, play roles in crosstalk among multiple hormone-mediated senescence pathways. Genes Genomics 33:373–381

53. Ueda H, Kusaba M (2015) Strigolactone regulates leaf senescence in concert with ethylene in Arabidopsis. Plant Physiol 169:138–147

54. Zhang DP (2014) Abscisic acid metabolism, transport and signaling. Springer, Dordrecht

55. He P, Osaki M, Takebe M et al (2005) Endogenous hormones and expression of senescence-related genes in different senescent types of maize. J Exp Bot 56:1117–1128

56. Gepstein S, Thimann KV (1980) Changes in the abscisic acid content of oat leaves during senescence. Proc Natl Acad Sci U S A 77:2050–2053

57. Gao S, Gao J, Zhu X et al (2016) ABF2, ABF3, and ABF4 promote ABA-mediated chlorophyll degradation and leaf senescence by transcriptional activation of chlorophyll catabolic genes and senescence-associated genes in Arabidopsis. Mol Plant 9:1272–1285

58. Yang JC, Zhang JH, Wang ZQ et al (2002) Abscisic acid and cytokinins in the root exudates and leaves and their relationship to senescence and remobilization of carbon reserves in rice subjected to water stress during grain filling. Planta 215:645–652

59. Lee IC, Hong SW, Whang SS et al (2011) Age-dependent action of an ABA-inducible receptor kinase, RPK1, as a positive regulator of senescence in Arabidopsis leaves. Plant Cell Physiol 52:651–662

60. Zhao Y, Chan Z, Gao J et al (2016) ABA receptor PYL9 promotes drought resistance and leaf senescence. Proc Natl Acad Sci U S A 113:1949–1954

61. Gudesblat GE, Russinova E (2011) Plants grow on brassinosteroids. Curr Opin Plant Biol 14:530–537

62. Gomes MDD, Netto AT, Campostrini E et al (2013) Brassinosteroid analogue affects the senescence in two papaya genotypes submitted to drought stress. Theor Exp Plant Physiol 25:186–195

63. He Y, Xu RJ, Zhao YJ (1996) Enhancement of senescence by epibrassinolide in leaves of mung bean seedling. Acta Phytophysiol Sin 22:58–62

64. Saglam-Cag S (2007) The effect of epibrassinolide on senescence in wheat leaves. Biotechnol Biotechnol Equip 21:63–65

65. Chory J, Nagpal P, Peto CA (1991) Phenotypic and genetic analysis of det2, a new mutant that affects light-regulated seedling development in Arabidopsis. Plant Cell 3:445–459

66. Hu Y, Jiang Y, Han X et al (2017) Jasmonate regulates leaf senescence and tolerance to cold stress: crosstalk with other phytohormones. J Exp Bot 68:1361–1369

67. He Y, Fukushige H, Hildebrand DF et al (2002) Evidence supporting a role of jas-

monic acid in Arabidopsis leaf senescence. Plant Physiol 128:876–884

68. Schommer C, Palatnik JF, Aggarwal P et al (2008) Control of jasmonate biosynthesis and senescence by miR319 targets. PLoS Biol 6:e230

69. Qi TC, Wang JJ, Huang H et al (2015) Regulation of Jasmonate-induced leaf senescence by antagonism between bHLH subgroup IIIe and IIId factors in Arabidopsis. Plant Cell 27:1634–1649

70. Jiang Y, Liang G, Yang S et al (2014) Arabidopsis WRKY57 functions as a node of convergence for jasmonic acid– and Auxin-mediated signaling in jasmonic acid–induced leaf senescence. Plant Cell 26:230–245

71. Yu J, Zhang Y, Di C et al (2016) JAZ7 negatively regulates dark-induced leaf senescence in Arabidopsis. J Exp Bot 67:751–762

72. Shan X, Wang J, Chua L et al (2011) The role of Arabidopsis Rubisco activase in jasmonate-induced leaf senescence. Plant Physiol 155:751–764

73. Vlot AC, Dempsey DMA, Klessig DF (2009) Salicylic acid, a multifaceted hormone to combat disease. Annu Rev Phytopathol 47:177–206

74. Abreu ME, Munne-Bosch S (2008) Salicylic acid may be involved in the regulation of drought-induced leaf senescence in perennials: a case study in field-grown Salvia officinalis L. plants. Environ Exp Bot 64:105–112

75. Zhang K, Halitschke R, Yin C et al (2013) Salicylic acid 3-hydroxylase regulates Arabidopsis leaf longevity by mediating salicylic acid catabolism. Proc Natl Acad Sci 110:14807–14812

76. Morris K, Mackerness SaH, Page T et al (2000) Salicylic acid has a role in regulating gene expression during leaf senescence. Plant J 23:677–685

77. Zhao X-Y, Wang J-G, Song S-J et al (2016) Precocious leaf senescence by functional loss of PROTEIN S-ACYL TRANSFERASE14 involves the NPR1-dependent salicylic acid signaling. Sci Rep 6:20309

78. Yamada Y, Umehara M (2015) Possible roles of strigolactones during leaf senescence. Plants 4:664–677

79. Woo HR, Chung KM, Park JH et al (2001) ORE9, an F-box protein that regulates leaf senescence in Arabidopsis. Plant Cell 13:1779–1790

80. Stirnberg P, Van De Sande K, Leyser HMO (2002) MAX1 and MAX2 control shoot lateral branching in Arabidopsis. Development 129:1131–1141

81. Yan H, Saika H, Maekawa M et al (2007) Rice tillering dwarf mutant *dwarf3* has increased leaf longevity during darkness-induced senescence or hydrogen peroxide-induced cell death. Genes Genet Syst 82:361–366

82. Snowden KC, Simkin AJ, Janssen BJ et al (2005) The decreased apical dominance1/Petunia hybrida CAROTENOID CLEAVAGE DIOXYGENASE8 gene affects branch production and plays a role in leaf senescence, root growth, and flower development. Plant Cell 17:746–759

83. Liu J, Novero M, Charnikhova T et al (2013) CAROTENOID CLEAVAGE DIOXYGENASE 7 modulates plant growth, reproduction, senescence, and determinate nodulation in the model legume Lotus japonicus. J Exp Bot 64:1967–1981

84. Yamada Y, Furusawa S, Nagasaka S et al (2014) Strigolactone signaling regulates rice leaf senescence in response to a phosphate deficiency. Planta 240:399–408

85. Gepstein S, Glick BR (2013) Strategies to ameliorate abiotic stress-induced plant senescence. Plant Mol Biol 82:623–633

86. Allu AD, Soja AM, Wu A et al (2014) Salt stress and senescence: identification of crosstalk regulatory components. J Exp Bot 65:3993–4008

87. Chen H-J, Lin Z-W, Huang G-J et al (2012) Sweet potato calmodulin SPCAM is involved in salt stress-mediated leaf senescence, H2O2 elevation and senescence-associated gene expression. J Plant Physiol 169:1892–1902

88. Xiao H-J, Liu K-K, Li D-W et al (2015) Cloning and characterization of the pepper CaPAO gene for defense responses to salt-induced leaf senescence. BMC Biotechnol 15:100

89. Khanna-Chopra R, Selote DS (2007) Acclimation to drought stress generates oxidative stress tolerance in drought-resistant than -susceptible wheat cultivar under field conditions. Environ Exp Bot 60:276–283

90. Chen D, Wang S, Xiong B et al (2015) Carbon/nitrogen imbalance associated with drought-induced leaf senescence in Sorghum bicolor. PLoS One 10:e0137026

91. Janack B, Sosoi P, Krupinska K et al (2016) Knockdown of WHIRLY1 affects drought stress-induced leaf senescence and histone modifications of the senescence-associated gene HvS40. Plants 5:37

92. Liu T, Longhurst AD, Talavera-Rauh F et al (2016) The Arabidopsis transcription factor ABIG1 relays ABA signaled growth inhibition and drought induced senescence. Elife 5:e13768

93. Bita CE, Gerats T (2013) Plant tolerance to high temperature in a changing environment:

scientific fundamentals and production of heat stress-tolerant crops. Front Plant Sci 4:273

94. Jespersen D, Zhang J, Huang B (2016) Chlorophyll loss associated with heat-induced senescence in bentgrass. Plant Sci 249:1–12

95. Jespersen D, Yu J, Huang B (2015) Metabolite responses to exogenous application of nitrogen, cytokinin, and ethylene inhibitors in relation to heat-induced senescence in creeping bentgrass. PLoS One 10:e0123744

96. Chen WQ, Provart NJ, Glazebrook J et al (2002) Expression profile matrix of Arabidopsis transcription factor genes suggests their putative functions in response to environmental stresses. Plant Cell 14:559–574

97. Guo Y, Gan SS (2012) Convergence and divergence in gene expression profiles induced by leaf senescence and 27 senescence-promoting hormonal, pathological and environmental stress treatments. Plant Cell Environ 35:644–655

98. He Y, Tang W, Swain JD et al (2001) Networking senescence-regulating pathways by using Arabidopsis enhancer trap lines. Plant Physiol 126:707–716

99. Guo Y (2013) Towards systems biological understanding of leaf senescence. Plant Mol Biol 82:519–528

100. Zentgraf U, Jobst J, Kolb D et al (2004) Senescence-related gene expression profiles of rosette leaves of Arabidopsis thaliana: leaf age versus plant age. Plant Biol (Stuttg) 6:178–183

101. Breeze E, Harrison E, Mchattie S et al (2011) High-resolution temporal profiling of transcripts during Arabidopsis leaf senescence reveals a distinct chronology of processes and regulation. Plant Cell 23:873–894

102. Woo HR, Koo HJ, Kim J et al (2016) Programming of plant leaf senescence with temporal and inter-organellar coordination of transcriptome in Arabidopsis. Plant Physiol 171:452–467

103. Guo Y, Cai Z, Gan S (2004) Transcriptome of Arabidopsis leaf senescence. Plant Cell Environ 27:521–549

104. Christiansen MW, Gregersen PL (2014) Members of the barley NAC transcription factor gene family show differential co-regulation with senescence-associated genes during senescence of flag leaves. J Exp Bot 65:4009–4022

105. Guo Y, Gan S (2006) AtNAP, a NAC family transcription factor, has an important role in leaf senescence. Plant J 46:601–612

106. Zhang K, Xia X, Zhang Y et al (2012) An ABA-regulated and Golgi-localized protein phosphatase controls water loss during leaf senescence in Arabidopsis. Plant J 69:667–678

107. Zhang K, Gan SS (2012) An abscisic acid-AtNAP transcription factor-SAG113 protein phosphatase 2C regulatory chain for controlling dehydration in senescing Arabidopsis leaves. Plant Physiol 158:961–969

108. Liang C, Wang Y, Zhu Y et al (2014) OsNAP connects abscisic acid and leaf senescence by fine-tuning abscisic acid biosynthesis and directly targeting senescence-associated genes in rice. Proc Natl Acad Sci U S A 111:10013–10018

109. Kim JH, Woo HR, Kim J et al (2009) Trifurcate feed-forward regulation of age-dependent cell death involving miR164 in Arabidopsis. Science 323:1053–1057

110. Kim HJ, Hong SH, Kim YW et al (2014) Gene regulatory cascade of senescence-associated NAC transcription factors activated by ETHYLENE-INSENSITIVE2-mediated leaf senescence signalling in Arabidopsis. J Exp Bot 65:4023–4036

111. Rushton DL, Tripathi P, Rabara RC et al (2012) WRKY transcription factors: key components in abscisic acid signalling. Plant Biotechnol J 10:2–11

112. Miao Y, Laun T, Zimmermann P et al (2004) Targets of the WRKY53 transcription factor and its role during leaf senescence in Arabidopsis. Plant Mol Biol 55:853–867

113. Hinderhofer K, Zentgraf U (2001) Identification of a transcription factor specifically expressed at the onset of leaf senescence. Planta 213:469–473

114. Solomos T (1988) Respiration in senescing plant organs: its nature, regulation, and physiological significance. In: Nooden LD, Leopold AC (eds) Senescence and aging in plants. Academic Press, San Diego, pp 111–145

115. Thomas H, Ougham H, Hortensteiner S (2001) Recent advances in the cell biology of chlorophyll catabolism. Adv Bot Res 35:1–52

116. Matile P (1992) Chloroplast senescence. In: Baker NR, Thomas H (eds) Crop photosynthesis:spatial and temporal determinants. Elsevier, Amsterdam, pp 413–440

117. Hörtensteiner S, Feller U (2002) Nitrogen metabolism and remobilization during senescence. J Exp Bot 53:927–937

118. Hörtensteiner S, Kräutler B (2011) Chlorophyll breakdown in higher plants. Biochim Biophys Acta 1807:977–988

119. Ren G, An K, Liao Y et al (2007) Identification of a novel chloroplast protein AtNYE1 regulating chlorophyll degradation during leaf

senescence in Arabidopsis. Plant Physiol 144:1429–1441

120. Wu S, Li Z, Yang L et al (2016) NON-YELLOWING2 (NYE2), a close paralog of NYE1, plays a positive role in chlorophyll degradation in Arabidopsis. Mol Plant 9:624–627

121. Rong H, Tang Y, Zhang H et al (2013) The stay-green rice like (SGRL) gene regulates chlorophyll degradation in rice. J Plant Physiol 170:1367–1373

122. Jiao BB, Wang JJ, Zhu XD et al (2012) A novel protein RLS1 with NB-ARM domains is involved in chloroplast degradation during leaf senescence in rice. Mol Plant 5:205–217

123. Thompson JE, Froese CD, Madey E et al (1998) Lipid metabolism during plant senescence. Prog Lipid Res 37:119–141

124. Watanabe M, Balazadeh S, Tohge T et al (2013) Comprehensive dissection of spatiotemporal metabolic shifts in primary, secondary, and lipid metabolism during developmental senescence in Arabidopsis. Plant Physiol 162:1290–1310

125. Kaup MT, Froese CD, Thompson JE (2002) A role for diacylglycerol acyltransferase during leaf senescence. Plant Physiol 129:1616–1626

126. Fan L, Zheng S, Wang X (1997) Antisense suppression of phospholipase D alpha retards abscisic acid- and ethylene-promoted senescence of postharvest Arabidopsis leaves. Plant Cell 9:2183–2196

127. He Y, Gan S (2002) A gene encoding an acyl hydrolase is involved in leaf senescence in Arabidopsis. Plant Cell 14:805–815

128. Diaz-Mendoza M, Velasco-Arroyo B, Santamaria ME et al (2016) Plant senescence and proteolysis: two processes with one destiny. Genet Mol Biol 39:329–338

129. Roberts IN, Caputo C, Criado MV et al (2012) Senescence-associated proteases in plants. Physiol Plant 145:130–139

130. Liu J, Wu YH, Yang JJ et al (2008) Protein degradation and nitrogen remobilization during leaf senescence. J Plant Biol 51:11–19

131. Zelisko A, Garcia-Lorenzo M, Jackowski G et al (2005) AtFtsH6 is involved in the degradation of the light-harvesting complex II during high-light acclimation and senescence. Proc Natl Acad Sci U S A 102:13699–13704

132. Martínez DE, Costa ML, Guiamet JJ (2008) Senescence-associated degradation of chloroplast proteins inside and outside the organelle. Plant Biol 10:15–22

133. Martinez DE, Costa ML, Gomez FM et al (2008) 'Senescence-associated vacuoles' are involved in the degradation of chloroplast

proteins in tobacco leaves. Plant J 56:196–206

134. Buchanan-Wollaston V, Earl S, Harrison E et al (2003) The molecular analysis of leaf senescence - a genomics approach. Plant Biotechnol J 1:3–22

135. Avila-Ospina L, Moison M, Yoshimoto K et al (2014) Autophagy, plant senescence, and nutrient recycling. J Exp Bot 65:3799–3811

136. Ono Y, Wada S, Izumi M et al (2013) Evidence for contribution of autophagy to rubisco degradation during leaf senescence in Arabidopsis thaliana. Plant Cell Environ 36:1147–1159

137. Xiong Y, Contento AL, Bassham DC (2005) AtATG18a is required for the formation of autophagosomes during nutrient stress and senescence in Arabidopsis thaliana. Plant J 42:535–546

138. Wada S, Ishida H, Izumi M et al (2009) Autophagy plays a role in chloroplast degradation during senescence in individually darkened leaves. Plant Physiol 149:885–893

139. Ishida H, Yoshimoto K, Izumi M et al (2008) Mobilization of rubisco and stroma-localized fluorescent proteins of chloroplasts to the vacuole by an ATG gene-dependent autophagic process. Plant Physiol 148:142–155

140. Masclaux C, Valadier M-H, Brugière N et al (2000) Characterization of the sink/source transition in tobacco (Nicotiana tabacum L.) shoots in relation to nitrogen management and leaf senescence. Planta 211:510–518

141. Gregersen PL, Holm PB, Krupinska K (2008) Leaf senescence and nutrient remobilisation in barley and wheat. Plant Biol 10:37–49

142. Guiboileau A, Yoshimoto K, Soulay F et al (2012) Autophagy machinery controls nitrogen remobilization at the whole-plant level under both limiting and ample nitrate conditions in Arabidopsis. New Phytol 194:732–740

143. Moschen S, Bengoa Luoni S, Di Rienzo JA et al (2016) Integrating transcriptomic and metabolomic analysis to understand natural leaf senescence in sunflower. Plant Biotechnol J 14:719–734

144. Cowan AK, Freeman M, Björkman P-O et al (2005) Effects of senescence-induced alteration in cytokinin metabolism on source-sink relationships and ontogenic and stress-induced transitions in tobacco. Planta 221:801–814

145. Cao ML, Zhou Z, Wang Z (2001) Performance of autoregulatory senescence delayed system in tobacco. Crop Res:26–28

146. Lin YJ, Cao ML, Xu CG et al (2002) Cultivating rice with delaying led-senescence

by P-SAG12-IPT gene transfonination. Acta Bot Sin 44:1333–1338

147. Swartzberg D, Dai N, Gan S et al (2006) Effects of cytokinin production under two SAG promoters on senescence and development of tomato plants. Plant Biol (Stuttg) 8:579–586

148. Rivero RM, Kojima M, Gepstein A et al (2007) Delayed leaf senescence induces extreme drought tolerance in a flowering plant. Proc Natl Acad Sci U S A 104:19631–19636

149. Liu L, Zhou Y, Szczerba MW et al (2010) Identification and application of a rice senescence-associated promoter. Plant Physiol 153:1239–1249

150. Qin H, Gu Q, Zhang J et al (2011) Regulated expression of an isopentenyltransferase gene (IPT) in peanut significantly improves drought tolerance and increases yield under field conditions. Plant Cell Physiol 52:1904–1914

151. Liu YD, Yin ZJ, Yu JW et al (2012) Improved salt tolerance and delayed leaf senescence in transgenic cotton expressing the Agrobacterium IPT gene. Biol Plant 56:237–246

152. Merewitz EB, Gianfagna T, Huang B (2011) Photosynthesis, water use, and root viability under water stress as affected by expression of SAG12-ipt controlling cytokinin synthesis in Agrostis stolonifera. J Exp Bot 62:383–395

153. Gan SS (2014) Leaf senescence as an important target for improving crop production. Adv Crop Sci Technol 2:e116

154. Chen Y, Qiu K, Kuai B et al (2011) Identification of an NAP-like transcription factor BeNAC1 regulating leaf senescence in

bamboo (Bambusa emeiensis 'Viridiflavus'). Physiol Plant 142:361–371

155. Fan K, Bibi N, Gan S et al (2015) A novel NAP member GhNAP is involved in leaf senescence in Gossypium hirsutum. J Exp Bot 66:4669–4682

156. Zhang Y, Cao Y, Shao Q et al (2012) Regulating effect of ZmNAP gene on anti-senescence and yield traits of maize. J Henan Agric Sci 41:19–24

157. Duguay J, Jamal S, Liu Z et al (2007) Leaf-specific suppression of deoxyhypusine synthase in Arabidopsis thaliana enhances growth without negative pleiotropic effects. J Plant Physiol 164:408–420

158. Wang T-W, Lu L, Zhang C-G et al (2003) Pleiotropic effects of suppressing deoxyhypusine synthase expression in Arabidopsis thaliana. Plant Mol Biol 52:1223–1235

159. Thompson JE, Hopkins MT, Taylor C et al (2004) Regulation of senescence by eukaryotic translation initiation factor 5A: implications for plant growth and development. Trends Plant Sci 9:174–179

160. Hopkins M, Taylor C, Liu Z et al (2007) Regulation and execution of molecular disassembly and catabolism during senescence. New Phytol 175:201–214

161. Wang TW, Zhang CG, Wu W et al (2005) Antisense suppression of deoxyhypusine synthase in tomato delays fruit softening and alters growth and development. Plant Physiol 138:1372–1382

162. Wang L, Xu C, Wang C et al (2012) Characterization of a eukaryotic translation initiation factor 5A homolog from Tamarix androssowii involved in plant abiotic stress tolerance. BMC Plant Biol 12:118

Part II

Phenotypic Analysis and Molecular Markers
of Plant Organ Senescence

Chapter 3

Phenotypic Analysis and Molecular Markers of Leaf Senescence

Liming Zhao, Yan Xia, Xiao-Yuan Wu, Jos H.M. Schippers, and Hai-Chun Jing

Abstract

The process of leaf senescence consists of the final stage of leaf development. It has evolved as a mechanism to degrade macromolecules and micronutrients and remobilize them to other developing parts of the plant; hence it plays a central role for the survival of plants and crop production. During senescence, a range of physiological, morphological, cellular, and molecular events occur, which are generally referred to as the senescence syndrome that includes several hallmarks such as visible yellowing, loss of chlorophyll and water content, increase of ion leakage and cell death, deformation of chloroplast and cell structure, as well as the upregulation of thousands of so-called senescence-associated genes (*SAGs*) and downregulation of photosynthesis-associated genes (*PAGs*). This chapter is devoted to methods characterizing the onset and progression of leaf senescence at the morphological, physiological, cellular, and molecular levels. Leaf senescence normally progresses in an age-dependent manner but is also induced prematurely by a variety of environmental stresses in plants. Focused on the hallmarks of the senescence syndrome, a series of protocols is described to asses quantitatively the senescence process caused by developmental cues or environmental perturbations. We first briefly describe the senescence process, the events associated with the senescence syndrome, and the theories and methods to phenotype senescence. Detailed protocols for monitoring senescence *in planta* and in vitro, using the whole plant and the detached leaf, respectively, are presented. For convenience, most of the protocols use the model plant species *Arabidopsis* and rice, but they can be easily extended to other plants.

Key words Leaf senescence, Visible yellowing, Chlorophyll, Ion leakage, Cell death, Senescence-associated genes (SAGs), *Arabidopsis*, Rice

1 Introduction

Leaf senescence represents the final stage of leaf development and is characterized by the visible color change of the leaf, the degradation of chlorophyll, and the activation of thousands of senescence-associated genes *(SAGs)* to initiate the recycling of the macromolecules and micronutrients of the senescing leaf [1]. Even for nonscientists, senescence is immediately apparent as it drives the development of amazing autumn colors and the ripening of

Yongfeng Guo (ed.), *Plant Senescence: Methods and Protocols*, Methods in Molecular Biology, vol. 1744,
https://doi.org/10.1007/978-1-4939-7672-0_3, © Springer Science+Business Media, LLC 2018

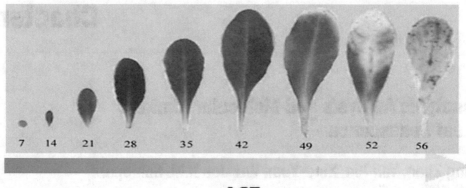

AGE

Fig. 1 Percentage of yellowing of first leaves during age-dependent senescence in *Arabidopsis*

crops and fruits [2], which can occur at a scale such that it transforms the appearance of the earth. This simple observation demonstrates the close relationship between senescence, crop yield, and shelf life [2, 3]. Indeed, it has been estimated that delaying leaf senescence in crops could increase the grain yield over 10% and substantially improve carbon and nitrogen economy.

As hinted in the previous section, leaf senescence can be perceived by changes in leaf appearance. As a matter of fact, dramatic changes at the morphological, physiological, cellular, and molecular levels occur following the onset and progression of senescence. These changes are commonly referred to as the senescence syndrome. The very first visible symptoms of leaf senescence are the changes of leaf color. After initial rapid expansion and reaching full maturation, a leaf starts the senescence process subjected to the control of leaf age/developmental stage or the induction by environmental stimuli. Under normal circumstances, first signs of typical senescence start with the yellowing of the leaf tip, which gradually develops inward into the leaf blade, and in the end the entire leaf blade becomes yellow, orange, or purple, and the leaf becomes dried and shrunken, leading to cell death (Fig. 1). Hence, characterization of the color change is one of the best morphological phenotyping for leaf senescence. Associated with such changes in leaf appearance, a range of physiological, biochemical, and cellular events occur in the leaf cells. The color changes are ascribed to the degradation of chlorophyll and the concomitant synthesis of carotenoids and anthocyanins. In a senescing cell, the metabolism shifts from anabolism to catabolism; macromolecules of various kinds including proteins, RNA, and DNA as well as polysaccharides are degraded into the basic unit compounds such as short peptides and amino acids, short stretches of nucleotides, sucrose, and glucose, which are then translocated to other parts via transport or activities. Subcellular organelles such as chloroplasts, vacuoles, nuclei, and cellular membranes are deformed. In the end, leaf cells remobilize all the essential carbon and nitrogen and ultimately die. Thus, physiological and cellular characterization of senescence can be achieved through quantifying the pigment content, nutrient content, ion leakage, and structure and morphology of subcellular organelles. The onset and progression of leaf senescence requires transcriptional activation. Indeed, many *SAG*s have been identified upregulated during the onset and progression of senescence. In *Arabidopsis*, a recent count shows that *SAG*s amount to over 3000 and the genes are involved in a wide range of biological processes including transcriptional regulation, molecule degradation, transportation, and stress tolerance. Over the years, many *SAG*s have been well characterized and used as molecular markers for various types of senescence. For instance, *SAG12*, encoding a cysteine proteinase, is well accepted by the senescence community as the marker for age-dependent/development-regulated leaf senescence, while *SAG13* is normally used for dark-induced leaf senescence. *EIN3*, a transcriptional factor essential for ethylene signaling, is used as the marker for ethylene-induced senescence. The development of these molecular makers allows the characterization of senescence at the molecular level, which can assess senescence prior to any visible and physiological changes.

Although the onset of senescence occurs inevitably due to aging of the leaf, the process can be initiated prematurely due to adverse environmental conditions and specific external factors. Previous studies have identified numerous genetic factors that modulate age- and/or environment-dependent senescence in the model plant species *Arabidopsis thaliana* and several crops [4–6]. Crucial to the identification of senescence mutants is the precise characterization of their phenotypes with quantitative methods. Over the years, many excellent experimental systems have been developed to characterize the senescence syndrome at the morphological, physiological, and molecular levels using *in planta* and in vitro assays. For *in planta* assays, whole plant senescence or leaves attached to the leaves at various positions are used as the objectives, whereas for in vitro assays, detached leaves are normally used. Initial phenotyping of the onset of leaf senescence is done by comparing the number of yellow leaves at a given time or developmental stage. Noteworthy, from previous senescence studies, it is clear that growth conditions (day length, light quality, humidity, soil type) have a major impact on the onset of leaf senescence. Furthermore, induction of senescence by external factors such as darkness, osmotic stress, or treatment of plant hormones shares many aspects with developmental senescence but also shows fundamental differences at the molecular level [7]. Therefore, subsequent experiments should aim at characterizing the molecular events associated with senescence by using molecular markers, such as the changes in the expression levels of senescence-associated genes.

Here we describe the basic methods to analyze the process of leaf senescence. Most importantly, the onset of senescence depends on the age or the developmental stage of the leaf examined, which should be kept in mind at all times during the preparation for each of the described experiments.

2   Materials

2.1   Plant Culture

1. Seeds: Of the plant species used (e.g., *Arabidopsis thaliana*, *Oryza sativa*).
2. 10% NaClO.
3. Pots: For culture in soil or hydroponic culture systems for growth in nutrient solution.
4. Soil and nutrient solution: Optimized for each plant species used (*see* **Note 1**).
5. Environmentally controlled plant growth chamber or greenhouse.
6. Trays and foil.
7. Watering system.

2.2   Detached Leaf Assays

1. Detached leaves: Detached mature first leaves from *Arabidopsis*, typically from 21-day-old plants. For rice, detached flag leaves from the main culm at the heading stage (*see* **Note 2**).
2. MES solution: 2.35 mM MES in water.
3. Petri dishes or 6-well multiwell plates.
4. Scissors.
5. Solutions for senescence induction: 500 mM sorbitol, 100 mM NaCl, 5 µM ABA, 10 µM ACC, 10 mM H_2O_2, etc.
6. Aluminum foil.

2.3   Leaf Pigment Extraction and Determination

1. Fine balance.
2. Extraction solution: 80% acetone.
3. Glass cuvettes.
4. Spectrophotometer (wavelengths 470, 647, and 663 nm).

2.4   Electrolyte Leakage Assay

1. Water bath.
2. Glass tubes.
3. Deionized water.
4. Conductivity meter.

2.5   Trypan Blue Staining of Leaves

1. Trypan blue solution: Dissolve 25 mg of trypan blue in 10 mL lactophenol and add 20 mL 96% ethanol.

2. Chloral hydrate solution: 25 g of chloral hydrate in 10 mL water.

3. 70% glycerol.

4. Stereomicroscope with camera.

2.6 Leaf Nutrient Content

1. Fine balance.

2. Scissors.

3. Deionized water.

4. Oven.

5. Mortar and pestle.

6. Mass spectrometer.

7. ICP-MS: Inductively coupled plasma-mass spectrometer.

2.7 Chloroplast Counting

1. 3.5% (v/v) glutaraldehyde.

2. 0.1 M Na_2EDTA: pH 9.0.

3. Water bath.

4. Light microscope.

2.8 Chloroplast Morphology by Transmission Electron Microscopy

1. Sorensen's phosphate buffer: 0.133 M Na_2HPO_4, 0.133 M KH_2PO_4, for 100 mL, mixed with 71.5 mL of Na_2HPO_4 and 28.5 mL of KH_2PO_4, pH 7.2.

2. 5% (v/v) glutaraldehyde: In Sorensen's phosphate buffer.

3. Ethanol: Absolute, 90, 70, 50, 30% (v/v).

4. Propylene oxide.

5. Araldite 502 resin.

6. 0.5% toluidine blue: In 1% borax.

7. Aqueous lead citrate: Saturated.

8. 0.5% uranyl acetate.

9. Refrigerator or cold room: 4 °C.

10. MT-2 Ultramicrotome.

11. Transmission electron microscope (TEM): Jeol 1200 Ex II.

2.9 RNA Isolation and Gene Expression Analysis

1. Plant RNA extraction kit or TRIzol.

2. UV spectrophotometer.

3. RNase-free DNase.

4. Reverse transcriptase.

5. SYBR Green Mastermix for qPCR.

6. Oligonucleotide primers (Table 1).

7. Real-time PCR system.

8. qPCR plates and optical foils.

Table 1
Oligonucleotide primer sequences for selected marker genes

Gene	Sequence forward primer	Sequence reverse primer
Arabidopsis thaliana		
SAG12 (At5g45890)	ACAAAGGCGAAGACGCTACTTG	ACCGGGACATCCTCATAACCTG
SAG13 (At5g51070)	TGCCATTGCTGAAGGACTAGCG	AGGGACATGATGCGTTTCGTCAAG
SAG14 (At5g14930)	AGCTCACGCCATGGAGTCTTCTTC	ACCACAAGCTTCCAAGTGCAG
SAG101 (At5g20230)	TTCGCTGCCGTTGTTGTCTTCG	AATTCGAGCTCGTCGCCTACAC
SAG113 (At5g13170)	TAAGCGCCGTTATGTGGTTCGC	ATCCCACCACGTTTGGAATCGC
CAB (At1g29930)	TGTTCCCGGAAAGTGAGCCAAG	TCTCCTCTCACACTCACGAAGC
Oryza sativa		
SGR (LOC_Os09g36200)	ACGCATGCAATGTCGCCAAATG	GAGCTGAGCTAAATGCCACTACG
RCCR1 (LOC_Os10g25030)	GCCGGGATCGACGATTGATTTC	ATGTCGATTGCGCCATTTGGG
Osh36 (LOC_Os05g39770)	ACACCGGCCAAAGTTGAGATCG	CAACCAGCACACCATGTCTTTC
Osh69 (LOC_Os08g38710)	AGATTTGCCTCGAGAGTGGTGAC	TCTGCAAGTGCCTCTCAACAGC

3   Methods

3.1   Plant Culture

1. Before sowing *Arabidopsis* seeds on soil, the seeds are placed in a micro-centrifuge tube covered with water and kept for 24 h at 4 °C. For rice, seeds are sterilized with 10% NaClO for 10 min and after five times washing with water placed on moisturized filter paper in a closed transparent plastic box. Subsequently, the boxes with rice seeds are placed for 6 days at 28 °C in the dark. Thereafter, the boxes with rice seedlings are placed under a 12/12 h day-night regime for another 6 days prior to transplanting them into soil. Or the plants can be grown in hydroponic systems (*see* **Note 1**).

2. Fill pots with appropriate soil/nutrient solution and place them into trays.

3. For *Arabidopsis*, dispense the seeds on soil with a suitable pipette, by placing three seeds in each pot. After sowing cover the trays with a transparent cover, and place them into a controlled plant growth chamber. For rice, transplant the seedlings into the soil, and place the pots into a plant growth chamber (*see* **Note 3**).

4. Once the *Arabidopsis* seedlings have reached stage 1.02 according to the Boyes et al. system [8] with two cotyledons and two visible leaves (around 7 days after sowing), remove the plastic cover. Subsequently, discard seedlings that are not of the correct developmental stage, and leave only one seedling per pot.

5. Depending on the type of analysis, 20–100 plants are needed for characterizing the senescence syndrome.

3.2   Phenotyping and Sampling of Leaf Material During Developmental/ Age-Dependent Senescence

1. For experiments with *Arabidopsis*, the first leaf pair will be visually monitored from day 14 until day 46. Under controlled growth conditions, the onset of senescence for the wild type occurs within a predictable time window [9]. For rice plants, age-dependent senescence is best studied by following the progression of this process in the leaves of the main culm [6]. As rice and other cereals display reproductive senescence, the yellowing of the upper five leaves of the main culm should be documented from the time point of panicle emergence (heading stage) over a period of 30–60 days.

2. Count the number of yellow leaves in 2-day intervals for *Arabidopsis* and 4-day intervals for rice. A leaf is considered to undergo senescence when more than 10% of the total leaf area has become yellow (*see* **Notes 4** and **5**).

3. To get more accurate estimate of yellow areas, we have developed a simple in-house method of quantifying leaf yellowing

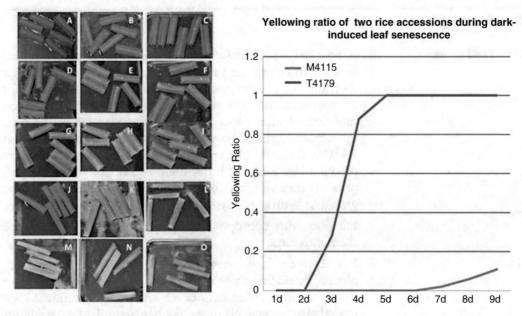

Fig. 2 An example showing the quantification of leaf yellowing areas as the indicator for the onset and progression of leaf senescence. The flag leaves of two rice accessions were harvested at the heading stage and incubated in the deionized water in darkness for up to 9 days and the photos taken. JavaScripts were then used to analyze the leaf blade and the yellow areas in pixels. The data were then converted into ratios and plotted. On the left panel, (**a–i**) represent the photos of 1- to 9-day dark-incubated leaves of rice Accession M4115, respectively, while (**j–o**) represent those of 1- to 6-day dark-incubated leaves of rice Accession T4179, respectively. The right panel shows the change of yellowing areas in the two accessions 1 to 9 day(s) after dark incubation

areas using the imaging analysis approach. Briefly, a high-resolution digital camera is used to take photos of the leaves under observation at various time points. The photos are then processed with JavaScripts (RGB.java and RGB_shape.java) to define the leaf blade and extract the pixels of G (green), R (Red), and B (Blue) channels, which are then used to calculate the yellow area.

4. Calculate for each genotype the percentage of yellow leaves per time point. Subsequently, plot the percentage of yellow leaves against time. As an illustration of the procedure, results for the comparison of two rice genotypes in a detached leaf assay are shown in Fig. 2.

5. The plotted data will reveal if a genotype has an early or late senescence phenotype as compared to the wild type. Based on this information, five time points are selected for sampling leaf material for subsequent biochemical and/or molecular analysis.

3.3 Senescence Phenotyping Using Detached Leaf Assays

1. Fill 6-well plates or Petri dishes with MES solution. Next to that, additional plates can be filled with solutions that contain senescence-inducing agents (*see* **Note 2** and Subheading 2.2).

2. Detach the first leaves of 21-day-old *Arabidopsis*, and immediately place them onto the solution in the 6-well plate with the adaxial side up. For rice, collect fully expanded flag leaves from the main culm, and cut them into pieces of 3 cm, and place them onto the solution with the adaxial side up. In case a biochemical and/or molecular analysis will be done on the leaf material, directly collect material for the zero-hour treatment.

3. Seal the Petri dishes and plates with Parafilm and place them into the appropriate environmental conditions. Commonly, plates are placed under constant light or dark regimes to stimulate the onset of leaf senescence (*see* **Note 6**).

4. Record the progression of senescence by taking pictures of the plates containing the detached leaves at days 0, 1, 3, 5, and 7.

5. Depending on the downstream analysis, harvest leaf material at days 0, 1, 3, 5, and 7. Quickly rinse the leaves twice with water, and directly freeze the material in liquid nitrogen for expression or metabolite analysis. Alternatively, record the fresh weight and proceed with the determination of chlorophyll and ion leakage, or stain the leaves with cell viability markers.

3.4 Chlorophyll and Carotenoid Measurements (See Note 7)

1. Place 50–100 mg fresh leaves in a 2.0 mL micro-centrifuge tubes and add 1.0 mL of 80% acetone.

2. Incubate leaves in the dark for at least 16 h at 4 °C.

3. Transfer the extracted pigments to a 1.0 mL cuvette, and determine the absorption against 80% acetone in a spectrophotometer at 470, 647, and 663 nm.

4. Determine the concentration of pigments as follows:
   Chlorophyll a (μg/mL) = $12.25 \times A_{663.2} - 2.79 \times A_{646.8}$
   Chlorophyll b (μg/mL) = $21.5 \times A_{646.8} - 5.1 \times A_{663.2}$
   Total carotenoids (μg/mL) = $(1000 \times A_{470} - 1.82 \times C_a - 85.02 \times C_b)/198$

5. Plot the measured pigment concentration against time.

3.5 Electrolyte Leakage Assay (See Note 8)

1. Prepare glass tubes with 10 mL deionized water.

2. Fill six tubes with five excised first leaves of *Arabidopsis* or five leaf pieces of rice per genotype and time point.

3. Place the glass tubes in a water bath of 25 °C and incubate for 30 min.

4. Measure the conductivity and record the data.

5. Place the glass tubes into a water bath of 96 °C and boil the leaves for 15 min.

6. After boiling let the glass tubes cool down to 25 °C and measure the conductivity.

7. Calculate the relative electrical conductivity as percentage of the first measurement over the second.

8. Plot the calculated relative conductivity against time.

3.6 Detection of Dead Cells with Trypan Blue (See Note 9)

1. Place excised leaves in 2.0 mL micro-centrifuge tubes and cover them with trypan blue staining solution.

2. Place the tubes at 96 °C for 1 min.

3. Leave the tubes at room temperature for 1–24 h.

4. Remove the staining solution (can be reused several times), and immerse the leaves in chloral hydrate solution for 2–4 h.

5. Replace the chloral hydrate solution and destain the leaves overnight.

6. Remove the chloral hydrate solution and place the leaves in 70% glycerol.

7. Place the destained leaves on a microscope slide and examine the staining with a stereomicroscope.

8. Document the staining by taking pictures.

3.7 Nutrient Measurement (See Note 10)

1. Harvest rosette leaves at various developmental stages or at different intervals of treatment.

2. Dry the samples in oven to constant weight.

3. Ground the samples into fine powder with a mortar and pestle.

4. The samples of 1000–2000 μg were carefully weighed in tin capsules.

5. Use a continuous-flow mass spectrometer to determine total N content.

6. Use ICP-MS (inductively coupled plasma-mass spectrometer) to determine the total content of micronutrients (e.g., Zn and Fe).

3.8 Chloroplast Number per Cell Area (See Note 11)

1. Leaf samples are fixed in 3.5% (v/v) glutaraldehyde for 1 h in the dark.

2. The samples are then washed once by 0.1 M Na_2EDTA, pH 9.0 buffer, and kept in it.

3. Incubate samples in a shaking water bath at 60 °C for 2.5 h, or store at 4 °C until counting.

4. Take pictures by light microscopy. Chloroplasts are counted in two separate planes for each cell to estimate chloroplast at the top and bottom surface of each cell.

5. Cell area and the number of chloroplast are quantified by ImageJ software.

3.9 Chloroplast Morphology by Transmission Electron Microscopy (See Note 12)

1. Leaves are fixed in 5% (v/v) glutaraldehyde in Sorensen's phosphate buffer, pH 7.2, and then sectioned into 1 mm-wide strips while submerged in glutaraldehyde.

2. Samples are dehydrated in 30, 50, 75, 95, and 100% ethanol.

3. Leaf sections are washed four times for 15 min with propylene oxide.

4. The samples are then embedded in Araldite 502 resin and are sectioned in 400 nm slides with a Porter-Blum MT-2 ultramicrotome. The samples are then stained with 0.5% toluidine blue in 1% borax. Thin sections, obtained using an Ultratome Nova, are stained with saturated aqueous lead citrate and uranyl acetate.

5. Transmission electron micrographs are obtained using a Jeol 1200 Ex II electron microscope.

6. Score 50 individual chloroplasts from different senescence stages to determine the distribution of chloroplast types at different stages of senescence.

3.10 Gene Expression Analysis (See Note 13)

1. Isolate total RNA with a Plant RNA extraction kit or TRIzol reagent according to the manufacturer's recommendations. We use 100 mg of fresh weight material for each sample.

2. Use a UV spectrophotometer to determine the RNA concentration by measuring the absorption at 260 nm. Calculate c (mg/mL) = $E_{260} \times 40 \times$ dilution factor.

3. After the RNA isolation, the contaminating DNA needs to be removed by treatment with an RNase-free DNase. After this step, repeat **step 2**.

4. Reverse transcribe 2 µg of total RNA into cDNA using an oligo(dT)18 primer.

5. Run the quantitative real-time PCR. Typical reaction mix (5 µL): 2.5 µL 2× SYBR Green Mastermix, 2.0 µL of 0.5 mM primer pairs, 0.5 µL cDNA (optional, dilute the cDNA prior to qRT-PCR). Cycler conditions: Initial denaturation for 10 min at 95 °C, followed by 45 cycles, comprising denaturation for 10 s at 95 °C, annealing for 20 s at 60 °C, elongation for 30 s at 72 °C, followed by detection.

6. Calculate the relative change in expression by using the comparative threshold cycle (Ct) method [10].

7. Plot the obtained relative expression data against time.

4 Notes

1. Plants can be grown in soil and in hydroponic culture. For rice hydroponic culture, we directly germinate the seeds on the medium after sterilization [11]. One consideration is that the

nutrient composition of the soil or the culture medium has a major effect on the timing of the senescence process [12].

2. For detached leaf assays, it is extremely important to take leaves of the same developmental stage. Leaves that are selected for the assay should not show any sign of senescence. Leaves that are still expanding are more resilient against senescence induction and cannot be used for these assays, especially as the effect of senescence-inducing agents is dependent strongly on the age of the leaf [1, 9].

3. In general we grow *Arabidopsis* plants under long-day conditions (16 h light/8 h dark, 22 °C/18 °C) and rice under even light and dark conditions (12 h light/12 h dark, 26 °C/22 °C).

4. During the counting of the number of yellow leaves, also take note of the total number of leaves, bolting time, and flowering time. The onset of senescence is related to the development of the whole plant, for instance, early senescence mutants often show an early flowering phenotype, while late mutants are often delayed in this aspect.

5. A leaf is often arbitrarily considered as a senescing leaf when the yellowing area is over 10% of the total leaf blade. This normally works fine when different genotypes or time points are compared. One problem with such scoring is that it is qualitative and cannot quantify the differences of yellowing areas. Furthermore, more accurate calculation of yellow areas is needed when subtle differences are compared. Quantification of the yellowing areas at different time points would also generate leaf yellowing progression curves and allow the comparison of the kinetics of senescence progression for different genotypes.

6. In our experience, we have noticed that constant light or darkness stimulates the onset of senescence in some mutants [9], while others remain unaffected. These simple experiments indicate to which extent the onset of leaf senescence is affected in a mutant or genotype.

7. The determination of chlorophyll and carotenoid levels is done as previously described [13]. The amount of each pigment is calculated relative to the fresh weight, implying that the weight of the material needs to be exactly determined.

8. During leaf senescence, the integrity of cellular membranes decreases which is reflected by an increase in electrolyte leakage.

9. Histological staining of leaves is used to reveal cellular and biochemical changes during the progression of the senescence syndrome. For instance, as cell death is the final result of senescence, a staining with trypan blue reveals if senescence progresses still sequentially or if mutants show precocious

activation of the cell death program. Next to trypan blue stains, we commonly stain for the accumulation of ROS, like superoxide (NBT staining) or hydrogen peroxide (DAB staining), as their accumulation is a sign for cell viability and stress.

10. During the senescence processes, proteins are degraded, and nutrients are remobilized from senescing leaves to other developing organs [14].

11. During the senescing process, degradation of chloroplast protein or dismantling of photosynthetic apparatus is very important for nitrogen remobilization. Several studies have shown that during leaf senescence, there is a decrease in the number of chloroplasts per cell, and thylakoid membranes differ totally from those in the green leaf [15, 16].

12. Green leaves have swirling and highly folded thylakoid membranes, while senescing leaves have loosened folding of thylakoid membranes. Furthermore, senescing leaves are often with lots of plastoglobuli [15].

13. For gene expression analysis in *Arabidopsis*, it is important to realize that the favored marker for age-dependent senescence is *SAG12*, whereas *SAG13* and *SAG14* are commonly used for monitoring stress-induced senescence. For rice, so far mainly genes involved in the breakdown of chlorophyll are commonly used (*SG* and *RCCR1*) together with several uncharacterized *SAGs* [6, 17].

References

1. Schippers JHM, Jing HC, Hille J et al (2007) Developmental and hormonal control of leaf senescence. In: Gan S (ed) Senescence processes in plants, vol 26. Blackwell Publishing, Oxford, pp 145–170

2. Graham LE, Schippers JHM, Dijkwel PP et al (2012) Ethylene and senescence processes. In: McManus MT (ed) Annual plant reviews, vol 44. Blackwell Publishing, Oxford, pp 305–341

3. Wu XY, Kuai BK, Jia JZ et al (2012) Regulation of leaf senescence and crop genetic improvement. J Integr Plant Biol 54:936–352

4. Kim JH, Woo HR, Kim J et al (2009) Trifurcate feed-forward regulation of age-dependent cell death involving miR164 in Arabidopsis. Science 323:1053–1057

5. Jing HC, Schippers JHM, Hille J et al (2005) Ethylene-induced leaf senescence depends on age-related changes and OLD genes in Arabidopsis. J Exp Bot 56:2915–2923

6. Liang C, Wang Y, Zhu Y et al (2014) OsNAP connects abscisic acid and leaf senescence by fine-tuning abscisic acid biosynthesis and directly targeting senescence-associated genes in rice. Proc Natl Acad Sci U S A 111:10013–10018

7. Buchanan-Wollaston V, Page T, Harrison E et al (2005) Comparative transcriptome analysis reveals significant differences in gene expression and signalling pathways between developmental and dark/starvation-induced senescence in Arabidopsis. Plant J 42:567–585

8. Boyes DC, Zayed AM, Ascenzi R et al (2001) Growth stage-based phenotypic analysis of Arabidopsis: a model for high throughput functional genomics in plants. Plant Cell 13:1499–1510

9. Jing HC, Sturre MJ, Hille J et al (2002) Arabidopsis onset of leaf death mutants identify a regulatory pathway controlling leaf senescence. Plant J 32:51–63

10. Livak KJ, Schmittgen TD (2008) Analysis of relative gene expression data using real-time quantitative PCR and the 2(-Delta Delta C(T)) method. Methods 25:402–408

11. Schmidt R, Mieulet D, Hubberten HM et al (2013) Salt-responsive ERF1 regulates reactive oxygen species-dependent signaling during the initial response to salt stress in rice. Plant Cell 25:2115–2131

12. Guiboileau A, Yoshimoto K, Soulay F et al (2012) Autophagy machinery controls nitrogen remobilization at the whole-plant level under both limiting and ample nitrate conditions in Arabidopsis. New Phytol 194:732–740

13. Wellburn AR (1994) The spectral determination of chlorophylls a and b, as well as total carotenoids, using various solvents with spectrophotometers of different resolution. J Plant Physiol 144:307–313

14. Uauy C, Distelfeld A, Fahima T, Blechl A, Dubcovsky J (2006) A NAC gene regulating senescence improves grain protein, zinc, and iron content in wheat. Science 314: 1298–1301

15. Evans IM, Rus AM, Belanger EM, Kimoto M, Brusslan JA (2010) Dismantling of Arabidopsis thaliana mesophyll cell chloroplasts during natural leaf senescence. Plant Biol 12:1–12

16. Pyke KA, Leech RM (1991) Rapid image-analysis screening-procedure for identifying chloroplast number mutants in mesophyll-cells of Arabidopsis-Thaliana (L) Heynh. Plant Physiol 96:1193–1195

17. Zhou Q, Yu Q, Wang Z et al (2013) Knockdown of GDCH gene reveals reactive oxygen species-induced leaf senescence in rice. Plant Cell Environ 36:1476–1489

Chapter 4

Investigation of Petal Senescence by TRV-Mediated Virus-Induced Gene Silencing in Rose

Chenxia Cheng, Junping Gao, and Nan Ma

Abstract

The classic reverse genetic screening, such as EMS-induced or T-DNA-mediated mutation, is a powerful tool to identify senescence-related genes in many model plants. For most non-model plants, however, this strategy is hard to achieve. Even for model plants, construction of a mutant library is usually labor and time-consuming. Virus-induced gene silencing (VIGS) provides an alternative to characterize gene function in a wide spectrum of plants through transient gene expression. To date, more than a dozen of VIGS vector systems have been developed from different RNA and DNA viruses, while *Tobacco rattle virus* (TRV) system might be one of the most used due to its wide host range and ease of use. Here, we describe a modified TRV vector, TRV-GFP, in which a green fluorescent protein (GFP) is fused to 3'-end of the coat protein (CP) gene in the TRV2 vector. Since the GFP-tagged CP protein could be traced under UV light *in planta*, identification of TRV-GFP-infected plants is easy. Application of this system in identifying genes regulating petal senescence in rose is described.

Key words Virus-induced gene silencing, Petal senescence, Rose, Functional analysis, pTRV, pTRV-GFP, Vacuum infiltration

1 Introduction

Senescence is an irreversible and highly programmed process in plants. For all horticultural crops, investigation of the senescence mechanism of economic organs is crucial for reducing postharvest loss. Screening and functional identification of senescence-related genes is an efficient way to understand the mechanism of organ senescence. However, classic genetic methods, such as genetic mapping and construction of mutant library, are usually more difficult to conduct in horticultural crops. In the last decade, the virus-induced gene silencing (VIGS) system has been broadly used in investigating gene function in many plant species.

VIGS is a powerful tool based on RNA-mediated innate plant defense system related to posttranscriptional gene silencing (PTGS) that has been widely used to study gene function in many plants

Yongfeng Guo (ed.), *Plant Senescence: Methods and Protocols*, Methods in Molecular Biology, vol. 1744,
https://doi.org/10.1007/978-1-4939-7672-0_4, © Springer Science+Business Media, LLC 2018

[1–5]. The PTGS process results in downregulation of a gene at the RNA level (i.e., after transcription) [6]. PTGS can be triggered by viral (double-stranded RNA) dsRNA, an intermediate step in viral replication [7]. After the virus infects the host plant, viral ssRNA is converted to dsRNA by an endogenous RNA-dependent RNA polymerase (RdRP) [4]. The dsRNA is then cleaved into 21–25 nt small-interfering RNA (siRNA) by DICER-like enzymes [4, 8]. The resulting short double-stranded products enter a RISC assembly pathway, and only one of the two strands is loaded into RISC to direct target recognition and degradation via sequence complementary [6, 9].

To date, several VIGS vectors have been developed for gene silencing in various plant species [7, 10], including vectors using *Apple latent spherical virus* (ALSV) [11], *Barley stripe mosaic virus* (BSMV) [12], *Potato virus X* (PVX) [13], and *Tobacco rattle virus* (TRV) [5, 14]. Among these, TRV provides the most robust results in terms of efficiency, ease of application, and milder virus-infected symptoms of plants [7]. TRV is a RNA virus with wide host range including *Nicotiana benthamiana* [7, 15], *Arabidopsis* [16], tomato [14], pepper [17], petunia [18], rose [19], *Aquilegia* [20], *Thalictrum* [21], strawberry [22], *Eschscholzia californica* [23], *Jatropha curcas* [24], *Mirabilis jalapa* [25], cotton [26], and *Paeonia lactiflora* [27]. Moreover, TRV can infect a variety of tissues and organs of plant including leaves [7, 15], flowers [18, 19, 21], roots [28], and fruits [14, 22].

TRV is a positive-strand RNA virus with a bipartite genome (RNA1 and 2). RNA1 is composed of 134 and 194 kDa replicase proteins, a 29 kDa movement protein, and a 16 kDa cysteine-rich protein whose function is not fully understood [7, 29]. RNA1 is fundamental for viral movement and replication in plants. RNA2 consists of the coat protein (CP) and nonstructural proteins (29.4 kDa protein and 32.8 kDa protein) [7]. The nonstructural proteins are not essential for plant infection in laboratory conditions. In the engineered TRV vector system, RNA1 and RNA2 are used to create two plasmids, namely, pTRV1 and pTRV2, both of which are required for plant infection [29, 30]. In the pTRV2 construct, the nonessential genes are replaced with a multiple cloning site (MCS) to carry the target fragment [1]. For the VIGS approach, a reporter gene is needed to indicate if the VIGS is working and whether the target gene is silenced. In the last decade, the *phytoene desaturase* (*PDS*) gene has been widely used as a reporter gene [14, 15]. *PDS* encodes a rate-limiting enzyme in carotene biosynthesis, and silencing of *PDS* produces characteristic photobleaching symptoms of host plants due to inhibition of carotene biosynthesis. However, silencing *PDS* also causes chloroplast damage, which substantially inhibits photosynthesis and impairs normal growth and development of host plants. In addition, it takes at least 7–10 days after TRV inoculation to observe the typical photobleaching symptoms. In some non-Solanaceae plants, such as rose (*Rosa* sp.), it needs more than 40 days to obtain similar photobleaching symptoms [5].

Therefore, we recently developed a modified pTRV vector, pTRV-GFP, by tagging a *GFP* CDS to the 3′-end of *CP* gene in pTRV2. GFP-tagged TRV virus could be traced by noninvasive monitoring with a portable UV lamp. Although movement of GFP-tagged TRV virus is somehow slower than the original TRV virus, the efficiency of TRV-GFP is the same as the original one. Furthermore, the GFP indicator does not affect normal growth and development of the host plants. Using the pTRV-GFP vector, we have successfully silenced several endogenous genes in a variety of plant species including *N. benthamiana*, *A. thaliana*, rose, strawberry (*Fragaria ananassa*), and chrysanthemum (*Dendranthema grandiflorum*) [5]. In this chapter, we describe the modified protocol for pTRV-GFP-mediated VIGS in functional analysis of senescence-related genes in rose plantlets.

2 Materials

2.1 Plant Cultivation

1. Rose (*R. hybrida* cv. Samantha) plantlets are propagated by in vitro cultivation.

2. Shoots propagation medium: Murashige and Skoog (MS) medium containing 1.0 mg/L 6-benzylaminopurine (6-BA), 0.05 mg/L α-naphthalene acetic acid (NAA), and 1.0 mg/L gibberellic acid (GA$_3$).

3. Rooting medium: 1/2-strength MS containing 0.1 mg/L NAA.

4. Plastic pots containing the mixture of 1:1 (v/v) peat/vermiculite.

2.2 Cloning and Vector Construction

1. For amplification of target fragment: Phanta® Max Super-Fidelity DNA Polymerase (Vazyme Biotech Co., Ltd., cat. no. P505, Nanjing, Jiangsu, China), specific forward and reverse primers (10 mM for each), dNTP mixture (10 mM for each), and Milli-Q water.

2. For DNA cleanup: PCR Purification Combo Kit (BioTeke, cat. no. DP1502, Beijing, China), Milli-Q water, appropriate percentage of agarose in TAE 1× (40 mM Tris-acetate and 1 mM Na 2EDTA).

3. For DNA concentration test: NanoDrop spectrophotometer.

4. For construction of VIGS vector: TRV-based silencing vectors pTRV1, pTRV2 (kindly provided by Dr. Yule Liu at Tsinghua University), and modified pTRV2-GFP (Fig. 1) [5], restriction enzymes, DNA Ligation Kit.

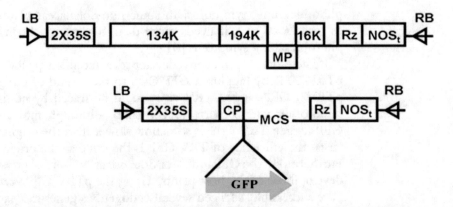

Fig. 1 Construction of the pTRV2-GFP vector. (**a**) Schematic diagram of the pTRV-GFP vector. The pTRV 5′- and 3′-untranslated regions (UTRs) are indicated by lines. Open boxes, TRV 2 × 35S promoter, Rz (self-cleaving ribozyme), nopaline synthase terminator (NOSt), and CP. Green arrow, GFP cistron. The figure is from Tian et al. (2014) [5] with modification

5. For *Escherichia coli* transformation: Trans5α Chemically Competent Cell (TransGen Biotech Co., Ltd., cat. no. CD201-01, Beijing, China), sterile (Luria-Bertani) LB liquid medium (1% tryptone, 0.5% yeast extract, 1% NaCl), and solid LB agar plates (1% tryptone, 0.5% yeast extract, 1% mM NaCl, 1% agar) containing 50 μg/mL kanamycin antibiotics for VIGS vectors.

6. For plasmid extraction from *E. coli*: AxyPrep™ Plasmid Miniprep Kit (Axygen®, cat. no. AP-MN-P-250, Union, CA, USA).

7. For stocking of *E. coli*: Sterile Eppendorf tubes, 50% sterile glycerol, liquid nitrogen, and −80 °C refrigerator.

2.3 Agrobacterium Transformation and Infiltration Preparation

1. For Agrobacterium transformation: Eppendorf tubes containing 40 μL of *A. tumefaciens* strain GV3101 competent cells stored at −80 °C (*see* **Note 1**), DK-S22 water bath. Sterile LB liquid medium and LB agar plates containing 50 μg/mL kanamycin and 25 μg/mL rifampicin antibiotics.

2. For Agrobacterium glycerol stocks: Sterile Eppendorf tubes, 50% sterile glycerol, liquid nitrogen, and −80 °C refrigerator.

3. For Agrobacterium GV3101 cultivation: 500 mL beaker flasks.

4. Spectrophotometer.

2.4 Agroinfiltration

1. LB liquid medium with kanamycin (50 μg/mL) and rifampicin (25 μg/mL).

2. Subculture medium: LB liquid medium containing 25 μg/mL rifampicin, 50 μg/mL kanamycin, 10 mM MES, and 20 mM acetosyringone.

3. Infiltration buffer (*see* **Notes 2** and **3**): 10 mM 2-(N-morpholino) ethanesulfonic acid (MES), 10 mM MgCl$_2$, 200 μM aceto-syringone, pH 5.6 (adjusted with 1 M KOH), 0.01% (v/v) Silwet L-77.

4. Rotary-vane vacuum pump.

5. Vacuum dryer.

2.5 Validation of Gene Silencing

1. For total RNA extraction: Extraction buffer (200 mM sodium tetraborate decahydrate, 30 mM EGTA, 1% deoxycholic acid sodium salt, 10 mM DTT, 2% PVP 40, 1% NP-40), Proteinase K, 2 M KCl$_2$, 8 M LiCl, 2 M LiCl, 1 M TRIS-HCl (pH 7.5), 2 M K-acetate (pH 5.5), ethanol, and RNA-free Milli-Q water.

2. For DNA digestion: Recombinant DNase I (RNase-free) (Takara Biomedical Technology Co., Ltd., cat. no. 2270A, Beijing, China) and Recombinant RNase Inhibitor (Takara Biomedical Technology Co., Ltd., cat. no. 2313A, Beijing, China).

3. For RT-PCR: Reverse Transcriptase M-MLV (RNase H) (Takara Biomedical Technology Co., Ltd., cat. no. 2641A, Beijing, China), 2× Taq MasterMix (Beijing ComWin Biotech Co., Ltd., cat. no. CW0682A, Beijing, China), AlphaImager 2200.

4. For Western blotting: GFP-specific antibody (Abcam Inc.), protein extraction buffer (100 mM TRIS, pH 6.8, 2.5% SDS, 100 mM dithiothreitol (DTT), 100 mM NaCl, 10% glycerol), polyvinylidene difluoride membrane, chemiluminescent substrate, X-ray films.

2.6 GFP Imaging

1. 100 W hand-held long-wave ultraviolet lamp (UV products, Upland, CA, USA; Black Ray model B 100AP/R).

2. Kodak wratten filter 15.

2.7 Phenotyping of Silenced Plants

1. Digital camera.

2. Colorimetric card.

2.8 Anthocyanin Analysis

1. Methanol: Containing 2% formic acid.

2. 0.22 μm reinforced nylon membrane filter (ANPEL, Shanghai, China).

3. Dionex system: Equipped with an UltiMate3000 HPLC pump, an UltiMate 3000 autosampler, a TCC-100 thermo-stated column compartment, and a PDA100 photodiode array detector (Thermo Fisher Scientific Inc., Waltham, MA, USA).

4. ODS-80Ts QA C18 column (250 mm length × 4.6 mm inner diameter, Tosoh Bioscience Shanghai Co., LTD., Shanghai, China).

5. Eluent A: 1% formic acid aqueous solution.

6. Eluent B: 15% methanol in acetonitrile, cyanidin 3-glucoside.

7. Anthocyanin standard: Cyanidin 3-glucoside (Sigma).

2.9 Senescence Marker Gene Analysis

1. For total RNA extraction: *see* Subheading 2.5, **item 1**.

2. For DNA digestion: *see* Subheading 2.5, **item 2**.

3. For qRT-PCR: Reverse Transcriptase M-MLV (RNase H-) (Takara Biomedical Technology Co., Ltd., cat. no. 2641A, Beijing, China), GoTaq® qPCR Master Mix (Promega Biotech Co., Ltd., cat. no. A6001, Beijing, China), Step One Plus™ real-time PCR system.

2.10 Electrolyte Leakage Measurement

1. Hole punch.

2. 0.4 M mannitol.

3. Conductivity meter.

3 Methods

3.1 Plant Growth

Rose (*R. hybrida* cv. Samantha) plantlets are propagated by tissue culture.

1. Rose stem fragments (at least 2 cm in length, including 1 node) are cultured on shoot propagation medium for 30 days to get plantlets.

2. Transfer the plantlets to rooting medium for 30 days.

3. After vacuum infiltration, infected plants are planted in plastic pots containing a mixture of 1:1 (v/v) peat/vermiculite at 23 °C, relative humidity 60–70%, 16 h/8 h photoperiod.

3.2 Vector Construction

1. Design primers for targeting the gene of interest (GOI) (*see* **Note 4**). Typically, three to four nucleotides are added to the 5′-end of the restriction enzyme site in the primers for efficient digestion. The optimal fragment length is from 200 to 400 bp [31], and the possibility of off-targets during PTGS could be predicted using the siRNA scan program (http://plantgrn. noble.org/pssRNAit/) [7] (*see* **Notes 5** and **6**).

2. Amplify the GOI fragment by PCR reaction, separate the PCR product by agarose gel electrophoresis, and cut the interested band from the gel under UV light with a clean, sharp scalpel. Purify the DNA fragment using PCR Purification Combo Kit according to the manufacturer's instruction. Quantify the DNA by NanoDrop spectrophotometry.

3. Make the pTRV2-*GOI*/pTRV2-GFP-*GOI* constructs. The DNA fragments from pTRV2 or pTRV2-GFP are released by restriction enzyme digestion (*see* **Note 7**). Stop the digestion

reactions at 85 °C for 10 min. For ligation reactions, prepare a new PCR tube and mix the digested *GOI* fragment and pTRV2/pTRV2-GFP vector, and add the ligation mix according to the manufacturer's instruction. Incubate at 25 °C for 5 min.

4. Transform *E. coli* with the pTRV2-*GOI*/pTRV2-GFP-*GOI* vectors. Take a tube of chemically competent *E. coli* Trans5α cells from −80 °C and thaw it on ice, add 5 μL of ligation products of pTRV2-*GOI*/pTRV2-GFP-*GOI* to 35 μL competent cells, incubate without shaking, and perform transformation following manufacturer's instruction. Pipette 50 μL transformed bacterial culture, and spread on a LB plate containing 50 μg/mL kanamycin. Incubate the plate overnight at 37 °C.

5. Validate the pTRV2-*GOI*/pTRV2-GFP-*GOI* vectors. Pick four to six single and isolated colonies from the overnight plate, and use each colony to inoculate 5 mL of LB broth with 50 μL/mL kanamycin in a 50 mL test tube. Incubate at 37 °C overnight with shaking at 180 rpm. Centrifuge the cells at 5000 × *g* for 5 min and pour off the supernatant. The pellet is used to extract plasmid DNA using the AxyPrep™ Plasmid Miniprep Kit. Validate the pTRV2-*GOI*/pTRV2-GFP-*GOI* vector by restriction enzyme digestion analysis and by DNA sequencing using pTRV2 specific primers (*see* **Note 8**). Once validated, a colony with correct insertions is used to generate a glycerol stock for future use (*see* Subheading 2.2, **item 7**).

3.3 Agrobacterium Transformation

1. Prepare plasmid DNA for pTRV2-*GOI*/pTRV2-GFP-*GOI* (silencing construct) or pTRV2 (vector control) and pTRV1 constructs using the AxyPrep™ Plasmid Miniprep Kit.

2. Transform Agrobacterium with individual pTRV construct. Use 50 μL Agrobacterium GV3101 competent cells for each pTRV construct. Thaw competent cells on ice. Add 5 μL of DNA plasmid to each tube. Mix by gently tapping with your index finger. Freeze the mixture in liquid nitrogen for 5 min, and then thaw them for 5 min at 37 °C in water bath. Chill the mixture for 5 min on ice. Add 500 μL of LB broth to each tube, and incubate for ~3 h at 28 °C with shaking at 180 rpm. Then, collect the bacteria by spinning at 5000 × *g* for 5 min. Pour off the supernatant, and resuspend the pellet in 100 μL fresh LB broth. Spread each culture on LB agar plates containing rifampicin (25 μg/mL) and kanamycin (50 μg/mL). Incubate the plates for at least 48 h at 28 °C in dark.

3. Validate the Agrobacterium clones. For each construct, pick four colonies from selective plates, and use them to inoculate 5 mL of LB broth containing kanamycin (50 μg/mL) and rifampicin (25 μg/mL). Incubate at 28 °C with shaking at 180 rpm for 24 h. Take 1 μL culture from each tube as template to perform PCR to identify positive colonies. Centrifuge

the cells at 5000 × g for 5 min and pour off the supernatant. The pellet is used to extract plasmid DNA. The vectors are validated by restriction digestion and DNA sequencing. Once validated, a correct colony is used to generate a glycerol stock for future use (*see* **Note 9**).

3.4 Agroinfiltration

1. Grow the Agrobacterium strain GV3101 containing pTRV2-*GOI*/pTRV2-GFP-*GOI* or pTRV1 vectors. Take the bacterial stocks from −80 °C, and put each stock into a 50 mL plastic tube containing 5 mL of LB liquid medium with kanamycin (50 μg/mL) and rifampicin (25 μg/mL). Grow them with shaking at 28 °C for ~ 24 h.

2. Put 1 mL of the bacterial cultures into 100 mL subculture medium, and incubate at 28 °C with shaking at 180 rpm for 10~12 h. Collect the Agrobacterium cells by centrifugation at 4000 × g for 15 min. Discard the supernatant and resuspend the pellet gently with ~3 mL infiltration buffer (*see* **Note 10**). Measure the absorbance of resulted solution with a spectrophotometer. Adjust the absorbance of each culture with infiltration buffer to the same reading of OD600, which may vary for different infection methods (*see* **Note 11**). Mix the Agrobacterium cultures containing pTRV1 and pTRV2/pTRV2-GFP or its derivatives in a ratio of 1:1 (v/v), and adjust the final OD600 to 0.8–1.5 (*see* **Note 11**). The mixture of Agrobacterium cultures is placed at room temperature in dark for 4–6 h before inoculation.

3. Take 30-day-old rooted rose plantlets from the rooting medium. Rinse the roots with distilled water to remove the medium, and briefly dry the plantlets with paper towel. Submerge whole plantlets in infiltration buffer containing pTRV1 and pTRV2/pTRV2-GFP or its derivatives (OD600 = ~1.0), and subject to vacuum at −0.8 kg/cm² twice, each for 60 s. Briefly wash the plantlets with distilled water, then transplant them into plastic pot containing a mixture of 1:1 (v/v) peat/vermiculite, and grow them at 23 °C, relative humidity 60–70%, and 16 h/8 h photoperiod. Cover the pots with plastic film to keep humidity for 5–7 days (*see* **Note 12**).

4. Grow the inoculated plants in the greenhouse at 21 ± 2 °C, relative humidity of ~60%, and 16 h day/8 h night until the observation period is reached (*see* **Note 13**). Higher temperature (above 25 °C) is harmful for VIGS.

5. Check phenotypes of the plants starting from 35 days after inoculation.

3.5 Evaluation of Gene Silencing

1. Collect top young leaves from the agroinfiltrated plants 30 days postinoculation.

2. Transfer the plant samples quickly to 2 mL capped tubes and freeze in liquid nitrogen. Store at −80 °C until use.

3. Extract total RNA from rose samples using the hot borate method as previously described [32]. Preheat the extraction buffer to 86 °C in a water bath before use. Grind the samples up in a mortar with liquid nitrogen to obtain a fine frozen powder. Add hot extraction buffer quickly to the samples in Eppendorf tubes and vortex it completely. Incubate the mixture on ice for 30 min. Add approximately 2–10 mg/mL Proteinase K into each tube. Incubate the tubes at 42 °C with gentle shaking in a shaking water bath for 1.5–2.5 h. Add 2 M KCl to a final concentration of 160 mM. Incubate on ice for 1 h or more. Precipitate RNA using 1/3 volume of 8 M LiCl to a final concentration of 2 M. Incubate the tubes on ice at 4 °C overnight. Rewash the pellet twice using 2 M LiCl, and then dissolve it in 1 M TRIS-HCl (pH 7.5). Precipitate the RNA with ethanol at −80 °C for 2 h. Dissolve the RNA pellet in 30 μL RNA-free water.

4. Synthesize first-strand cDNA using 1 μg total RNA with oligo (dT) (for target genes) or random primer (for *CP-GFP*).

5. Conduct semiquantitative reverse transcription-PCR (RT-PCR) for each sample. The PCR products are separated on a 1.2% agarose gel, and the images are scanned and analyzed using an AlphaImager 2200.

6. GFP-specific antibodies are used to detect the CP-GFP fused protein. Total protein is extracted from leaves of individual rose plant with 300 μL protein extraction buffer. Quantify protein content by Bradford assays. Proteins (10 mg from each sample) are separated by 10% SDS-PAGE and then transferred to a polyvinylidene difluoride membrane. CP-GFP is detected after an overnight incubation at room temperature with a 1:10,000 dilution of the anti-GFP antibodies conjugated to alkaline phosphatase. Detect alkaline phosphatase using a chemiluminescent substrate and expose to X-ray films (Fig. 2).

3.6 GFP Imaging

For whole-mount visualization, the plants are illuminated with a 100 W portable long-wave ultraviolet lamp and are photographed with a Kodak wratten filter 15 (Fig. 2).

3.7 Phenotyping of Target Gene-Silenced Plants

1. Record the flower image daily with a digital camera.

2. Color fading is an important sign for petal senescence. Use the colorimetric card to trace the color changes (Fig. 3).

3.8 Anthocyanin Analysis

1. Extract anthocyanins from petal materials using 5 mL methanol containing 2% formic acid as described [33].

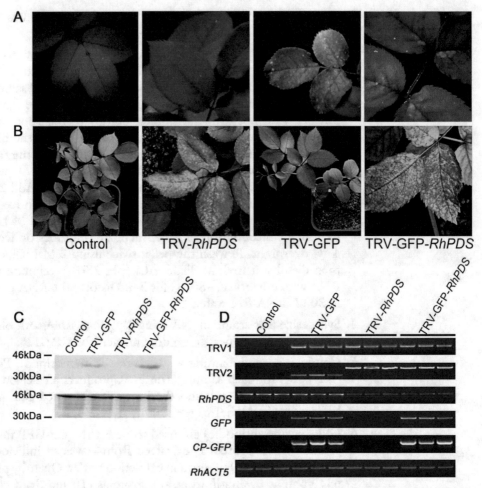

Fig. 2 Validation of the pTRV-GFP vector in rose plants. (**a, b**) pTRV- and pTRV-GFP- infiltrated rose plants. The plants were photographed under UV illumination (upper) and normal light (bottom) at 30 and 45 d after inoculation, respectively. (**c**) CP-GFP protein levels in the upper leaves 30 d after inoculation. Coomassie blue staining was used for confirmation of equal loading in each lane. (**d**) Semiquantitative RT-PCR of TRV1, TRV2, *RhPDS*, GFP, and CP-GFP in control and inoculated plants. *RhACT5* was used as an internal control. The figure is from Tian et al. (2014) [5] with modification

2. Filter the extract through a 0.22 μm reinforced nylon membrane filter for HPLC-diode array detection and HPLC-MS analyses. Replicates of each sample are analyzed in triplicate.

3. The chromatographic separation is performed on a Dionex system equipped with an UltiMate3000 HPLC pump, an UltiMate 3000 autosampler, a TCC-100 thermostated column compartment, and a PDA100 photodiode array detector. A 20 μL aliquot of extract is injected and analyzed in an ODS-80Ts QA C18 column protected with a C18 guard cartridge. A gradient elution with the following composition is used: 5%

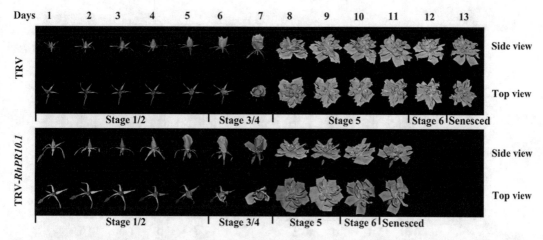

Fig. 3 Flower senescence phenotypes of *RhPR10.1*-silenced rose plants. Flowers were recorded every day. The figure is from Wu et al. (2016) [35] with modification

Eluent B at 0 min, 35% Eluent B at 25 min, and 5% Eluent B at 30 min. The flow rate is 0.8 mL min-1, and chromatograms are acquired at 525 nm for anthocyanins, with DAD data being recorded from 200 to 800 nm.

4. The total content of anthocyanins in each sample is measured semiquantitatively by linear regression of the commercially available standard cyanidin 3-glucoside (Fig. 4).

3.9 Analysis of the Expression of Senescence Maker Genes

1. Isolate total RNA and do reverse transcription as in Subheading 3.5, **step 3**, and Subheading 3.5, **step 4**.

2. For qRT-PCR of *RhSAG12*: qRT-PCR reactions are performed using 1 mL cDNA as template and the Step One Plus real-time PCR system with GoTaq® qPCR Master Mix. *RhACT5* is used as an internal control [34] (*see* **Note 14**).

3.10 Measurement of Electrolyte Leakage Rates

1. Collect rose petals from the same layer of control plants and gene-silenced plants, respectively. Excise 1 cm diameter discs from the center of the petals using a hole punch.

2. Membrane leakage is determined by measurement of electrolytes leaking from rose petal discs [35]. Immerse sixteen rose petal discs from each treatment in 15 mL of 0.4 M mannitol at room temperature with gentle shaking for 3 h, and measure initial conductivity of the solution with a conductivity meter. Total conductivity was determined after sample incubation at 85 °C for 20 min.

3. Calculate the electrolyte leakage rates as the percentage of initial conductivity divided by total conductivity (Fig. 5).

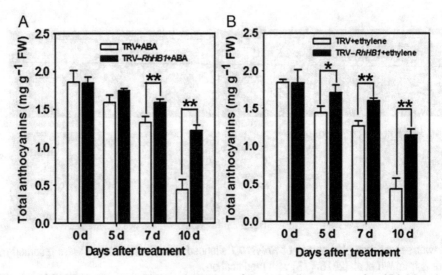

Fig. 4 Silencing of *RhHB1* increases anthocyanin content of the petal discs in rose after ABA (**a**) and ethylene (**b**) treatments. *RhHB1* was silenced in petal discs by VIGS, and the discs were treated with 100 μM abscisic acid (ABA) or 10 μL/L ethylene for 24 h. Values are means SD ($n = 3$). Asterisks indicate statistically significant differences (*$p < 0.05$, **$p < 0.01$, student's t test). The figure is from Lü et al. (2014) [33] with modification

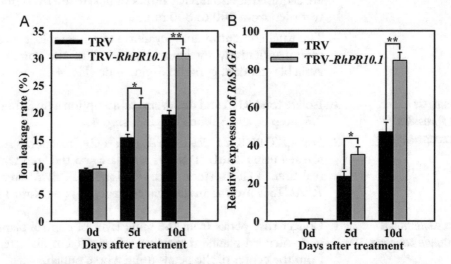

Fig. 5 Silencing of *RhPR10.1* promotes flower senescence in rose. (**a**) Ion leakage rate and (**b**) expression of *RhSAG12* in *RhPR10.1*-silenced and pTRV control petals. The results are the means of three biological replicates with standard deviations. Asterisks indicate significant differences (student's t test, *$p < 0.05$, **$p < 0.01$). The figure is from Wu et al. (2016) [35] with modification

4   Notes

1. Other disarmed Agrobacterium strains, such as EHA105 and LBA4404, can also be used.

2. The infiltration buffer should be freshly prepared using stock solutions. (a) 1 M MES (100×) pH 6.3: Dissolve MES in ster-

ile water and adjust the pH with KOH 1 M. Require sterilization by filtration and split in 10 mL aliquots and store at
−20 °C. (b) 1 M MgCl$_2$ (100×): Dissolve in sterile water, and
sterilize by filtration. (c) 50 mM acetosyringone solution:
Dissolve 78.48 mg acetosyringone in 2 mL of dimethyl sulfoxide (DMSO) and filter-sterilize. Split in 200 µL aliquots and
store at −20 °C. *Important*: DMSO should be handled in a
fume cupboard, and protective clothing should be worn when
handling it. Use pipette tips that are safe to handle solvents or
use glass pipettes.

3. MES and acetosyringone are photosensitive and need to be
   stored under darkness. Tubes containing MES or acetosyringone stocks should be wrapped with aluminum foil.

4. The target fragment could be designed in a conserved coding
   region (for silencing multiple homogenous genes) or in an
   untranslated region (UTR) (for silencing a certain gene specifically) of the gene of interest.

5. The optimal length of target fragment for efficient VIGS is
   200–400 bp with minimal off-target effects. Insert lengths
   should be in the range of ~200 to ~1300 bp [31]. Inserts
   larger than 400 bp would increase the chance of off-target
   silencing [31]. Homopolymeric region (i.e., poly(A/T) tail)
   sequences should be avoided.

6. It is critical to confirm that the selected fragment could not
   produce similar 21 nt siRNAs that can match other genes. This
   can be tested by using the Basic Local Alignment Search Tool
   (BLAST) or other siRNA prediction software.

7. We use SnapGene® 2.3.2 (2017© GSL Biotech LLC, Carlsbad,
   CA, USA) program to choose the suitable restriction enzymes.

8. pTRV2 vector forward primer, TGGGAGATGATACGCTGTT;
   pTRV2 vector reverse primer, CCTAAAACTTCAGACACG.

9. Individual bacterial colony carrying each construct can be
   streaked on an LB agar plate containing antibiotics and grown
   as described in this step. After bacterial growth, these plates
   can be stored for 1 week at 4 °C. Liquid cultures of bacteria
   can be stored as glycerol stocks up to 1 year at −80 °C. pTRV1
   stock is less stable in storage and make sure you validate the
   glycerol stock before use.

10. It is crucial to resuspend the cells completely with the infiltra-
    tion buffer. Avoid using dead cells by harvesting the cells at the
    logarithmic growth phase of the bacteria.

11. OD600 of the final infiltration mixture need to be optimized
    for each single experiment, depending on the target plant spe-
    cies and infiltration methods. For rose, the recommended
    range for the vacuum infiltration is ~1.0.

12. During the period of covering with films, make sure the inside temperature is not higher than 25 °C to avoid reducing of the VIGS efficiency.

13. The efficiency of VIGS is reduced at higher (above 25 °C) temperature. Temperature higher than 27 °C is not suitable for VIGS experiments [36].

14. *SAG12* is a typical senescence-related gene in plants. The primers for qRT-PCR of *RhSAG12* are forward 5'-AGCGGAGAA GCCTTTCAGTC-3' and reverse 5'-CAGCATGGT TCAGGC TGGTA-3'; the primers for *RhACT5* are forward 5'-GAGC GTTTCAGATGCCC AGA-3' and reverse 5'-TGGTGGGGC AACCACCTTA-3'.

Acknowledgments

We would thank Dr. Yule Liu (Tsinghua University) for providing the original pTRV vector and for his excellent advices. We would thank Dr. Daqi Fu (China Agricultural University) for his kind and valuable help for the improvement of VIGS approach.

References

1. Ratcliff F, Martin-Hernandez AM, Baulcombe DC (2001) Technical advance: tobacco rattle virus as a vector for analysis of gene function by silencing. Plant J 25:237–245

2. Shao Y, Zhu H, Tian H et al (2008) Virus-induced gene silencing in plant species. Russ J Plant Physiol 55:168–174

3. Baulcombe D (1999) Viruses and gene silencing in plants. In: 100 years of virology. Springer, New York, pp 189–201

4. Pflieger S, Richard MM, Blanchet S et al (2013) VIGS technology: an attractive tool for functional genomics studies in legumes. Funct Plant Biol 40:1234–1248

5. Tian J, Pei H, Zhang S et al (2014) TRV-GFP: a modified *Tobacco rattle virus* vector for efficient and visualizable analysis of gene function. J Exp Bot 65:311–322

6. Vaucheret H, Béclin C, Fagard M (2001) Post-transcriptional gene silencing in plants. J Cell Sci 114:3083–3091

7. Senthil-Kumar M, Mysore KS (2014) Tobacco rattle virus-based virus-induced gene silencing in *Nicotiana benthamiana*. Nat Protoc 9: 1549–1562

8. Ding S-W, Voinnet O (2007) Antiviral immunity directed by small RNAs. Cell 130:413–426

9. Baulcombe D (2004) RNA silencing in plants. Nature 431:356–363

10. Becker A, Lange M (2010) VIGS-genomics goes functional. Trends Plant Sci 15:1–4

11. Igarashi A, Yamagata K, Sugai T et al (2009) Apple latent spherical virus vectors for reliable and effective virus-induced gene silencing among a broad range of plants including tobacco, tomato, *Arabidopsis thaliana*, cucurbits, and legumes. Virology 386:407–416

12. Scofield SR, Huang L, Brandt AS et al (2005) Development of a virus-induced gene-silencing system for hexaploid wheat and its use in functional analysis of the *Lr21*-mediated leaf rust resistance pathway. Plant Physiol 138: 2165–2173

13. Faivre-Rampant O, Gilroy EM, Hrubikova K et al (2004) Potato virus X-induced gene silencing in leaves and tubers of potato. Plant Physiol 134:1308–1316

14. Liu Y, Schiff M, Dinesh-Kumar S (2002) Virus-induced gene silencing in tomato. Plant J 31:777–786

15. Liu Y, Schiff M, Marathe R et al (2002) Tobacco *Rar1*, *EDS1* and *NPR1/NIM1* like genes are required for N-mediated resistance to tobacco mosaic virus. Plant J 30:415–429

16. Wang C, Cai X, Wang X et al (2006) Optimisation of tobacco rattle virus-induced gene silencing in *Arabidopsis*. Funct Plant Biol 33:347–355

17. Chung E, Seong E, Kim YC et al (2004) A method of high frequency virus induced gene silencing in chili pepper (*Capsicum annuum* L. cv. Bukang). Mol Cells 17:377–380

18. Chen J-C, Jiang C-Z, Gookin T et al (2004) Chalcone synthase as a reporter in virus-induced gene silencing studies of flower senescence. Plant Mol Biol 55:521–530

19. Ma N, Xue J, Li Y et al (2008) *Rh-PIP2; 1*, a rose aquaporin gene, is involved in ethylene-regulated petal expansion. Plant Physiol 148:894–907

20. Gould B, Kramer EM (2007) Virus-induced gene silencing as a tool for functional analyses in the emerging model plant *Aquilegia* (columbine, Ranunculaceae). Plant Methods 3:6

21. Di Stilio VS, Kumar RA, Oddone AM et al (2010) Virus-induced gene silencing as a tool for comparative functional studies in *Thalictrum*. PLoS One 5:e12064

22. Jia H, Chai Y, Li C et al (2011) Abscisic acid plays an important role in the regulation of strawberry fruit ripening. Plant Physiol 157:188–199

23. Wege S, Scholz A, Gleissberg S et al (2007) Highly efficient virus-induced gene silencing (VIGS) in California poppy (*Eschscholzia californica*): an evaluation of VIGS as a strategy to obtain functional data from non-model plants. Ann Bot 100:641–649

24. Ye J, Qu J, Bui HTN et al (2009) Rapid analysis of *Jatropha curcas* gene functions by virus-induced gene silencing. Plant Biotechnol J 7:964–976

25. Singh A, Liang Y-C, Kumar P et al (2012) Co-silencing of the *Mirabilis* antiviral protein (MAP) permits virus-induced gene silencing (VIGS) of other genes in four O'Clock plants (*Mirabilis jalapa*). J Hortic Sci Biotech 87:334–340

26. Gao X, Britt RC Jr, Shan L et al (2011) Agrobacterium-mediated virus-induced gene silencing assay in cotton. J Vis Exp 54:2938

27. He Z, Chen C (2016) First report of tobacco rattle virus infecting chinese herbaceous peony (*Paeonia lactiflora* Pall.) in China. Plant Dis 100:2543–2543

28. Valentine T, Shaw J, Blok VC et al (2004) Efficient virus-induced gene silencing in roots using a modified tobacco rattle virus vector. Plant Physiol 136:3999–4009

29. Macfarlane SA (1999) Molecular biology of the tobraviruses. J Gen Virol 80:2799–2807

30. Lange M, Yellina AL, Orashakova S et al. (2013) Virus-induced gene silencing (VIGS) in plants: an overview of target species and the virus-derived vector systems. Virus-Induced Gene Silencing: Methods and Protocols, 1–14

31. Liu E, Page JE (2008) Optimized cDNA libraries for virus-induced gene silencing (VIGS) using tobacco rattle virus. Plant Methods 4:5

32. Xue J, Li Y, Tan H et al (2008) Expression of ethylene biosynthetic and receptor genes in rose floral tissues during ethylene-enhanced flower opening. J Exp Bot 59:2161–2169

33. Lü P, Zhang C, Liu J et al (2014) RhHB1 mediates the antagonism of gibberellins to ABA and ethylene during rose (Rosa Hybrida) petal senescence. Plant J 78:578–590

34. Pei H, Ma N, Tian J et al (2013) An NAC transcription factor controls ethylene-regulated cell expansion in flower petals. Plant Physiol 163:775–791

35. Wu L, Ma N, Jia Y et al (2016) An ethylene-induced regulatory module delays flower senescence by regulating cytokinin content. Plant Physiol 173:11–21

36. Senthil-Kumar M, Mysore KS (2011) Virus-induced gene silencing can persist for more than 2 years and also be transmitted to progeny seedlings in *Nicotiana benthamiana* and tomato. Plant Biotechnol J 9:797–806

Chapter 5

Phenotypic Analysis and Molecular Markers of Plant Nodule Senescence

Xiuying Xia

Abstract

Leguminous crops can form nodules to fix atmospheric nitrogen (N_2). Senescence of nodules is associated with a rapid decline in N fixation. During the process of nodule senescence, a number of visible or detectable changes on morphology, biochemistry, and physiology occur. Here we describe several methods for examining the senescing phenotypes of nodules, including rhizobium inoculation, nitrogenase activity determination with the acetylene reduction assay, leghemoglobin content determination, and apoptotic cell identification with TdT-mediated dUTP-biotin nick end-labeling (TUNEL) staining.

Key words Nodule, Senescence, Nitrogen fixation, Acetylene reduction assay (ARA), TdT-mediated dUTP-biotin nick end labeling (TUNEL), Leghemoglobin (LHb)

1 Introduction

Leguminous crops can form a symbiosis with N_2-fixing bacteria, leading to the development of root nodules, in which atmospheric nitrogen (N_2) can be reduced into ammonia under limited nitrogen conditions. The natural aging process of nodules, known as nodule senescence, is defined as a biochemical and physiological event observed at the final stage of maturation that requires the transcription of new genes and results in death of the nodule [1]. During nodule senescence, many visible or detectable internal and external changes occur.

The first symptoms of senescence are changes in color and loss of turgidity in the early formed nodules. The pink color due to the presence of leghemoglobin (LHb) darkens or becomes green. At the cellular level, ultrastructural changes also take place: the cytoplasm becomes low electron dense, and vesicles and ghost membranes appear as a result of host and symbiosome membrane disintegration. The final outcome of nodule senescence is apoptosis or death of both bacteroids and nodule cells. These phenotypic changes can be detected by micromorphology analysis,

Yongfeng Guo (ed.), *Plant Senescence: Methods and Protocols*, Methods in Molecular Biology, vol. 1744, https://doi.org/10.1007/978-1-4939-7672-0_5, © Springer Science+Business Media, LLC 2018

TdT-mediated dUTP-biotin nick end-labeling (TUNEL) staining assay, and in situ live/dead staining assay. TUNEL staining is one of the most widely used methods for detecting DNA damage (DNA fragmentation) or identifying apoptotic cells in situ. The assay relies on the ability of the terminal deoxynucleotidyl transferase (TdT) to incorporate labeled dUTP into free 3′-OH termini of the genomic DNA fragment. The dUTPs can be modified by fluorophores or haptens, including biotin or bromine, and can be detected directly in the case of a fluorescently modified nucleotide, or indirectly with streptavidin or antibodies, if biotin-dUTP or BrdUTP are used, respectively [2]. In situ live/dead analysis could be carried out by utilizing mixtures of SYTO 9 green-fluorescent nucleic acid stain and the red-fluorescent nucleic acid stain, propidium iodide. These stains differ both in their spectral characteristics and in their ability to penetrate healthy bacterial cells. The SYTO 9 stain generally labels all bacteria in a population. Propidium iodide penetrates only bacteria with damaged membranes. When both dyes are present, SYTO 9 stain fluorescence will be reduced. Thus, with an appropriate mixture of the SYTO 9 and propidium iodide stains, bacteria with intact cell membranes stain fluorescent green, whereas bacteria with damaged membranes stain fluorescent red. This method has been found to work well with a broad spectrum of bacterial types including some rhizobia [3, 4].

Nodule senescence is characterized by a rapid decline in the biological fixation of N_2 and leghemoglobin content. The cyanmethemoglobin method is adopted from the standard procedure for the determination of hemoglobin (Hb) in blood. This procedure is based on the oxidation of hemoglobin and its derivatives to methemoglobin in the presence of alkaline potassium ferricyanide. Methemoglobin reacts with potassium cyanide to form cyanmethemoglobin, which has maximum absorption at 540 nm. The color intensity measured at 540 nm is proportional to the total hemoglobin concentration. Leghemoglobin (LHb) is structurally and chemically similar to blood Hb. The cyanmethemoglobin method has been proved to be a rapid, quantitative method for the determination of LHb in plant tissue [5, 6]. Nodule senescence is easy to monitor using the acetylene reduction assay, which measures declines in the activities of nitrogenase, the enzyme that carries out nitrogen fixation. ARA is based on the nitrogenase-catalyzed reduction of C_2H_2–C_2H_4, gas chromatographic isolation, and quantitative measurement of C_2H_2 and C_2H_4 with a H_2-flame analyzer. The assay was successfully applied to measurements of nitrogenase activities because of its high sensitivity, speed, low cost, and nondestructive nature [7].

Understanding how genes operate during senescence and identifying molecular markers is crucial to senescence control. A hallmark of nodule senescence is the triggering of a wide range of proteolytic activities that cause large-scale protein degradation [8].

In particular, a group of upregulated cysteine proteases are considered to be excellent markers for developmental nodule senescence [9]. Besides these markers, in a transcriptome analysis, several tags identified are upregulated during developmental nodule senescence [10] and have been considered as potential molecular markers of nodule senescence. Quantitative RT-PCR is the mature and ideal method to quantitatively detect gene expression. In situ hybridization is used to reveal the location of specific nucleic acid sequences on chromosomes or in tissues and is a crucial step for understanding the organization, regulation, and function of genes. *Agrobacterium rhizogenes*-mediated transformation can produce composite plants with a transformed root but untransformed shoot. Recently, this method has been used as an effective tool to study the function of individual genes involved in root and nodule biology [11, 12].

In this chapter we describe a number of protocols commonly used in phenotypic and molecular analyses of nodule senescence. This includes nodulation and *Agrobacterium rhizogenes*-mediated transformation, micromorphology analysis and leghemoglobin content determination, nitrogenase activity measurement, apoptosis detection in tissue sections, and expression analysis of nodule senescence marker genes.

2 Materials

2.1 *Bradyrhizobium japonicum* Inoculation and Nodulation

1. NH_4Cl solution: 16 g/100 mL.

2. $FeCl_3$ solution: 0.67 g/100 mL.

3. $CaCl_2$ solution: 1.5 g/100 mL.

4. $MgSO_4$ solution: 18 g/100 mL.

5. Modified arabinose gluconate (MAG) medium: For 1 L, add 1.3 g HEPES, 1.1 g MES, 1.0 g yeast extract, 1.0 g arabinose, 1.0 g gluconic acid, 0.22 g KH_2PO_4, 0.25 g Na_2SO_4, 2 mL NH_4Cl solution, 2 mL $FeCl_3$ solution, 2 mL $CaCl_2$ solution, 2 mL $MgSO_4$ solution, and adjust to pH 6.6 with KOH. Autoclave 30 min at 120 °C. Add 18 g agar per liter for solid medium. Store at 4 °C.

6. 70% ethanol.

7. 2% sodium hypochlorite (v/v).

8. Nitrogen-free MS-modified nutrient solution: PhytoTechnology Laboratories.

9. Vermiculite: Size 2–4 mm. Autoclave 45 min at 120 °C before use.

10. Metro mix soil.

2.2 Agrobacterium rhizogenes-Mediated Transformation

1. *Agrobacterium rhizogenes*: K599 or other strains harboring constructed vector carrying genes of interest.

2. YEP medium: To make 1 L, add 10 g Bacto Peptone, 5 g yeast extract, 5 g NaCl, 12 g agar for solid medium, pH 7.0.

3. Murashige and Skoog basal (MS) medium: Sigma.

4. 100 mM acetosyringone solution: Dissolve 196.2 mg of aceto-syringone in 10 mL dimethyl sulfoxide (DMSO).

5. 200 mg/mL cefotaxime solution: Dissolve 2 g cefotaxime in 10 mL H_2O, filter sterilize with 0.22 μm filters, and store at −20°C.

2.3 Leghemoglobin (LHb) Determination and Acetylene Reduction Assay

1. LHb extract solution: Drabkin's reagent, 1:2 (w/v), Sigma.

2. Human hemoglobin: Sigma.

3. Acetylene (C_2H_2).

4. Ethylene (C_2H_4).

5. Spectrophotometer.

6. Gas chromatography: Hewlett-Packard 5890 series II with a Porapak-T column.

2.4 In Situ Live/Dead Staining

1. 6% (w/v) agarose: Weight 6 g low melting temperature aga-rose in 100 mL distilled water, allow cooling to 37 °C, and remove bubbles before agarose-embedding on ice.

2. LIVE/DEAD® BacLight Bacterial Viability Kits: Invitrogen.

3. Live/dead staining solution: Combine 5 mM SYTO 9 and 30 mM PI in the 50 mM Tris–HCl, pH 7.0 buffer, and mix thoroughly.

4. Embedding cassettes.

5. Leica VT1200S vibratome.

6. Confocal laser scanning microscope: Leica TCS SP2.

2.5 Micromorphology Analysis

1. FAA solution: Mix 10 mL 35–40% formaldehyde, 5 mL acetic acid glacial, and 85 mL 70% ethanol.

2. Xylene.

3. Ethanol series: 100, 95, 85, 70, 50, and 30% ethanol.

4. Xylene/Ethanol solution, 1:1.

5. Paraffin.

6. Paraplast Plus tissue-embedding medium: Sigma.

7. Embedding cassettes.

8. Microtome.

9. Poly L-lysine.

10. 1% toluidine blue solution: Dissolve 1 g toluidine blue in saturated sodium borate ($Na_2B_4O_7 \cdot 10H_2O$), and store in darkness.

11. Permount® Mounting medium: Fisher.

2.6 TUNEL Assay

1. DeadEnd™ Colorimetric TUNEL System: Promega.

2. 1× PBS (phosphate-buffered saline): Dissolve 8 g NaCl, 0.2 g KCl, 1.44 g Na_2HPO_4, and 0.23 g KH_2PO_4 in distilled water, adjust volume to 1000 mL, and adjust pH to 7.4 with HCl.

3. 0.85% NaCl solution.

4. 4% formaldehyde: Add 4 g paraformaldehyde in 100 mL PBS in a fume hood. Dissolve by heating the closed bottle in a water bath at 65 °C for 2 h. Store at 4 °C.

5. 20 μg/mL Proteinase K solution: Dilute 10 mg/mL Proteinase K stock solution (from the DeadEnd™ Colorimetric TUNEL System) in PBS by 1:500.

6. Equilibration buffer: 200 mM potassium cacodylate (pH 6.6), 25 mM Tris–HCl (pH 6.6), 0.2 mM DTT, 0.25 mg/mL BSA, 2.5 mM cobalt chloride, from the DeadEnd™ Colorimetric TUNEL System.

7. rTdT reaction buffer: 98 μL equilibration buffer, 1 μL biotinylated nucleotide mix, 1 μL rTdT enzyme, from the DeadEnd™ Colorimetric TUNEL System.

8. 0.3% hydrogen peroxide: Prepare fresh from hydrogen peroxide reagent stock prior to use.

9. Streptavidin HRP: 1:500 Dilute horseradish peroxidase-labeled streptavidin (0.5 mg/mL) in PBS.

10. RNase-free DNase: Ambion Turbo DNA-free Kit™.

11. 20× SSC buffer: Dissolve 87.7 g NaCl and 44.1 g sodium citrate in 400 mL deionized water, adjust pH to 7.2 with 10 N NaOH, and bring volume to 500 mL. Dilute to 2× with deionized water before use.

12. Diaminobenzidine (DAB) staining solution: Add 50 μL of the DAB substrate buffer (20×) to 950 μL deionized water. Then add 50 μL of the DAB chromogen (20×) and 50 μL of hydrogen peroxide solutions (20×). Keep the DAB solutions away from light and use within 30 min.

13. Permount® Mounting Medium: Fisher.

14. Plastic Coverslips: Fisher.

2.7 qRT-PCR Analysis

1. Plant RNeasy® Kit (Qiagen).

2. RNase-free DNase: Ambion Turbo DNA-*free* Kit™.

3. Stop solution: 20 mM EGTA (pH 8.0)

4. iScript™ cDNA Synthesis Kit (Bio-Rad)

5. iQ™ SYBR® Green Supermix (Bio-Rad).

6. NanoDrop® ND-1000 spectrophotometer.

7. Bio-Rad® iCycler iQ thermocycler.

2.8 In Situ Hybridization

1. DIG RNA Labeling Kit (SP6/T7): Sigma.

2. DEPC water: 1 mL DEPC (Sigma), 1 L sterile ultrapure water. Autoclave.

3. 1× PBS: 8 g NaCl, 0.2 g KCl, 1.44 g Na_2HPO_4, 0.23 g KH_2PO_4, 800 mL DEPC water. Adjust pH to 7.4 with HCl. Adjust volume with DEPC water to 1000 mL. Autoclave.

4. PBSw (100 mL): 100 μL Tween 20, 100 mL 1× PBS. Filter sterilize.

5. 4% PFA: Add 4 g paraformaldehyde in 100 mL PBS in a fume hood. Dissolve by heating the closed bottle in a water bath at 65 °C for 2 h. Store at 4 °C.

6. 20× SSC (1 L): 175.3 g NaCl, 88.2 g sodium citrate, 700 mL DEPC water. Adjust pH to 7.0 with HCl or NaOH. Adjust volume to 1 L with DEPC water. Autoclave.

7. Proteinase K solution (2.5 mg/mL): Make a solution of 1 part Proteinase K stock (Sigma, 20 mg/mL)/7 parts PBSw.

8. Denhardt's solution (100×): 10 g Ficoll, 10 g PVP (polyvinylpyrrolidone), 10 g bovine serum albumin (BSA), 500 mL sterile dH_2O.

9. Prehybridization solution (RNase-free): 500 μL of 50% formamide, 60 μL of 5 M NaCl stock, 10 μL of 1 M Tris–HCl (pH 7.5) stock, 2 μL of 0.5 M EDTA stock, 50 μL of 100× Denhardt's solution, 100 μL of 10% dextran sulfate, 1 μL 100 mg/mL heparin, adjust volume with DEPC water to 1 mL.

10. RNase A buffer: 2.922 g NaCl, 1.2114 g Tris, 100 μL Tween 20. Adjust pH to 7.5 with HCl or NaOH. Adjust volume to 100 mL with DEPC water and filter sterilize.

11. MABT stock (5×): Dissolve 58 g maleic acid, 43.5 g NaCl, 55 g Tween 20 in 900 mL water. Bring the pH to 7.5 by adding Tris base. Bring final volume to 1 L.

12. Blocking buffer: 1× MABT with 0.1% Triton X-100 and 2% sheep serum.

13. Pre-staining buffer: 100 mM Tris–HCl pH 9.5, 100 mM NaCl, 10 mM $MgCl_2$.

14. Alkaline phosphatase buffer (NTMT) (with 5 mM levamisole): For 80 mL, mix 8 mL 1 M Tris–HCl, pH 9.5, 4 mL 1 M $MgCl_2$, 1.6 mL 5 M NaCl, 80 μL Tween 20, 63.4 mL ddH_2O. Add levamisole to a final concentration of 5 mM.

3   Methods

3.1 Bradyrhizobium japonicum Inoculation and Nodulation of Soybean

1. Streak out the *B. japonicum* strain USDA 110, and incubate it in MAG agar medium at 25 °C for 7 days (*see* **Note 1**).

2. Transfer a single colony into MAG broth, and incubate at 25 °C on a rotary shaker running at 180 rpm until the cell density reach an optical density of OD_{600} 0.5. The suspension is then used for soybean seed inoculation.

3. Surface sterilize healthy soybean seeds with 70% ethanol for 2 min and 2% sodium hypochlorite (v/v) for 10 min, then wash thoroughly with sterile distilled water, and place the seeds on a piece of sterile moist filter paper in a Petri dish to germinate in darkness.

4. When the hypocotyls are 1 cm long, seeds are rinsed in the suspension of *B. japonicum* (OD_{600} 0.5) for 2 h, then placed in bottles filled with moist autoclaved vermiculite, and incubated in a growth chamber under the condition of 16 h light at 25 °C and 8 h darkness at 23 °C.

5. Seedlings receive 20 mL of nitrogen-free MS-modified nutrient solution every 3 days (*see* **Note 2**).

6. Three weeks later, select uniform plants with nodules, and transfer them to plastic pots (25×18 cm^2) containing 10 L of Metro mix soil (*see* **Note 3**).

7. Grow the soybean plants in a greenhouse under a 12 h photoperiod at 27 °C, a night temperature of 21 °C, and 70% relative humidity. Plants are managed routinely. 500 mL nitrogen-free MS solution is supplied to a single plant twice monthly.

8. Harvest the plants for analysis. Wash the roots with distilled water. The nodules should be kept in humid conditions throughout their processing.

3.2 Agrobacterium rhizogenes-Mediated Transformation and Nodulation

1. Construct recombinant plasmids carrying target genes. Transform the plasmids to *Agrobacterium rhizogenes* via heat shocking.

2. Streak *Agrobacteria* onto solid YEP medium containing antibiotics. After approximately 2 days, a single colony is inoculated in 50 mL of liquid YEP medium with antibiotics and incubated at 28 °C and 180 rpm until an OD_{600} 0.5 is reached (*see* **Note 4**).

3. Centrifuge the activated bacterial cultures for 10 min, collect the bacterial cells, and resuspend in 50 mL MS liquid medium (without antibiotics) containing 100 µM acetosyringone. Shake the bacterial cultures for 4 h at 28 °C in a rotatory shaker at 120 rpm before cocultivation (*see* **Note 5**).

4. Seeds are sterilized and germinated as described in Subheading 3.1. When the hypocotyls are 1 cm long, the seeds are wounded by carefully stabbing with a scalpel blade into the hypocotyls perpendicularly, and then immerse the explants into the bacterial culture for 30 min (*see* **Note 6**).

5. Blot the seeds with autoclaved filter papers, place in 50-mm-diameter Petri dishes with MS medium, and grow them in darkness at 28 °C for cocultivation.

6. Three days later, infected explants are subsequently rinsed thoroughly in sterile H_2O supplied with 300 mg/L cefotaxime (*see* **Note 7**) and then transferred to autoclaved vermiculite moistened with MS nutrient solution supplied with 200 mg/L cefotaxime. The plantlets are grown under the conditions of 16 h light at 25 °C and 8 h darkness at 23 °C.

7. Three weeks later, remove the taproot, leave only lateral transgenic roots to develop. Then transfer the plants with well-developed hairy roots to containers filled with autoclaved soil. Placed the plants in a growth chamber in a 14/10 h photoperiod at 25/21 °C.

8. Water the plants with nitrogen-free MS nutrient solution daily for 3 days to starve the plant.

9. Three days later, inject each plant root with 20 mL 0.2 OD_{600} rhizobia suspension to inoculate. (*B. japonicum* culture as described in Subheading 3.1, with the OD_{600} of 0.2).

10. Transgenic and nodulated plants are grown under a 12 h photoperiod at 25 °C, a night temperature of 21 °C, and 70% relative humidity. Plants are watered daily. 500 mL nitrogen-free MS solution is supplied to single plant once a week.

11. Nodules are collected at various senescing stages for paraffin-embedded section slides and nitrogenase activity detection.

3.3 Leghemoglobin (LHb) Content Determination

1. Excise nodules at different development stages from taproots (or transformed hair roots) (*see* **Notes 8** and **9**), wash them with running water, and blot up quickly, weight, and put into a 5 mL centrifuge tube.

2. Crush and grind the nodule samples in cooled LHb extract solution.

3. Centrifuge the mixture at 500 *g* for 15 min to throw down the large particles of nodule tissue, and transfer the supernatant to another centrifuge tube.

4. Extract the nodule tissue twice more and combine the supernatants.

5. Centrifuge the supernatants at 20,000 *g* for 30 min.

6. Read the absorbance of the cleared supernatant at 540 nm against Drabkin's solution using a spectrophotometer.

7. Use human hemoglobin (HB) as a standard to plot a calibration curve of absorbance values versus the HB concentration (mg/mL). The curve is linear, passing through the origin.

8. Determine the total LHb concentration (mg/mL) of each test directly from the calibration curve. Values are expressed as milligrams per gram of nodule mass (fresh weight).

3.4 Acetylene Reduction Assay for Nitrogenase Activity Measurement

1. Harvest soybean plants at different senescence stages. Plants are excavated with roots and shoot, immediately transported to the laboratory, and assay within 2 h of collection (*see* **Note 10**).

2. Excise and collect the nodules from the upper 5 cm of the roots.

3. Nodules are incubated in 50 mL serum bottles (*see* **Note 11**). A 10% volume of air is then removed and replaced with an equal volume of acetylene. The bottles are kept flat to increase the contact area with acetylene.

4. Incubate the nodules at room temperature for 1 h, then withdraw 0.5 mL gas sample, and analyze ethylene formation by gas chromatography with a Porapak-T column using pure ethylene as internal standards.

5. The ethylene content of the sample is calculated from the observed peak height by reference to a standard curve of log ethylene concentration versus log peak height.

3.5 In Situ Live/Dead Staining

1. Excise the nodules from the upper 5 cm of the roots at different senescence stages. Place the nodules in embedding cassettes, and embed them in 6% (w/v) agarose on ice immediately.

2. Slice nodules transversally by 70 μm with a Leica VT1200S vibratome (*see* **Note 12**).

3. Incubate the sections in the live/dead staining solution at room temperature in the dark for 20 min (*see* **Note 13**).

4. Remove the sections from the staining solution and rinse in distilled water.

5. Observe the sections and acquire images with a confocal laser scanning microscope. The excitation/emission maxima are about 480/500 nm for SYTO 9 stain and 490/635 nm for propidium iodide.

3.6 Micromorphology Analysis

1. Excise 15–20 crown nodules from the upper 5 cm of the roots at different senescence stages, immerse nodules immediately in FAA solution with 5–10 times of the nodule volume, vacuum for 10 min, and fix for 24 h at room temperature.

2. Fixed nodules are dehydrated in an increasing ethanol series (70, 85, 95, and 100%), followed by xylene/ethanol solution for 30 min and then three times in xylene 30 min each.

3. The nodules are infiltrated with several changes of paraffin at 60 °C over a 3-day time period.

4. Place the nodules in embedding cassettes, and embed them in Paraplast Plus tissue-embedding medium.

5. The Paraplast Plus-embedded tissues are sectioned transversally (5–10 μm thick) with a microtome (*see* **Note 12**).

6. Sections obtained from each nodule are placed on poly L-lysine-treated slides, and floated in distilled water to flat. Allow sections to air-dry for 10 min, and then bake at 45 °C in an oven overnight.

7. Deparaffinize sections in three changes of xylene, 5 min each.

8. Rehydrated in descending concentrations of ethanol (100, 95, 70, 50, and 30%) for 3 min each, and then rinse in distilled water.

9. Stain nodule sections in 0.5% toluidine blue solution for 10 min.

10. Rinse slides in distilled water briefly and air-dry. Mount slides with Permount® Mounting Medium.

11. Stained sections are examined and photographed under bright field using a microscope. See examples shown in Fig. 1.

3.7 TUNEL Assay for Identifying Apoptotic Cells in Senescing Nodules

Carry out all the procedures at room temperature unless otherwise specified.

1. Nodules from different senescence stages are fixed, embedded in paraffin, and cut transversally. The slides are deparaffinated in xylene and rehydrated in ethanol series as described in Subheading 3.6.

2. Wash the slides successively in 0.85% NaCl and PBS for 5 min each.

3. Fix the slides in 4% paraformaldehyde for 15 min. Then wash the slides in PBS for 5 min twice.

4. Blot away excess water carefully, and pipette 100 μL 20 μg/mL Proteinase K buffer to cover sections. Incubate 15 min (*see* **Note 14**). Then wash slides with PBS for 5 min twice.

5. Refix the tissue sections by immersing the slides in 4% paraformaldehyde solution for 5 min. Wash slides in PBS for 5 min twice.

6. Carefully blot away excess water, and then cover the slides with 100 μL of equilibration buffer for 10 min.

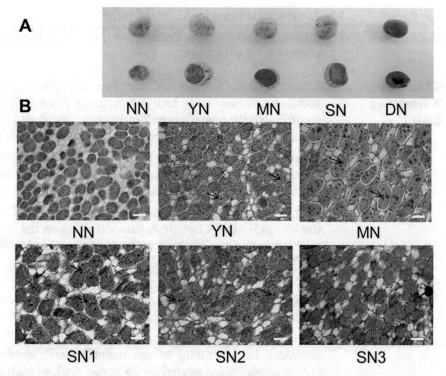

Fig. 1 Changes in nodule structure of soybean during developmental senescence. (**a**) Visible appearance of the nodule harvested from difference age, showing color changes in inner tissues. (**b**) Light micrographs of transverse median sections from nodules of different age, showing morphological changes coinciding with the nodule development process. *NN* newly formed nodule, *YN* young nodule, *MN* mature nodule, *SN1–SN3* senescing nodule, *DN* disintegrated nodule. Arrows show bacteroids. Bar, 40 μm

7. Remove equilibration buffer, and cover the sections with 100 μL of rTdT reaction buffer (*see* **Note 15**). Cover the sections with plastic coverslips to ensure even distribution of the reagent. Incubate slides in a humidified chamber at 37 °C for 60 min to allow the end-labeling reaction to occur.

8. Remove the plastic coverslips, and incubate slides in 2× SSC buffer for 15 min to terminate the end-labeling reaction.

9. Wash the slides in PBS for 5 min twice to remove unincorporated biotinylated nucleotides.

10. Immerse the slides in 0.3% hydrogen peroxide in PBS for 5 min to block the endogenous peroxidases.

11. Wash the slides in PBS for 5 min twice.

12. Incubate the slides with 100 μL streptavidin HRP (diluted 1: 500 in PBS) at 25 °C for 30 min.

13. Wash slides for 5 min in PBS twice. Stain with 100 μL DAB staining solution for 10 min in the dark (*see* **Note 16**). Monitor color development until there is a light brown background.

14. Rinse several times in deionized water.

15. Mount the slides with Permount® Mounting Medium, and observe with a microscope.

3.8 qPCR Analysis of Nodule Senescence Marker Genes

1. Extract total RNAs from nodules (from the upper 5 cm of the root) of different senescence stages using Plant RNeasy® Kit according to the manufacturer's instruction (*see* **Note 17**). RNA is quantified with a NanoDrop® ND-1000 spectrophotometer.

2. Treat RNAs with RNase-free DNase at 37 °C for 30 min, using 1 unit of DNase per μg of total RNA. Add 1 μL stop solution, stay at 65 °C for 10 min, and put on ice or −80 °C.

3. First strand cDNA is synthesized from 1 μg of total RNA with the iScript™ cDNA Synthesis Kit, according to the manufacturer's instruction. Thermal cycler is programmed as follows: 25 °C, 5 min, 1 cycle; 42 °C, 30 min, 1 cycle; 85 °C, 5 min, 1 cycle; 4 °C, hold.

4. Design and synthesize sense and antisense primers according to target genes sequence (*see* **Note 18**).

5. Quantitative PCR is performed with the iQ™ SYBR® Green Supermix Kit according to the manufacturer's instructions. cDNA products are diluted by 25 times. 5 μL of each diluted sample is used as a template, and 100 nM of sense and antisense primers are used in a 25 μL reaction for real-time PCR amplification.

6. PCR reactions are run and analyzed on a Bio-Rad® iCycler iQ thermocycler. The PCR parameters are initial denaturation, 1 cycle at 95 °C for 3 min; amplification and quantification, 40 cycles at 95 °C for 30 s, 55 °C for 30 s, and 72 °C for 1 min; cycle threshold (Ct) values were determined by IQ-5 Bio-Rad software assuming 100% primer efficiency.

7. Relative transcript levels are calculated as the ratio of normalized gene expression against the housekeeping gene (*see* **Note 18**).

8. The reactions are performed in triplicate.

3.9 Nodule Senescence-Associated Gene Detection by Digoxigenin (DIG)-Labeled RNA Probe In Situ Hybridization

1. Extract total RNA of nodules. Amplify the cDNA by RT-PCR using special primers designed on target gene as described in Subheading 3.8.

2. Sense and antisense probes are synthesized following the protocol of the DIG RNA Labeling Kit.

3. Nodules from different senescence stages are fixed in 4% (w/v) paraformaldehyde for 4 h, dehydrated in a graded ethanol series, embedded in paraffin, and sectioned at 6 μm thickness. Sections are fixed on slides.

4. The slides are deparaffinated and rehydrated as described in Subheading 3.6.

5. Wash the slides in PBS for 5 min.

6. Cover the sections with 20 μg/mL proteinase K solution (pre-warm for 10 min at 37 °C), and incubate for 30 min.

7. Stop digestion by incubating the slides with 0.2% glycine in PBS for 5 min at room temperature and rinse in PBS.

8. Immerse slides in ice-cold 20% (v/v) acetic acid for 20 s. Immerse the slides in 0.25% acetic anhydride for 10 min. Wash slides twice in PBS.

9. Dehydrate the slides by washing for approximately 1 min per wash in 70% ethanol, 95% ethanol, and 100% ethanol, and then air-dry.

10. Add 100 μL of prehybridization solution to each slide and incubate the slides for 1 h in a humidified hybridization chamber at 60 °C.

11. Add 6 μL DIG-labeled probe to 100 μL prehybridization. Heat at 95 °C for 2 min in a PCR block to denature the RNA probe. Chill on ice immediately to prevent reannealing.

12. Drain off the prehybridization solution. Add 100 μL pre-warmed (42 °C) hybridization solution per slide. Cover the sample with a cover slip to prevent evaporation. Incubate in the humidified hybridization chamber at 42 °C overnight.

13. Wash the slides in 2× SSC/50% formamide for 5 min twice at 42 °C.

14. Wash the slides in 1× SSC for 5 min at 42 °C.

15. The slides are treated with RNaseA buffer for 30 min at 37 °C to eliminate nonspecific background.

16. Wash the slides in MABT for 30 min at 37 °C. Repeat the MABT wash at room temperature. Dry the slides.

17. Add 200 μL blocking buffer to each section. Block for 1–2 h at room temperature.

18. Drain off the blocking buffer. Add 1 μL anti-label antibody to blocking buffer (1:2000). Incubate overnight for 1–2 h at room temperature.

19. Wash the slides with MABT for 10 min three times at room temperature.

20. Wash the slides 2 × 10 min each at room temperature with pre-staining buffer.

21. Immerse the slides in freshly made alkaline phosphatase buffer (NTMT) (with 5 mM levamisole) for 5 min twice.

22. Add 4.5 μL/mL NBT and 3.5 μL/mL BCIP in NTMT, incubate in darkness for 12 h.

23. Wash with 1× PBS for 5 min twice. Dehydrate with gradual ethanol and xylene. Mount the slide and observe under a microscope.

4   Notes

1. *B. japonicum* will not grow on TY or LB medium. Temperatures above 30 °C will inhibit *Bradyrhizobium* growth. You should be able to see colonies after 7 days.

2. Use nitrogen-free nutrient solution to grow plants as nodulation will be inhibited by nitrogen.

3. In order to obtain uniform nodulated plants to minimize systematic error, and to reduce the frequency of secondary nodulation, all the plants were transferred to soil after nodulation.

4. A single colony is not necessary. You can also active the *A. rhizogenes* using a stock stored in glycerol as inoculums.

5. The supplementation of MS medium with acetosyringone, used prior cocultivation, can increase the transformation efficiency in soybean. The acetosyringone can also be supplied to cocultivation medium.

6. The *A. rhizogenes*-mediated transformation can also be carried out by injecting directly into the hypocotyls or the cotyledonary nodes with a syringe delivering approximately 5–10 µL of inoculum into the wound.

7. To inhibit *Agrobacterium* overgrowth, you can adopt rinsing of the seedling in water supplied with cefotaxime for 2 days on a shaker.

8. To obtain the age-matched uniform nodule, it is necessary to collect nodules from the taproot or the upper 5 cm of the hair root.

9. If you want to determinate LHb in nodules of different senescent stages synchronously, the nodule samples can be frozen at −80 °C, and the LHb content will not change.

10. Nitrogenase activities in closed systems can be affected by assay conditions, but for comparative purposes and by using a short assay time this method provides reliable data, so the assay should be finished within 2 h. You can use attached nodules on roots as samples. Nitrogen fixation can also be measured on intact plants using acetylene reduction assay as described by Pfeiffer [13]. The values of N_2 fixation abilities obtained from attached nodules are higher than that of the excised nodules.

11. The nodules should be divided into several equal bottles if their weight exceeds 1 g. The volume or weight of the nodule in every bottle is ensued at almost the same value at each test. Ensure that the volume of nodule samples should not exceed 1/3 of the container.

12. Nodules are classified as 'indeterminate' and 'determinate' according to their mode of development. Indeterminate nodules, such as those of pea, clover and alfalfa have a persistent

apical meristem. Indeterminate nodules comprise of a number of easily distinguished zones. During nodule development, the senescence zone gradually moves from the proximal part of the nodule to the distal part, until it reaches the apex. Determinate nodules such as those of soybean have no active meristem. There is a radial gradient of development with senescence beginning at the center and spreading outward. For the purpose of examining senescence status in different areas or zones in nodules, the indeterminate nodules should be cut longitudinally while the determinate nodule should be cut transversally.

13. Propidium iodide and SYTO 9 stain both bind to nucleic acids. Propidium iodide is a potential mutagen. DMSO is known to facilitate the entry of organic molecules into tissues. So all the reagents in this test should be used with appropriate care. To optimize the staining, the proportions of the two dyes should be adjusted for optimal discrimination by experimenting with a range of concentrations of SYTO 9 dye combination with a range of propidium iodide concentrations.

14. Prolonged incubation with Proteinase K may cause sections to come off the slide. 15 min incubation for 5–10 μm paraffin sections and 25 min incubation for 50–70 μm vibratome sections can obtain ideal results.

15. Use 1 μL of autoclaved, deionized water instead of rTdT enzyme for preparing control reaction buffer.

16. The staining time may need to be optimized according to different samples.

17. Total RNAs can also be isolated by phenol/chloroform extraction from samples grounded to a fine powder and allowed to thaw into five volumes of a buffer containing 0.1 M NaCl, 2% SDS, 50 mM Tris–HCl pH 9.0, 10 mM EDTA, and 20 mM β-mercaptoethanol.

18. Potential molecular markers available for nodule senescence: TC100437 (MTR_5g022560), cysteine proteinase gene, nodule senescence marker in *Medicago truncatula* [10]; *MTD1* (identical to an EST clone accession number AW559774) and *MTD2* (highly homologous to an *Arabidopsis thaliana* gene accession number AL133292) from *M. truncatula* are preferentially and specifically upregulated in nodules following simple defoliation [14]; *DD15* (accession number AY196194) identified from soybean maybe involved in remobilization of reserves during nodule senescence [15]; *CP* EST in soybean (GenBank accession number, AI973914) [16]; *senescence-associated nodulin1 (SAN1)* (encoding 2-oxoglutarate-dependent dioxygenases) multigene family member from *Glycine max* (*SAN1A* is downregulated during induced senescence, while *SAN1B* is upregulated during induced senescence) [17]; *Asnodf32*, encoding a nodule-specific cysteine proteinase in

Astragalus sinicus, reference primers for qPCR, 5′-TGATGCT AGTGGCTCAGACTTTC-3' and 5′-ATGTATCCTTCTTCT CCCCAGC-3′) [18]; *Tp-nsr1* (GenBank accession number, AM411056); and *Tp-nse1* (GenBank accession number, AM411035), identified from red clover. *Tp-nsr1* is identical to an existing marker that encodes a hypothetical protein with an aspartyl protease domain. *Tp-nse1* has no known function and is clearly upregulated in senescing nodules [19]. 18s rRNA could be used as housekeeping gene in soybean, reference primers: 5′-GTATGGTCGCAAGGCTGAAAC-3′ and 5′-TTA GCAGGCTGAGGTCTCGT-3′.

References

1. Evans PJ, Gallesi D, Mathieu C et al (1999) Oxidative stress occurs during soybean nodule senescence. Planta 208:73–79

2. Gavrieli Y, Sherman Y, Ben-Sasson SA (1992) Identification of programmed cell death in situ via specific labeling of nuclear DNA fragmentation. J Cell Biol 119:493–501

3. Haag AF, Baloban M, Sani M et al (2011) Protection of *Sinorhizobium* against host cysteine-rich antimicrobial peptides is critical for symbiosis. PLoS Biol 9:e1001169

4. Bourcy M, Brocard L, Pislariu CI et al (2013) *Medicago truncatula* DNF2 is a PI-PLC-XD-containing protein required for bacteroid persistence and prevention of nodule early senescence and defense-like reactions. New Phytol 197:1250–1261

5. Wilson DO, Reisenauer HM (1963) Determination of leghemoglobin in legume nodules. Anal Biochem 6:27–30

6. Schiffmann J, Löbel R (1970) Haemoglobin determination and its value as an early indication of peanut rhizobium efficiency. Plant and Soil 33:501–512

7. Hardy RW, Holsten RD, Jackson EK et al (1968) The acetylene-ethylene assay for N_2 fixation: laboratory and field evaluation. Plant Physiol 43(8):1185–1207

8. Pladys D, Rigaud J (1985) Senescence in french-bean nodules: occurrence of different proteolytic activities. Physiol Plant 63:43–48

9. Pérez Guerra JC, Coussens G, De Keyser A et al (2010) Comparison of developmental and stress-induced nodule senescence in *Medicago truncatula*. Plant Physiol 152(3):1574–1584

10. Van de Velde W, Guerra JC, De Keyser A et al (2006) Aging in legume symbiosis. A molecular view on nodule senescence in *Medicago truncatula*. Plant Physiol 141:711–720

11. Estrada-Navarrete G, Alvarado-Affantranger X, Elías Olivares J et al (2006) Agrobacterium rhizogenes transformation of the *Phaseolus* spp.: a tool for functional genomics. Mol Plant Microbe Interact 19:1385–1393

12. Jian B, Hou W, Wu C et al (2009) *Agrobacterium rhizogenes*-mediated transformation of super root-derived *Lotus corniculatus* plants: a valuable tool for functional genomics. BMC Plant Biol 9:78–93

13. Pfeiffer NE, Malik NSA, Wagner FW (1983) Reversible dark-induced senescence of soybean root nodules. Plant Physiol 71:393–399

14. Curioni PMG, Reidy B, Flura T, Vögeli-Lange R, Nösberger J, Hartwig UA (2000) Increased abundance of MTD1 and MTD2 mRNAs in nodules of decapitated *Medicago truncatula*. Plant Mol Biol 44:477–485

15. Alesandrini F, Frendo P, Puppo A et al (2003a) Isolation of a molecular marker of soybean nodule senescence. Plant Physiol Biochem 41:727–732

16. Alesandrini F, René M, Van de Sype G et al (2003) Possible roles for a cysteine protease and hydrogen peroxide in soybean nodule development and senescence. New Phytol 158:131–138

17. Webb CJ, Weiher CC, Johnson DA (2008) Isolation of a novel family of genes related to 2-oxoglutarate-dependent dioxygenases from soybean and analysis of their expression during root nodule senescence. Plant Physiol 165(16):1736–1744

18. Li Y, Zhou L, Li Y et al (2008) A nodule-specific plant cysteine proteinase, AsNODF32, is involved in nodule senescence and nitrogen fixation activity of the green manure legume *Astragalus sinicus*. New Phytol 180:185–192

19. Webb KJ, Jensen EF, Heywood S et al (2010) Gene expression and nitrogen loss in senescing root systems of red clover (*Trifolium pratense*). J Agric Sci 148:579–591

Chapter 6

Quantitative Analysis of Floral Organ Abscission in *Arabidopsis* Via a Petal Breakstrength Assay

Chun-Lin Shi and Melinka A. Butenko

Abstract

Petal breakstrength (pBS) is a method to study floral organ abscission by quantitating the force required to pull a petal from the receptacle. However, it is only well established in some labs and used in a subset of abscission studies. Here, we describe the mechanism and operation of the pBS meter, as well as detailed measurement and further data analysis. We show that it is a powerful tool to detect early or delayed floral organ abscission in mutant or transgenic plants, which is not easily detected by phenotypic investigation.

   Key words Abscission, Floral organ abscission, Cell separation, Petal breakstrength, Abscission zone, Flower position

1 Introduction

Abscission is a regulated process of dropping plant organs, such as leaves, fruit, and flower petals. At the cellular level, the presence of an abscission zone (AZ) is a prerequisite for abscission to take place [1]. During the cell separation process, cell wall-modifying and cell wall-hydrolyzing enzymes are involved in degrading the middle lamella between two adjacent cell files (reviewed by [2–8]). It has also been discovered that expression of these enzymes is regulated by different factors and pathways, such as the peptide ligand IDA [9], receptor-like kinases HAE and HSL2 [10, 11], KNOX HD proteins [12], ethylene [13, 14], and auxin [15].

Petal breakstrength (pBS) assay measures the force required to remove a petal from the receptacle (in gram equivalents), which provides a quantitative insight into the physical integrity of the petal abscission zone. The pBS assay to study floral abscission was first described in Patterson 1998 [16] and was measured using a stress transducer developed by A. B. Bleecker and E. P. Spalding (University of Wisconsin–Madison). Compared to methods broadly used to study abscission in plants, such as visualization of AZ by scanning electron microscope (SEM) (*see* Chapter 26) and

Yongfeng Guo (ed.), *Plant Senescence: Methods and Protocols*, Methods in Molecular Biology, vol. 1744,
https://doi.org/10.1007/978-1-4939-7672-0_6, © Springer Science+Business Media, LLC 2018

investigation of expression of AZ-associated reporters, pBS has only been used in a subset of abscission studies [9, 14, 17]. This is mainly due to the lack of a commercial product that can be purchased. After that the first pBS meter was developed by Lease (2006) [18]; it has been successfully used to characterize abscission phenotype of mutants and transgenic plants in several labs [1, 10–13, 15, 19–24]. We describe here a modified pBS meter and detail the steps for both parameter settings and measurements. We also show here some examples of pBS patterns, which correlated with specific AZ cellular modifications.

2   Materials

The pBS meter mainly consists of a petal gripper/clamp, sensor/stress transducer, electronic circuit, and personal computer (PC). The meter described here is built after the description in Lease et al. (2006) [18], with some modifications (Fig. 1).

1. An ultralight miniature clamp: Instead of using a petal gripper and an aluminum strip, an ultralight miniature clamp is suspended (arrow in Fig. 1b).

2. A load transducer (AD Instruments MLT050/D, range 0–50 g): The transducer's output is repeatedly sampled at a rate of 21 ksps (Fig. 1b) (*see* **Note 1**).

3. Lightweight nylon thread: The clamp and the load transducer are connected with lightweight nylon thread (Fig. 1b).

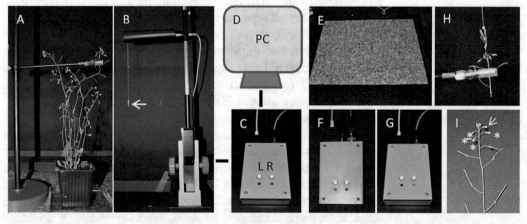

Fig. 1 The pBS meter. (**a**) The plant holder. (**b**) The load transducer and a petal clamp (arrow). (**c**) The user interface with microchips sitting inside, push buttons are indicated with L and R. (**d**) The PC. (**e**) The stable surface. (**f**) The pBS meter is initiated, indicated by the red LED light. (**g**) The pBS meter is ready to use, indicated by the green LED light. (**h**) A single petal from interested position is captured by the petal clamp. (**i**) Main inflorescence with interested positions, position 1 is indicated by arrow, and the following positions are marked with stars

4. A differential amplifier (Linear Technology LT1013): The output from the transducer is amplified by the differential amplifier.

5. The signal is digitized by a microcontroller's 10-bit AD converter (Microchip 12F675).

6. A second microcontroller (Microchip 16F676) is used to handle the user interface push buttons and indicator LEDs (Fig. 1c).

7. A PC: To collect and store data from pBS measurements (Fig. 1d).

8. A plant holder with clamp (Fig. 1a) (*see* **Note 2**).

9. A table with a stable surface: The whole pBS meter is sitting on it preventing vibrations from the surroundings (Fig. 1e) (*see* **Note 3**).

3 Methods

Arabidopsis thaliana plants are grown in soil in growth chambers at 22 °C under long days (16 h day/8 h dark) at a light intensity of 100 mE·m22·s21 (*see* **Note 4**).

3.1 pBS Meter Setting Up

Switch on the pBS meter and PC; the meter needs about 10 min to initiate and stabilize, indicated by the red LED light (Fig. 1f). The setting steps may differ from individual meters used in different labs, while the main parameters used for measurement should be similar.

1. Open hyperterminal (*see* **Note 5**).

2. Give a name.

3. Choose COM1 for connect using, click OK.

4. Restore defaults for port setting, 9600 bits per second, 8 data bits, none parity, 1 stop bits, none flow control, and click OK.

5. Click file and choose properties.

6. Click settings, ASCII setup.

7. Cross for append line feeds to incoming line ends for ASCII receiving.

8. Click transfer and capture text, and save the text file under your folder.

9. When the meter is ready, indicated by the green LED light (Fig. 1g), measure the empty control (*see* **Note 6**).

10. Measure the standard 1 g control (*see* **Note 7**).

3.2 pBS Measurement

Carry out pBS measurements on adult plants with 18–20 flower positions (*see* **Note 8**).

1. Take a plant and fix the main stem to the plant holder (Fig. 1a, h) (*see* **Note 9**).

2. Roll the clamp down, and use it carefully to catch a single petal of flower at position 2 (Fig. 1h, i) (*see* **Note 10**).

3. Press button R at the user interface (Fig. 1c).

4. Roll the clamp up and pull the petal off (*see* **Note 11**).

5. Press button L at the user interface, and the value of measurement will appear on the PC screen (*see* **Note 12**).

6. Remove the petal from clamp, and repeat the measurement with another petal sitting opposite of the same flower (*see* **Note 13**).

7. Continue the measurement with flower position 4, 6, 8, and so forth, if applicable (*see* **Note 14**).

8. Measure for a minimum of 15 plants and obtain a minimum of 24 measurements at each position (*see* **Note 15**).

9. Import raw data into Microsoft Excel, and normalize raw data with standard control if necessary (*see* **Note 7**).

10. Measure the empty and standard control again before starting with a new plant line (*see* **Note 16**).

11. Analyze data and calculate average values and the standard deviation (SD) at each position (*see* **Note 17**). Do student's t-test between mutant or transgenic plants and control WT plants (*see* **Notes 18** and **19**). Generate a graph in Excel (Fig. 2a).

3.3 Analysis of pBS Pattern

So far, three different pBS profiles have been described which give a good indication of where the abscission process is affected (Fig. 2b). For mutants which have either a precocious or delayed floral abscission, a profile similar to WT is observed with the lowest pBS measurements being earlier or later than for that floral position observed in WT plants. For mutants deficient in abscission, the profile depends on whether the mutant develops a functional abscission zone or is capable of inducing the abscission process or whether there is no differentiation of AZ cells. An *ida*-like profile is an example of the former, while a *bop*-like profile is an example of the latter patterns (Fig. 2b).

4 Notes

1. This is to facilitate pBS measurements where the force required to remove the petal approached values of 0.2 g equivalents. A different pBS meter which measures the force with N has been described [15, 20].

2. This is to handle the flowers easily.

3. This provides the best balance for the pBS meter.

4. Generally 20 plants are required to be grown in soil, and 15 healthy plants in good growing condition are used for pBS measurement.

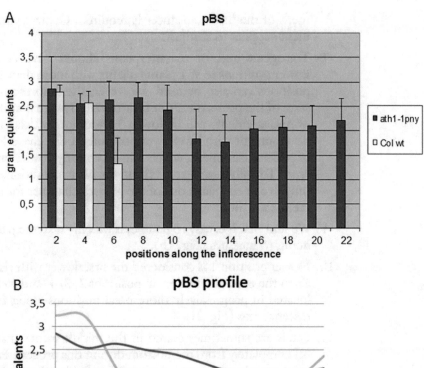

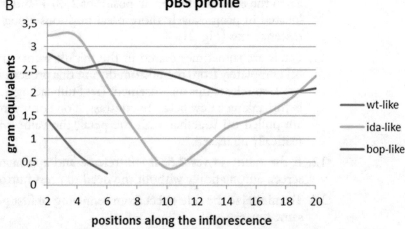

Fig. 2 Examples of pBS measurements of plants of different genotypes. (**a**) pBS measured from positions 2 to 22 along the inflorescence of 15 Col WT and *ath1pny* plants. Standard deviations are shown as thin lines at the top of the columns. (**b**) pBS profiles of WT-like, ida-like, and bop-like plants

5. The hyperterminal is a program to calculate load transducer values to gram equivalents.

6. Empty control should give a value of 0 or close to 0. Otherwise, the pBS meter should be restarted.

7. The 1 g standard control should give a value of 1. It may be lower than 1 (e.g., 0.9) depending on environmental and pBS meter conditions, such as room temperature or longtime continuous measurements. This small difference can be normalized by dividing all the values of measurements with that of the standard control (e.g., 0.9). A value less than 0.85 for standard control is not acceptable on our hands, in which case readjust-

ment of the load transducer is required. Contact your equipment supplier to readjust.

8. Transgenic and/or mutant plants should have comparable flower positions to WT plants. Plants with more than 20 flower positions can also be used. However, the inflorescence meristem of the plants must not be arrested. Beware that the values of measurements will be lower. In addition, a delay in flower development with less flower positions could give a low value of pBS measurement, which mimics an early abscission phenotype. To minimize stress effects on plant condition, we take only five or six plants out of the growth chamber for measurement each time.

9. We find that it is easy to handle, if we only take the part we are going to measure (Fig. 1h).

10. Flower position 1 is considered the first flower with visible petals to the eyes, and flowers in positions 2, 3, 4, and so forth are located in progressively more basal positions along the inflorescence axis (Fig. 1i).

11. Petals are sometimes teared in the middle rather than pulled off completely from the bottom during this process. Press button L to release this measurement (*see* Subheading 3.2, **step 5**) before taking a new petal. In contrast, if one or several stamens are pulled off together with the petal, the value will be dramatically increased.

12. If the value is over 2.5, it will release and appear on the PC screen automatically without the need to press button L.

13. To minimize the side effect from removing of first petal in the same flower.

14. In our lab, we do the measurements at every other flower positions starting from position 2. Depending on the research purpose, each position or every position within a certain developmental process is applied in other labs [10, 15, 19].

15. The plant number is to consider individual variations; the minimum number of measurements is for performing statistical analysis. For the flower position prior to the abscising flower, the force is too low to measure, and petals are dropped during handling of the plant. The minimum of 24 measurements is not applied for these positions.

16. We find that the load transducer can be less sensitive after long-time continuous measurements, which gives lower values (*see* **Note 7**). It is important to measure both empty and standard control when switching to a new plant line.

17. To minimize experimental mistake, we find that it is useful to exclude two maximum and two minimum values at each position before data analysis.

Table 1
pBS measurement of WT plants in different ecotypes

	Position 2	Position 4	Position 6	Position 8
Col	2.78 ± 0.15	2.56 ± 0.24	1.32 ± 0.53	N.A.
C24	2.41 ± 0.64	2.17 ± 0.41	0.48 ± 0.26	0.02 ± 0.05
Ler	1.57 ± 0.12	1.53 ± 0.13	0.36 ± 0.19	N.A.

N.A. not applied

Table 2
Averages of pBS measurements of Col WT plants in different labs

Position 2	Position 4	Position 6	Ref.
1.7	1.1	0.2	[14]
2.7	2.5	1.2	[11]
2.8	2.8	1.7	[10]
2.4	1.1	0.3	[24]
0.8	1.6[a]	N.A.	[19]

[a]Position 3 rather than position 4 is measured, and measurements of position 4 and so forth are N.A.
N.A. not applied

18. Mutant or transgenic plants must be in the same ecotype as WT control. This is because WT plants in different ecotypes shed their petals at different flower positions, which will consequently affect the values of pBS measurements (Table 1). Avoid using plant hybrids from different ecotypes. In this case caution should be taken that a slight delay or premature abscission may not be consistent due to big individual variations contributed by the mixed genome of different ecotypes.

19. The values of measurements from different labs can be quite different, mainly due to the specific pBS meters, setting of meters used in different labs, as well as plant growing conditions (Table 2).

Acknowledgments

This work was supported by Grants 13785/F20 to M.A.B and 348256/F20 to C-L.S from the Research Council of Norway.

References

1. Mckim SM, Stenvik GE, Butenko MA et al (2008) The BLADE-ON-PETIOLE genes are essential for abscission zone formation in Arabidopsis. Development 135:1537–1546

2. Aalen RB, Butenko MA, Stenvik G-E et al (2006) Genetic control of floral abscission. In: de Silva JT (ed) Floriculture, ornamental and plant biotechnology: advances and topical issues. Global Science Books, London, pp 101–108

3. Bleecker A, Patterson SE (1997) Last exit: senescence, abscission, and meristem arrest in *Arabidopsis*. Plant Cell 9:1169–1179

4. Estornell LH, Agusti J, Merelo P et al (2013) Elucidating mechanisms underlying organ abscission. Plant Sci 199-200:48–60

5. Lewis MW, Leslie ME, Liljegren SJ (2005) Plant separation: 50 ways to leave your mother. Curr Opin Plant Biol 9:59–65

6. Liljegren A (2013) Strategic collaboration councils in the mental health services: what are they working with? Int J integr Care 13:e009

7. Patterson SE (2001) Cutting loose. Abscission and Dehiscence in Arabidopsis. Plant Physiol 126:494–500

8. Roberts JA, Elliot KA, Gonzalez-Carranza ZH (2002) Abscission, dehiscence and other cell separation processes. Ann Rev Plant Biol 53:131–158

9. Butenko MA, Patterson SE, Grini PE et al (2003) *INFLORESCENCE DEFICIENT IN ABSCISSION* controls floral organ abscission in Arabidopsis and identifies a novel family of putative ligands in plants. Plant Cell 15:2296–2307

10. Cho SK, Larue CT, Chevalier D et al (2008) Regulation of floral organ abscission in Arabidopsis thaliana. Proc Natl Acad Sci U S A 105:15629–15634

11. Stenvik GE, Tandstad NM, Guo Y et al (2008) The EPIP peptide of INFLORESCENCE DEFICIENT IN ABSCISSION is sufficient to induce abscission in Arabidopsis through the receptor-like kinases HAESA and HAESA-LIKE2. Plant Cell 20:1805–1817

12. Shi CL, Stenvik GE, Vie AK et al (2011) Arabidopsis class I KNOTTED-like homeobox proteins act downstream in the IDA-HAE/HSL2 floral abscission signaling pathway. Plant Cell 23:2553–2567

13. Patterson S, Butenko M, Kim J (2007) Ethylene responses in abscission and other processes of cell separation in Arabidopsis. In: *Advances in plant ethylene research*. Springer, New York, pp 271–461. https://doi.org/10.1007/978-1-4020-6014-4_60

14. Patterson SE, Bleecker AB (2004) Ethylene-dependent and -independent processes associated with floral organ abscission in Arabidopsis. Plant Physiol 134:194–203

15. Basu MM, Gonzalez-Carranza ZH, Azam-Ali S et al (2013) The manipulation of auxin in the abscission zone cells of Arabidopsis flowers reveals that indoleacetic acid signaling is a prerequisite for organ shedding. Plant Physiol 162:96–106

16. Patterson SE (1998) Characterization of delayed floral organ abscission and cell separation in Arabidopsis. University of Wisconsin, Madison, Madison

17. Fernandez DE, Heck GR, Perry SE et al (2000) The embryo MADS domain factor AGL15 acts postembryonically. Inhibition of perianth senescence and abscission via constitutive expression. Plant Cell 12:183–198

18. Lease KA, Cho SK, Walker JC (2006) A petal breakstrength meter for Arabidopsis abscission studies. Plant Methods 2:2

19. Chen MK, Hsu WH, Lee PF et al (2011) The MADS box gene, FOREVER YOUNG FLOWER, acts as a repressor controlling floral organ senescence and abscission in Arabidopsis. Plant J 68:168–185

20. Gonzalez-Carranza ZH, Shahid AA, Zhang L et al (2012) A novel approach to dissect the abscission process in Arabidopsis. Plant Physiol 160:1342–1356

21. Kim J, Dotson B, Rey C et al (2013) New clothes for the jasmonic acid receptor COI1: delayed abscission, meristem arrest and apical dominance. PLoS One 8:e60505

22. Liu B, Butenko MA, Shi CL et al (2013) NEVERSHED and INFLORESCENCE DEFICIENT IN ABSCISSION are differentially required for cell expansion and cell separation during floral organ abscission in Arabidopsis thaliana. J Exp Bot 64:5345–5357

23. Sane AP, Tripathi SK, Nath P (2007) Petal abscission in rose (Rosa bourboniana var Gruss an Teplitz) is associated with the enhanced expression of an alpha expansin gene, RbEXPA1. Plant Sci 172:481–487

24. Wei PC, Tan F, Gao XQ et al (2010) Overexpression of AtDOF4.7, an Arabidopsis DOF family transcription factor, induces floral organ abscission deficiency in Arabidopsis. Plant Physiol 153:1031–1045

Chapter 7

Characterization of Climacteric and Non-Climacteric Fruit Ripening

Xiaohong Kou and Mengshi Wu

Abstract

Senescence is the terminal stage of plant development. It is a strategic and tactical response to seasonal and unpredictable stresses. As an important part of plant senescence, fruit ripening is normally viewed distinctly as climacteric or non-climacteric. In this chapter we describe protocols for the determination of a number of parameters that have been used in characterizing the ripening behavior of fruits. These include changes in respiratory rate, ethylene, flesh firmness, sugar, acidity, starch, pectin, enzymes, aroma volatiles, and expression of ripening-related genes during fruit ripening and senescence.

Key words Climacteric, Non-climacteric, Senescence, Ethylene, Ripening

1 Introduction

Senescence, which is part of a cloud of terms referring generally to the process or condition of growing old, has a specialized meaning in plant biology. Current physiological understanding of the senescence condition and its positive roles in plant growth, differentiation, adaptation, survival, and reproduction supports a definition that acknowledges senescence to be a phase of development that follows the completion of growth, which is absolutely dependent on cell viability and may or may not be succeeded by death [1]. Fruit ripening plays an important role in plant senescence and has received considerable attention because of the dramatic changes in a wide range of metabolic processes that occur before and after this event and because it has relevance for basic as well as applied research. Climacteric ripening is characterized by an upsurge in the respiration rate accompanying an autocatalytic ethylene production peak during fruit ripening. In contrast, non-climacteric fruit ripening involves cyanide-insensitive respiration to a lesser extent than in climacteric fruits, and the upsurge of respiration and ethylene is not observed or is transitory after ethylene application [2]. The patterns of ethylene and CO_2 production have been commonly used as criteria to identify climacteric fruits.

Yongfeng Guo (ed.), *Plant Senescence: Methods and Protocols*, Methods in Molecular Biology, vol. 1744,
https://doi.org/10.1007/978-1-4939-7672-0_7, © Springer Science+Business Media, LLC 2018

The ripening of fleshy fruits represents the unique coordination of developmental and biochemical pathways leading to changes in color, texture, aroma, and nutritional quality of mature seed-bearing plant organs. The gaseous plant hormone ethylene plays a key regulatory role in ripening of many fruits, including some representing important contributors of nutrition and fiber to the diets of humans. Examples include banana, apple, pear, most stone fruits, melons, squash, and tomato. Molecular exploration of the role of ethylene in fruit ripening has led to the affirmation that mechanisms of ethylene perception and response defined in the model system *Arabidopsis thaliana* are largely conserved in fruit crop species, although sometimes with modifications in gene family size and regulation. Positional cloning of genes defined by ripening defect mutations in the model fruit system tomato has recently led to the identification of both novel components of ethylene signal transduction and unique transcription factor functions influencing ripening-related ethylene production [3]. Kou et al. reported that the *Arabidopsis* silique has characteristics of a climacteric fruit and that *AtNAP*, a NAC family transcription factor gene whose expression is increased with the progression of silique senescence, plays an important role in its senescence. Silique senescence was delayed for 4–5 days in the *atnap* knockout mutant plants. The ethylene climacteric was delayed for 2 days in the *atnap* silique, and the associated respiratory climacteric was suppressed [4].

Climacteric fruits show a dramatic increase in the rate of respiration during ripening and this is referred to as the climacteric rise. The rise in respiration occurs either simultaneously with the rise in ethylene production or follows soon afterward. The increase in the rate of ethylene production is usually logarithmic [5]. Nonclimacteric fruits, on the other hand, ripen without ethylene and respiration bursts. During respiration plants consume O_2 and release CO_2; thus respiration rate can be determined by monitoring the change of CO_2 concentration in a sealed container over a certain period of time. Ethylene production is usually evaluated via gas chromatography. A number of other parameters have also been used to assist with the judgment of harvest maturity of fruits. Changes in fruit firmness, sugar, acidity, starch, pectin, enzymes, and aroma volatiles can be monitored to determine their relationship with harvest maturity and applicability as markers of ripening. Flesh firmness is usually assayed on pared surfaces from opposite sides of the fruit using a fruit penetrometer. Sugars can be measured using the HPLC system, and titratable acidity of the fruit can be measured with sodium hydroxide. For determination of starch content, sugars are removed by ethanol extraction, followed by solubilization of starch with dilute perchloric acid, after which total cell carbohydrates are measured through colorimetric determination by means of the anthrone reaction [6]. Pectin content can be estimated by measuring galacturonic acid through the carbazole and sulfuric acid spectrophotometric determination method [7].

In some cases ripening in non-climacteric fruits resembles the climacteric pattern of ripening, for which more specific analytic methods are needed to distinguish climacteric from non-climacteric fruits. Aroma volatiles produced by fruit and expression of ripening-related genes, for example, may be used in the determination of ripening pattern of fruits.

2 Materials

2.1 Respiratory Rate

1. Infrared CO_2 gas analyzer.
2. 0.2 L chambers.

2.2 Ethylene Production

1. A gas chromatograph with a flame ionization detector and an activated alumina column (model 6890 N, Agilent, Santa Clara, CA, USA).
2. A chromatograph column loaded with HP-5 (5% phenylmethylsiloxane) and a 30 m capillary alumina column (Agilent 19091J-413).
3. 1 L glass jars.
4. N_2 gas.

2.3 Application of Ethylene

1. Azetil: Containing 5.0% ethylene at the concentration of 1000 μL/L.
2. 0.19 m^3 chambers.

2.4 Application of 1-MCP

1. SmartFresh wettable powder: Containing 0.14% 1-MCP active ingredient at the concentration of 300 nL/L.
2. Flasks and chambers.

2.5 Determination of Flesh Firmness

1. GY-4 fruit penetrometer (Digital Force Gauge, Shanghai, China).

2.6 Determination of Sugar Content

1. Alliance e2695 HPLC system (Waters Co., USA) equipped with an evaporative light scattering detector (ELSD, Waters 2424, USA).
2. 80% ethanol.
3. D(−) fructose, D(+) galactose, D(+) glucose, D(+) sucrose, b-lactose, D(+) maltose, and D(+) raffinose in water (HPLC-grade) for six different concentration levels, viz., 50, 100, 500, 1000, 2500, and 5000 ppm.
4. 5 mL syringe filters.
5. 0.2 μL membrane filters.

2.7 Assessment of Fruit Acidity

1. 0.1 M sodium hydroxide: Dissolve 4 g sodium hydroxide in 1 L deionized water.

2.8 Determination of Fruit Starch [6]

1. Glucose standard: Dissolve 0.100 g of anhydrous glucose in 100 mL water containing 0.10% benzoate. Dilute 10 mL of the stock solution to 1000 mL, and use 5 mL of the diluted solution as the standard for 50 μg glucose. Prepare the dilute standard fresh.

2. Anthrone-sulfuric acid: Dissolve 2 g anthrone in 1 L of cold 95% sulfuric acid (*see* **Note 1**).

3. Ethyl alcohol-water: Dilute 1680 mL of 95% ethyl alcohol with water to make 2 L of 80% alcohol.

4. 52% perchloric acid: Add 270 mL of 72% perchloric acid to 100 mL water. Store in glass-stoppered containers.

5. 25 × 250 mm borosilicate glass tubes.

6. Heaters.

7. Centrifuge.

8. Flasks.

9. Spectrophotometer.

2.9 Determination of Water-Soluble Pectin [7]

1. Centrifuge.

2. 95% ethanol.

3. Sulfuric acid.

4. 1.5 g/L carbazole-ethanol solution: Weigh 0.150 g carbazole, and dissolve in 100 mL distilled water.

5. Galacturonic acid standard solution: Add 100 mg galacturonic acid to about 40 mL water and make up to 100 mL in a volumetric flask. Dilute the 1 mg/mL galacturonic acid to 0, 10, 20, 40, 60, 80 μg/mL, respectively.

6. Spectrophotometer.

2.10 Antioxidant Enzyme Activities

1. 1 M sodium phosphate buffer (pH 7.8): Prepare 1 M Na_2HPO_4 and 1 M NaH_2PO_4, and mix them to get 1 M sodium phosphate buffer (pH 7.8).

2. Extraction buffer: 0.2 M sodium phosphate buffer, pH 7.8, 1 mM phenylmethylsulfonyl fluoride (PMSF), 2 mM EDTA, and 5% insoluble PVPP.

3. 0.1 M ascorbate.

4. 0.1 M methionine.

5. 57 μM NBT: Dissolve nitrotetrazolium blue chloride (NBT) in 70% dimethylformamide (DMF).

6. 1 M Hepes-KOH (pH 7.6): Dissolve Hepes in water and use potassium hydroxide pellets to adjust pH to 7.5.

7. 0.1 M H_2O_2: Dissolve 30% H_2O_2 in water.

8. 1 M sodium phosphate buffer (pH 7.0): Prepare 1 M Na_2HPO_4 and 1 M NaH_2PO_4, and mix them to get 1 M sodium phosphate buffer (pH 7.0).

9. 0.1 M guaiacol: In ethanol.

10. 1.3 µM riboflavin.

11. 30% (v/v) Triton X-100:In 0.01 M PBS (pH 7.3).

12. 0.5 M EDTA-Na_2.

2.11 Volatile (Aroma) Analysis

1. Glass vials.

2. Electric analytical balance.

3. Vortex.

4. Glass beakers.

5. Volumetric flasks.

6. Saturated calcium chloride solution (20 °C): Weigh 74.5 g $CaCl_2$ to a glass beaker, and add 100 mL water to dissolve (*see* **Note 2**).

7. Phenylethyl alcohol: 0.1% (v/v).

8. n-Alkane standards (C8–C20) (HPLC-grade): Octane, nonane, decane, undecane, dodecane, tridecane, tetradecane, pentadecane, hexadecane, heptadecane, octadecane, nonadecane, and eicosane.

9. Crimp-top caps with TFE/silicone septa seals (Alltech Associates, Inc., Deerfield, IL).

10. Gas chromatograph (GC) (6890N, Agilent Technologies,Wilmington, DEL).

11. 1-cm-long SPME fibers containing 50/30 µm divinylbenzene/carboxene on polydimethylsiloxane coating (57329-U DVB/CarboxenTM/PDMS Stable FlexTM Fiber, Supelco, Bellefonte, PA).

12. Capillary column: 30 m × 0.25 mm i.d. × 0.25 µm thick (HP-5MS, Agilent Technol.).

13. 0.1% (v/v) phenylethyl alcohol: Pipette 100 µL phenylethyl alcohol (HPLC-grade) to a 100 mL glass beaker containing about 40 mL water and make up to 100 mL in a volumetric flask.

14. Chem-Station software (G1791CA, Version C.00.00, Agilent Technol.).

2.12 Expression of Ripening-Related Genes

1. Fruits of uniform size sampled at different developmental stages.

2. Tissues from the fruits cut into small cubes, then quickly frozen in liquid nitrogen, and stored at −80 °C until use.

3. Column Plant RNAOUT kit (Tiandz#71203, China).

4. RNase-Free DNase.

5. Agarose.

6. Spectrophotometer.

7. iScrip™ cDNA Synthesis Kit (Bio-Rad Product #170–8890).

8. iQ™ SYBR Green Supermix (Bio-Rad Product #170–8880).

9. Bio-Rad® IQ-5 thermocycler.

3   Methods

3.1   Respiratory Rate

Samples of air (1.0 mL) exiting from chambers are taken to determine rates of CO_2 on the day after harvest and subsequently at 24 h intervals [8] to evaluate respiratory rate. Carbon dioxide concentration is determined by pulse infrared gas analysis. Measurements are standardized using calibrated compressed gas mixtures purchased from the British Oxygen Co., Australia. The CO_2 concentration in the flasks was determined by infrared method.

1. For each sample, three or four fruits are weighed and enclosed singly in 0.2 L chambers at 20 °C.

2. Connect two vents of airtight specimen bottle to form a closed-circuit circulation, and record the CO_2 level (W_1) in the vessel when the portable infrared CO_2 gas analyzer show a stable digital of CO_2 level.

3. Fruits with known masses are then placed in the bottle for 1 h at room temperature [9], and record the CO_2 level (W_2). Results are expressed as mg $CO_2 \cdot kg^{-1} \cdot h^{-1}$ [10].

4. Respiratory rate is calculated as follows:

$$\text{The respiratory rate} \left(\text{mg } CO_2 \cdot kg^{-1} \cdot h^{-1} \right)$$
$$= \left[V \times (W_2 - W_1)\% \times 1.9643 \times 10^{-3} \right] / (m \times t)$$
$$= (W_2 - W_1)\% \times 0.00294645$$

where

V: The volume of airtight specimen bottle

W_1: The level of CO_2 of blank control

W_2: The level of CO_2 after putting the fruit for about 1 h

m: Mass of fruit, kg

t: Measure time, h

3.2   Ethylene Production

1. Fresh fruits of each stage are selected and placed in 1 L glass jars for 1 h at 20 °C.

2. One milliliter of the jar headspace gas is withdrawn and injected into a gas chromatograph with a flame ionization detector and an activated alumina column.

3. The ethylene produced ($nL \cdot g^{-1}$ fresh weight [FW]$\cdot h^{-1}$) is analyzed using the following conditions: chromatograph column included HP-5 and a 30 m capillary alumina column, temperatures of 80 °C for the column and 150 °C for the detector are used, the rate of the carrier gas flow is N_2 with 40 $mL \cdot min^{-1}$, and hydrogen pressure is 0.6 $kg \cdot cm^{-2}$. The experiment is done with three replications [11].

3.3 Application of Ethylene

1. Fruits are treated with Azetil in hermetically sealed chambers (0.19 m^3) for 24 h at 23 °C.

2. After treatment, open the chambers and keep the fruits at 23 ± 1 °C and $85 \pm 5\%$ RH [10].

3.4 Application of 1-MCP

1. Predetermined amounts of SmartFresh WP are transferred to capped flasks.

2. Add 20 mL prewarmed (50 °C) distilled water to the flasks, and stir until complete dissolution of the powder.

3. Fruits are placed in hermetically sealed chambers, and the flasks are opened inside the chambers at 23 °C for 12 h (*see* **Note 3**).

4. After treatment, the chambers are opened and fruits kept under room conditions [10, 12].

3.5 Determination of Flesh Firmness

1. Flesh firmness is analyzed on two sides of each fruit using a GY-4 fruit penetrometer [11].

2. The strength of flesh firmness is recorded as N cm^{-2}. The experiment is done with three replications.

3.6 Determination of Sugar Content

1. Approximately 0.15 g of sample is dissolved in 20 mL of ethanol (80%) by stirring well (20 min) followed by sonication (10 min) and further stirring (5 min) [13].

2. The resulting mixture is centrifuged at 4 °C in $4000 \times g$ for 10 min.

3. The supernatant is injected into the 20 µL loop with 5 mL syringe filters containing 0.2 µL membrane filters.

4. A standard solution mixture is prepared by dissolving D(−) fructose, D(+) galactose, D(+) glucose, D(+) sucrose, b-lactose, D(+) maltose, and D(+) raffinose in water (HPLC-grade) for six different concentration levels, viz., 50, 100, 500, 1000, 2500, and 5000 ppm.

5. The HPLC analysis is performed with the prevailing carbohydrate column (S5 lm, 250 mm × 4.6 mm i.d.) isocratically following the HPLC conditions: flow rate = 1.0 µL/min; data rate = 1 pps; run time = 15 min; gain = 1; column heater temperature = 35 °C; sample temperature = 5 °C; pressure = 50 psi; nebulizer, heating

(90%) and injection volume= 10 μL. Acetonitrile/water mixture (70:30) is used as the mobile phase.

6. Sugar contents are calculated based on standard curves.

3.7 Assessment of Fruit Acidity

1. Fruits are carefully dissected into pulp, skin, and seeds. Then, the pulp fraction is immediately frozen in liquid nitrogen and stored at −80 °C until analysis.

2. The titratable acidity of the juice is measured with sodium hydroxide (0.1 M NaOH) up to pH 7.0 [14].

3.8 Determination of Fruit Starch [6]

3.8.1 Sample Preparation

1. Blend 100 g sample with equal weight of water in a mechanical blender for 5 min.

2. Weigh 5.00 g sample of this slurry into a 50 mL centrifuge tube, and extract 4 times with 30 mL hot 80% alcohol by centrifugation at $2000 \times g$ for 10 min and decantation.

3. After the final extraction, add water to the sugar-free residue to 10 mL.

4. Cool the mixture in an ice water bath, and stir while adding 13 mL of 52% perchloric acid reagent.

5. Stir occasionally for 15 min.

6. Add 20 mL of water, stir, and centrifuge at $2000 \times g$ for 10 min.

7. Pour the solution into a 100 mL volumetric flask.

8. To the residue add water to 5 mL, cool as before, and add 6.5 mL of perchloric acid reagent.

9. Solubilize the samples as above (**steps 4–6**) for 15 min (*see* **Note 4**), and wash the content of the tube into the 100 mL volumetric flask. Bring to the final volume and filter.

3.8.2 Determination of Starch

1. Dilute 5 mL of the filtered starch solution to 500 mL.

2. Pipette 5 mL of the diluted solution into a 25 × 250 mm borosilicate glass tube, cool in a water bath, and add 10 mL of fresh anthrone-sulfuric acid.

3. After the anthrone reagent has been added to all of a series of sample tubes cooled in water, mix each one thoroughly and heat them together for 7.5 min at 100 °C.

4. Cool the tubes rapidly to 25 °C in water bath, and determine the color intensities at 630 nm.

5. Prepare a standard curve, pipette 5 mL of the standard solution and repeat **steps 2–4**, and use this calibrated curve to obtain the yield of glucose from starch. Multiply glucose yield by 0.90 to convert to starch.

3.9 Determination of Water-Soluble Pectin [7]

3.9.1 Pectin Extraction

1. Weigh 1.00 g of fresh fruit pulp into a mortar and ground to slurry. Pour the slurry into a 50 mL centrifuge tube. Wash the content of the mortar twice into the tube with 95% ethanol.

2. Add 25 mL 95% ethanol to the tube and keep in boiling water bath for 30 min.

3. Cool the mixture to room temperature and centrifuge at $10,000 \times g$ for 10 min. Discard the supernatant.

4. Repeat **steps 1–3** twice.

5. Add 20 mL distilled water to the residue, and keep in 50 °C water bath for 30 min to dissolve pectin.

6. Cool the mixture in water to room temperature and centrifuge at $10,000 \times g$ for 10 min. Transfer the supernatant to a 100 mL volumetric flask.

7. Repeat **steps 5** and **6** twice. Rinse 2× with water into the flask and bring to the volume (*see* **Note 5**).

3.9.2 Pectin Determination

1. Pipette 1.0 mL pectin solution or the standard solution into a 10 mL test tube, carefully add 6.0 mL concentrated sulfuric acid, and thaw in the boiling water bath for 20 min.

2. Cool the mixture to room temperature, and add 0.2 mL 1.5 g/L carbazole-ethanol solution. Shake vigorously.

3. Measure the absorbance at the wavelength of 530 nm after standing in the dark for 30 min.

4. The total pectin concentration is calculated based on a standard curve.

3.10 Antioxidant Enzyme Activities [15]

3.10.1 Enzyme Extraction

1. One gram of frozen ground tissue is homogenized with a pre-cooled mortar and pestle in 1.5 mL extraction buffer. For extraction of APX, 5 mM ascorbate is added to the extraction buffer.

2. After homogenization, the extracts are centrifuged at $14,000 \times g$ at 4 °C for 30 min. The supernatant was collected and immediately used for enzyme assay.

3.10.2 SOD Activity

1. Make 1 mL of reaction mixture containing 0.1 M sodium phosphate buffer (pH 7.8), 10 mM methionine, 57 µM NBT, 1.3 µM riboflavin, 0.025% (v/v) Triton® X-100, and 0.11 µM EDTA and 20 µL enzyme extract (*see* **Note 6**).

2. The reaction is initiated by illuminating the reaction mixture with 2 of 15 W fluorescent lights. After 10 min illumination, the reaction is stopped by removing the light source. A blank tube with reaction mixture and extraction buffer is kept in the dark as a negative control. Sample tubes with reaction mixture and enzyme extracts are kept in light [15].

3. SOD activity is calculated by NBT reduction in light with extraction buffer (control tube) minus NBT reduction with sample extraction (sample tube). One unit of SOD activity is defined as the amount of enzyme required to cause 50% inhibition of the reduction of NBT as monitored at 560 nm.

3.10.3   APX Activity

1. The reaction mixture contains 50 mM Hepes-KOH (pH 7.6), 0.1 mM EDTA-Na$_2$, 0.5 mM ascorbate (*see* **Note 7**), and 50 μL enzyme extracts. The reaction is initiated by adding 0.5 mL 0.4 mM H$_2$O$_2$.

2. The absorbance rate is monitored at 290 nm over 3 min. APX activity is determined by monitoring the decrease in absorbance at 290 nm as ascorbate being oxidized.

3.10.4   CAT Activity

1. The reaction mixture contains 50 mM sodium phosphate buffer (pH 7.0) and 10 μL extracted sample.

2. The reaction is initiated by adding 0.5 mL 10 mM H$_2$O$_2$. And the absorbance is monitored at 240 nm over 5 min. A molar extinction coefficient of 39.4 mM^{-1} cm^{-1} is used to calculate CAT activity.

3.11   POX Activity

1. The reaction mixture contains 50 mM sodium phosphate buffer (pH 7.0), 10 mM H$_2$O$_2$, and 10 mM guaiacol.

2. The reaction is initiated by adding 10 μL extracted sample. And the absorbance is recorded by monitoring the increase in absorbance at 470 mm over 3 min due to guaiacol oxidation. A molar extinction coefficient of 26.8 mM^{-1} cm^{-1} is used to calculate POX activity, which is expressed as mol guaiacol min^{-1} mg^{-1} protein.

3.12   Volatile (Aroma) Analysis

3.12.1   Sample Preparation

1. Weigh 2.00 g of the fresh fruit sample, and homogenize with saturated calcium chloride solution [2, 16] in a vortex (*see* **Note 8**).

2. Transfer the mixture into a 10 mL glass vial, and add 10 μL of 0.1% (v/v) phenylethyl alcohol (internal standard) [2].

3. Seal the vial using crimp-top caps with TFE/silicone septa seals, and place in the heat tray of the GC at 60 °C for 20 min for the headspace to form [17].

3.12.2   Solid-Phase Microextraction (SPME) [2]

1. Enter 22 mm of the needle of the 1-cm-long SPME fiber (*see* **Note 9**) into the vial headspace and remain 20 min at 60 °C for absorbing the volatiles.

2. Set the gas chromatography-mass spectrometry (GC-MS) and inject the samples. Perform the analysis with the following parameters [2]:

    (a) The analyses are conducted with a MPS2 Gerstel Multipurpose sampler coupled to the GC-MS.

(b) The injection port is operated at 280 °C in the splitless mode and subjected to a pressure of 80 psi.

(c) Volatiles are separated on a 30 m × 0.25 mm i.d. × 0.25 μm thickness capillary column that contained 5% phenylmethyl silicone as a stationary phase.

(d) The carrier gas is helium with a flow rate of 1.5 mL·min^{-1}.

(e) The initial oven temperature is 35 °C, followed by a ramp of 2 °C min^{-1} up to 75 °C, and then at 50 °C min^{-1} to reach a final temperature of 250 °C, which is held for 5 min.

(f) The inlet liner used is a 2637505 SPME/direct (Supelco), 78.5 mm × 6.5 mm × 0.75 mm.

(g) Mass spectra are obtained by electron ionization (EI) at 70 eV, and a spectrum range of 40–450 m/z is used.

(h) The detector works at 230 °C and in full scan with data acquisition and ion mass captured between 30 and 300 a.m.u.

(i) Desorb the volatiles from the SPME fiber into the GC injection port set at 280 °C for 3 min.

3. Evaluate the chromatograms and mass spectra using the ChemStation software.

4. Identify the peaks by comparing the experimental spectra with those of the National Institute for Standards and Technology (NIST98, search version 2.0) data bank.

5. Calculate the Kovats indices for all identified compounds. The volatile compounds are positively identified by comparing their Kovats retention indices (*Is*) with previously reported *Is*. The *Is* of unknown compounds were determined via sample injection with a homologous series of alkanes (C_8–C_{20}). C_6 and C_7 are calculated from compounds present in the chromatograms.

6. Results from the volatile analyses are expressed as the percentage of each compound's integrated area relative to the total integration (*see* **Note 10**). The Chemical Abstract Service (CAS) numbers of the volatiles reported in the NIST98 database also found in the website, http://webbook.nist.gov/chemistry/nameser.html, are used for obtaining their corresponding IUPAC names by checking the website http://www.chemindustry.com/apps/chemicals.
The Kovats index is given by the equation:

$$ I = 100 \times \left[n + (N - n) \frac{\log\left(t'_{r(unknown)}\right) - \log\left(t'_{r(n)}\right)}{\log\left(t'_{r(N)}\right) - \log\left(t'_{r(n)}\right)} \right] $$

where

I: Kovats retention index

n: The number of carbon atoms in the smaller n-alkane

N: The number of carbon atoms in the larger n-alkane

t_r': The adjusted retention time

For temperature-programmed chromatography, the Kovats index is given by the equation:

$$I = 100 \times \left[n + (N - n) \frac{t_{r(\text{unknown})} - t_{r(n)}}{t_{r(N)} - t_{r(n)}} \right]$$

where

I: Kovats retention index

n: The number of carbon atoms in the smaller n-alkane

N: The number of carbon atoms in the larger n-alkane

t_r: The retention time

3.13 Expression of Ripening-Related Genes

3.13.1 RNA Isolation

1. Total RNA is extracted from fruit tissues using the Column Plant RNAOUT kit [18].

2. Remove genomic DNA by 15 min incubation at 37 °C with RNase-Free DNase.

3. Analyze the purity and integrity of the RNA by both agarose gel electrophoresis and the A_{260}:A_{230} and A_{260}:A_{280} ratios using a spectrophotometer [18].

3.13.2 cDNA Synthesis

1. cDNA is synthesized using the iScrip™ cDNA Synthesis Kit according to the manufacturer's protocol [4].

3.13.3 Identification of ACO, ACS, and ETR Isoforms

1. Full-length isoform sequences of tomato SAMS, ACO, ACS, ETR, and CTR and other ripening-related genes are first obtained from the NCBI database and are used to identify any related accessions using the BLASTn search in the NCBI database as well as an EST database of a given material [12].

3.13.4 qPCR Analysis

1. The cDNA obtained above is used as a template for amplification of ripening-related genes.

2. For each qPCR reaction, 1 µL of each diluted sample is used as a template in a 25 µL reaction containing 12.5 µL 2× SYBR Green Supermix, 8.5 µL ddH$_2$O, and 1 µL of each primer [4].

3. All qPCR reactions were performed on a Bio-Rad® IQ-5 thermocycler with 40 cycles and an annealing temperature of 55 °C. Cycle threshold (Ct) values were determined by the IQ-5 Bio-Rad software assuming 100% primer efficiency [4].

4. Relative gene expression was analyzed by qPCR and the $2^{-\triangle\triangle CT}$ method.

4   Notes

1. Store the reagent at around 0 °C and prepare fresh every 2 days. This reagent is unstable and gives high background and variable results when stored for too long.

2. Saturated calcium chloride solution can reduce solubility and improve volatilization of the analytes in aqueous solutions, thus making more volatile substances adsorbed to the fiber and enhancing the response values. Saturated $CaCl_2$ solution can also stops enzymatic reactions that could cause further changes in volatile levels [19].

3. The flasks containing SmartFresh wettable powder are opened inside the chambers, which are immediately sealed to avoid gas loss.

4. The solution should be cooled to room temperature when perchloric acid reagent is added.

5. Add 1–2 mL water to the tube and centrifuge. Pipette the supernatant to the flask.

6. For determination of enzyme activities, the reaction mixtures are freshly made before the reaction.

7. 1 mM ascorbate is only used for APX (EC 1.11.1.11) activity.

8. The fruit samples should be adequately mixed with saturated calcium chloride solution to make the aroma released well.

9. The fiber should be preconditioned in the injection port at 250 °C for 3 h when being used for the first time. The preconditioning should be 1 h if the fiber has been used before.

10. To quantify the volatile compounds, the samples are tested in triplicates, and the integrated areas based on the total ion chromatogram are normalized to the areas of the internal standard and averaged. The relative concentrations of volatile compounds in the samples are determined by comparison with the concentration of the internal standard (0.1% v/v phenyl-ethyl alcohol), assuming a response factor of 1.

Acknowledgments

This work was supported by the National Natural Science Foundation of China (project no. 31171769). We are very grateful to Qiong Chen, Chen Liu, Lihua Han, and Jiyun Wu in Tianjin University for helping in collecting and checking over documents.

References

1. Thomas H (2013) Senescence, ageing and death of the whole plant. New Phytol 197:696–711

2. Obando-Ulloa JM, Moreno E, García-Mas J et al (2008) Climacteric or non-climacteric behavior in melon fruit 1. Aroma volatiles. Postharvest Biol Technol 49:27–37

3. Barry CS, Giovannoni JJ (2007) Ethylene and fruit ripening. J Plant Growth Regul 26:143–115

4. Kou XH, Waltkins CB, Gan SS (2012) Arabidopsis AtNAP regulates fruit senescence. J Exp Bot 63:6139–6147

5. Paul V, Pandey R, Srivastava GC (2012) The fading distinctions between classical patterns of ripening in climacteric and non-climacteric fruit and the ubiquity of ethylene —an overview. J Food Sci Technol 49:1–21

6. McCready RM, Guggolz J, Silviera V et al (1950) Determination of starch and amylose in vegetables. Anal Chem 22:1156–1158

7. Ma WP, Ni ZJ, Ren X et al (2011) Effects of different storage temperatures on the quality and physiology of melon. J Anhui Agric Sci 39:14996–15000

8. Abdi N, Holford P, McGlasson WB et al (1997) Ripening behavior and responses to propylene in four cultivars of Japanese plums. Postharvest Biol Technol 12:21–34

9. Gupta A, Pal RK, Rajam MV (2013) Delayed ripening and improved fruit processing quality in tomato by RNAi-mediated silencing of three homologs of 1-aminopropane-1-carboxylate synthase gene. J Plant Physiol 170:987–995

10. Azzolini M, Jacimino AP, Bron IU et al (2005) Ripening of "Pedro Sato" guava: study on its climacteric or non-climacteric nature. Braz J Plant Physiol 17:299–306

11. Sun JH, Luo JJ, Tian L et al (2013) New evidence for the role of ethylene in strawberry fruit ripening. J Plant Growth Regul 32:461–470

12. Aizat WM, Able JA, Stangoulis JCR et al (2013) Characterisation of ethylene pathway components in non-climacteric capsicum. BMC Plant Biol 13:191–204

13. Shanmugavelan P, Kim SY, Kim JB et al (2013) Evaluation of sugar content and composition in commonly consumed Korean vegetables, fruits, cereals, seed plants, and leaves by HPLC-ELSD. Carbohydr Res 380:112–117

14. Chervin C, El-Kereamy A, Roustan JP et al (2004) Ethylene seems required for the berry development and ripening in grape, a non-climacteric fruit. Plant Sci 167:1301–1305

15. Lee J, Cheng LL, Rudel DR et al (2012) Antioxidant metabolism of 1-methylcyclopropene (1-MCP) treated 'empire' apples during controlled atmosphere storage. Postharvest Biol Technol 65:79–91

16. Aguiar MCS, Silvério FO, de Pinho GP et al (2014) Volatile compounds from fruits of Butiacapitata at different stages of maturity and storage. Food Res Int 62:1095–1099

17. Jung H, Lee SJ, Lim JH et al (2014) Chemical and sensory profiles of makgeolli, Korean commercial rice wine, from descriptive, chemical, and volatile compound analyses. Food Chem 152:624–632

18. Kou XH, Wang S, Wu MS et al (2014) Molecular characterization and expression analysis of NAC family transcription factors in tomato. Plant Mol Biol Rep 32:501–516

19. Berna AZ, Buysens S, Natale CD et al (2005) Relating sensory analysis with electronic nose and headspace fingerprint MS for tomato aroma profiling. Postharvest Biol Technol 36:143–155

Part III

Hormonal Control of Plant Senescence

Chapter 8

Ethylene Treatment in Studying Leaf Senescence in *Arabidopsis*

Zhonghai Li and Hongwei Guo

Abstract

Plant leaf senescence is the final process of leaf development that involves the mobilization of nutrients from old leaves to newly growing tissues. Leaf senescence involves a coordinated action at the cellular and organism levels under the control of a highly regulated genetic program. However, leaf senescence is also influenced by multiple internal and environmental signals that are integrated into the age information. Among these internal factors, the simple gaseous phytohormone ethylene is well-known to be an important endogenous modulator of plant leaf senescence and fruit ripening. Ethylene and its biosynthetic precursor ACC (1-aminocyclopropane-1-carboxylic acid) as well as inhibitors of ethylene biosynthesis aminoethoxyvinylglycine (AVG) and action (AgNO₃) or the inhibitor of ethylene perception 1-MCP (1-methylcyclopropene) have been widely used by the research community of plant leaf senescence. However, until now, no systemically experimental method about the usage of ethylene in studying the molecular mechanisms of leaf senescence in *Arabidopsis* is available. Here, we provide detailed methods for exogenous application of ethylene and its inhibitors in studying *Arabidopsis* leaf senescence, which has been successfully used in our laboratory.

Key words Leaf senescence, Ethylene, Inhibitor, *Arabidopsis*

1 Introduction

Ethylene is a simple gaseous phytohormone that regulates a wide variety of processes during plant development, including cell division, cell elongation, fruit ripening, abscission, and flower and leaf senescence, as well as biotic and abiotic stress responses [1]. Application of exogenous ethylene promotes leaf senescence, while treatments with inhibitors of ethylene biosynthesis or action retard senescence [2]. The role of ethylene in controlling the timing of developmental leaf senescence is further hinted by the ethylene-insensitive mutant *ein2* that was identified as the delayed senescence mutant *ore2/ore3* [3, 4]. EIN2 plays an essential role in ethylene signaling, and loss-of-function mutants such as *ein2-5* show complete ethylene insensitivity [5]. Transgenic plants overexpressing *EIN2* exhibit early flowering and early senescence

Yongfeng Guo (ed.), *Plant Senescence: Methods and Protocols*, Methods in Molecular Biology, vol. 1744,
https://doi.org/10.1007/978-1-4939-7672-0_8, © Springer Science+Business Media, LLC 2018

phenotypes compared with the *ein2-5* mutant [6]. By contrast, loss-of-function of EIN2 delays leaf senescence process [3]. These results demonstrate that EIN2 is a positive regulator of senescence, which has been deposited in the leaf senescence database (LSD) [6–8]. Kim et al. reported that EIN2 controls the leaf senescence process partly through regulating the expression level of *miR164* and one of its target genes, *oresara1* (*ORE1/NAC2*), which is a positive regulator of leaf senescence [4]. *miR164* progressively decreases during leaf aging, and this reduction is almost completely abolished by loss-of-function mutations in *EIN2* [4]. *miR164* functions as a "brake" against premature senescence caused by overexpression of *ORE1/NAC2* and fine-tunes the timing of senescence [4]. As a downstream component of EIN2, the transcription factor EIN3 was demonstrated to be a functional senescence-associated gene, which accelerates senescence by directly repressing *miR164* transcription [9]. These molecular and genetic studies bring about a wholly connected signaling cascade of *EIN2-EIN3-miR164-NAC2* [9]. Recently, Kim et al. reported that EIN3 directly binds the promoters of *ORE1/NAC2* and *AtNAP* and induces their transcription, which suggests that EIN3 controls leaf senescence process by regulating *ORE1/NAC2* and *AtNAP* [10], which have been previously reported as key regulators of leaf senescence [11]. Collectively, these studies suggest that EIN3, a downstream signaling molecule of EIN2, regulates leaf senescence via a gene regulatory network involving *miR164* and senescence-associated *NAC* family transcription factors [9, 11]. Recently, Qiu et al. reported that EIN3 could directly promote chlorophyll degradation by physically binding to the promoters of three major chlorophyll catabolic genes (CCGS), NYE1, NYC1, and PAO, to activate their expressions [12]. Meanwhile, ORE1/ NAC2, a direct target of EIN3 [11], also activates the expression of the similar set of chlorophyll catabolic genes directly [12]. This work reveals that EIN3, ORE1, and CCGs constitute a coherent feed-forward loop involving in the robust regulation of ethylene-mediated chlorophyll degradation during leaf senescence in *Arabidopsis* [12].

Although ethylene has been widely used by the research community of plant senescence worldwide [13], no systemically experimental methods is available. Here, we provide detailed methods for exogenous application of ethylene and its inhibitors in studying *Arabidopsis* leaf senescence. Ethylene-insensitive mutants such as *ein2*, *ein3 eil1*, or *etr1* can be used as positive controls. Ethylene biosynthesis inhibitor aminoethoxyvinylglycine (AVG) [14] and action inhibitor $AgNO_3$ [15] are also used to treat detached leaves under dark conditions.

2   Materials

1. Plant materials: All of the transgenic lines and mutants were derived from the wild-type *Arabidopsis thaliana* Columbia (Col-0) ecotype and cultivated in growth chambers under long-day conditions (LDs, 16 h light/8 h dark) at 22 °C under fluorescence illumination (100–150 $\mu E\ m^{-2}\ s^{-1}$). Seeds were sterilized and stratified in the dark at 4 °C for 3 days and germinated on Murashige and Skoog (MS) medium (pH 5.7) [16] supplemented with 1% sucrose and 0.8% (w/v) agar. The plants of *etr1-1*, *ein2-5*, *ein3-1 eil1-1*, *eto1*, *EIN3ox* (*35S:EIN3*) [17], *pEIN3:GUS/Col-0* [9] and *5xEBS:GUS/Col-0* [18] are used.

2. Murashige and Skoog (MS) medium: MS medium [16] is supplemented with 0.8% agar (w/v) and 1% sucrose (w/v) autoclaved at 121 °C for 20 min. Adjust the pH of the medium to 5.7–5.8 with a solution of 10 N KOH. After the medium is cooled down to 60 °C, non-autoclavable additives such as antibiotics and plant hormones can be added.

3. 100 μM ACC (1-aminocyclopropane-l-carboxylic acid) (*see* **Note 1**).

4. 50 μM AVG [(*S*)-*trans*-2-amino-4-(2-aminoethoxy)-3-butenoic acid hydrochloride] (*see* **Note 2**).

5. 50 μM $AgNO_3$ (*see* **Note 3**).

6. 50 μM MeJA (methyl jasmonate): 1.146 μL 95% MeJA is firstly dissolved in ethanol, and then diluted in 100 mL H_2O to a final concentration of 0.005% ethanol; this concentration of ethanol is included in all of the treatments, including the water controls.

7. X-Gluc (5-bromo-4-chloro-3-indolyl glucuronide) (*see* **Note 4**).

8. GUS staining buffer: 100 mM Na phosphate buffer, pH 7.0, 10 mM EDTA, 0.1% Triton X-100, 2 mM potassium ferricyanide, and 2 mM potassium ferrocyanide [19].

9. Seed sterilization solution: 75% (v/v) ethanol/water solution supplemented with 0.05% (v/v) Triton X-100.

10. Ethylene gas.

11. An airtight chamber with inlet and outlet ports.

12. A gas flow-through system [20].

13. A gas chromatograph equipped with a flame-ionization detector with a 30 m HP-PLOT column [21].

3 Methods

3.1 Surface Sterilization of Arabidopsis Seeds (See Note 5)

1. Put seeds in sterilization solution for 15 min, and agitate every 30 s. About 1 mL solution in a 1.5 mL centrifuge tube is suitable for 100 μL seeds.

2. Centrifuge briefly to settle the seeds, and then remove the solution.

3. Wash seeds with 1 mL ethanol as soon as possible.

4. Discard the ethanol solution, and dry the seeds in a clean bench for approximately 1 h. Vacuum concentrator can be used to dry the seeds if the machine is available in the laboratory (approximately 20 min).

3.2 ACC-Induced Leaf Senescence Under Dark Conditions

1. Detach the third or fourth rosette leaves from 4-week-old plants, such as Col-0, *ein2-5*, *ein3 eil1*, and *EIN3ox*.

2. Put the detached leaves with right-side up in Petri dishes containing 100 μM ACC (*see* **Note 6**).

3. Seal the Petri dishes with Parafilm tape, and wrap with double-layer aluminum foil, and then keep them at 22 °C.

4. Take pictures of leaves after treatment under darkness for 0, 3, and 4 days.

5. Measure senescence-related parameters, including chlorophyll content, Fv/Fm, and ion leakage.

6. Take and freeze the treated leaves immediately in liquid N_2. Store the leaf samples at -80 °C until use for RNA extraction and qRT-PCR analysis of senescence-associated genes.

7. Repeat the experiment three times.

3.3 Inhibition of Dark-Induced Leaf Senescence with AVG Treatments

1. Detach the third or fourth rosette leaves from 4-week-old plants, such as Col-0, *ein2-5*, as well as *ein3 eil1*.

2. Put the detached leaves with right-side up in Petri dishes containing 50 μM AVG (*see* **Note 6**).

3. Seal the Petri dishes with Parafilm tape, and wrap with double-layer aluminum foil, and keep the Petri dishes at 22 °C.

4. Take pictures of leaves after treatment under darkness for 0, 3, 4, and 5 days.

5. Measure senescence-related parameters, including chlorophyll content, Fv/Fm, and ion leakage.

6. Take and freeze the treated leaves immediately in liquid N_2. Store the leaf samples at −80 °C until use for RNA extraction and qRT-PCR analysis of senescence-associated genes.

7. Repeat the experiment three times.

3.4 Inhibition of Dark-Induced Leaf Senescence by Pretreatment with AgNO₃

1. Detach the third or fourth rosette leaves from 4-week-old plants, such as Col-0, *ein2-5*, as well as *ein3 eil1* double mutant.

2. Pretreat the detached leaves by floating on 50 μM AgNO₃ solution for 1 h (*see* **Note 7**).

3. Wash the leaves with water three times.

4. Transfer the leaves (right-side up) to Petri dishes containing test solutions without AgNO₃.

5. Seal the Petri dishes with Parafilm tape, and wrap with double-layer aluminum foil, and then keep the Petri dishes at 22 °C.

6. Take pictures of leaves after treatment under darkness for 5 days.

7. Repeat the experiment three times.

3.5 Study of Roles of the Ethylene Signal Pathway in MeJA-Induced Leaf Senescence Under Dark Conditions (See Note 8)

1. Detach the third or fourth rosette leaves from 4-week-old plants, including Col-0, *ein2-5*, *ein3 eil1*, as well as *coi1*.

2. Pretreat the detached leaves with or without 50 μM AgNO₃ solution.

3. Put the pretreated leaves with right-side up in Petri dishes containing 50 μM MeJA (*see* **Note 9**).

4. Seal the Petri dishes with Parafilm tape, and wrap with double-layer aluminum foil, and then keep the Petri dishes at 22 °C.

5. Take pictures of leaves after treatment under darkness for 3 days.

6. Repeat the experiment three times.

3.6 Study the Effects of Ethylene Signal Pathway Components on Natural Leaf Senescence

1. Sow the seeds of *etr1-1*, *ein2-5*, *ein3 eil1*, as well as *EIN3ox* in soil.

2. Grow the mutant plants in soil under long-day conditions (16 h light/8 h dark) along with wild-type controls [22].

3. Detach the fourth rosette leaf of individual plants 12, 16, 20, 24, 28, 32, and 36 days after emergence (DAE) (*see* **Note 10**).

4. Take pictures of leaves or whole plants.

5. Measure senescence-related parameters, including chlorophyll content, Fv/Fm, and ion leakage.

6. Freeze the treated leaves in liquid N_2, and store them at −80 °C until use for RNA extraction and qRT-PCR analysis of SAG gene expression.

3.7 Treatment of Whole Plants with Ethylene Gas (See Note 11)

1. Sow the seeds of *etr1-1*, *ein2-5*, *ein3 eil1*, as well as *EIN3ox* in soil.

2. Grow the mutant plants in soil pots under long-day conditions (16 h light/8 h dark) along with wild-type controls.

3. Put the pots with plants into an airtight chamber, and treat with either air or 100 µL L^{-1} ethylene gas for 48 h.

4. Take out the plants and take pictures.

5. Measure senescence-related parameters, including chlorophyll content, Fv/Fm, and ion leakage.

6. Repeat the experiment three times.

3.8 Measurement of EIN3 Transcription Levels and Protein Activities During Leaf Senescence

1. Grow the transgenic plants *pEIN3:GUS*/Col-0 and *5xEBS:GUS*/Col-0 in soil under long-day conditions (16 h light/8 h dark) at 22 °C [9].

2. Detach the rosette leaves of individual plants at DAE 12, 16, 20, 24, 28, 32, and 36, and then put the leaves in Petri dishes.

3. Immerse the leaves in GUS staining solution [19].

4. Incubate the leaves overnight in darkness at 37 °C.

5. Rinse leaves in 70% ethanol for at least 5 min after staining.

6. Clear the chlorophyll in fresh 70% ethanol for at least 4 h at room temperature or about 30 min at 65 °C.

7. Take pictures.

3.9 Measurement of Ethylene Production

1. Weigh ~1 g freshly collected leaves into 50 mL glass flasks with 10 mL liquid MS media under darkness for 1 day at 22 °C.

2. Take 100 µL air samples from each flask, and inject into the column of the gas chromatograph.

3. Record the observations from at least three replicating flasks.

4. Verify the retention time with standard samples of 0.1, 1, 10, and 50 ppm ethylene gas.

5. Quantify the amount of ethylene production.

4 Notes

1. ACC is a biosynthetic precursor of ethylene.

2. AVG is an inhibitor of ethylene biosynthesis.

3. AgNO$_3$ is an inhibitor of ethylene action.

4. X-Gluc is the most commonly used substrate for GUS histochemical staining in plants.

5. If seeds are to be germinated on MS medium plates, they must be sterilized before sowing.

6. The detached leaves are placed right-side up in Petri dishes.

7. The detached leaves are pretreated with 50 µM AgNO$_3$ solution for less than 1 h.

8. The plant hormones jasmonic acid (JA) and ethylene are the major endogenous low-molecular-weight signal molecules involved in regulating senescence in plants [23]. In *Arabidopsis thaliana*, an intact JA-ethylene signaling pathway is thought to be necessary for the normal initiation of senescence [4, 9, 23–26].

9. MeJA is firstly dissolved in ethanol and then diluted in H_2O.

10. DAE (day after emergence) 12 is approximately equal to DAS (day after sowing) 21.

11. To examine the stimulating effects of ethylene on leaf senescence, the entire rosettes are put in an airtight chamber for 48 h. The chamber is fitted with inlet and outlet ports and connected to a flow-through system [20] that passed air or air containing ethylene at 100 μL L^{-1}, at a rate of 100 mL min^{-1}.

Acknowledgments

This work was supported by grants from the National Natural Science Foundation of China (91017010 to H.G.), the National Basic Research Program of China (973 Program; 2012CB910902), the Ministry of Agriculture of China (2010ZX08010-002) to H.G., the China Postdoctoral Science Foundation-funded project (2014 M560015 and 2015 T80013), and the Postdoctoral Fellowship at Peking-Tsinghua Center for Life Sciences to Z. L.

References

1. Bleecker AB, Kende H (2000) Ethylene: a gaseous signal molecule in plants. Annu Rev Cell Dev Biol 16:1–18

2. Wang NN, Yang SF, Charng Y (2001) Differential expression of 1-aminocyclopropane-1-carboxylate synthase genes during orchid flower senescence induced by the protein phosphatase inhibitor okadaic acid. Plant Physiol 126:253–260

3. Oh SA, Park JH, Lee GI et al (1997) Identification of three genetic loci controlling leaf senescence in Arabidopsis thaliana. Plant J 12:527–535

4. Kim JH, Woo HR, Kim J et al (2009) Trifurcate feed-forward regulation of age-dependent cell death involving miR164 in Arabidopsis. Science 323:1053–1057

5. Alonso JM, Hirayama T, Roman G et al (1999) EIN2, a bifunctional transducer of ethylene and stress responses in Arabidopsis. Science 284:2148–2152

6. Li Z, Zhao Y, Liu X et al (2014) LSD 2.0: an update of the leaf senescence database. Nucleic Acids Res 42:D1200–D1205

7. Liu X, Li Z, Jiang Z et al (2011) LSD: a leaf senescence database. Nucleic Acids Res 39:D1103–D1107

8. Li Z, Zhao Y, Liu X et al (2017) Construction of the leaf senescence database and functional assessment of senescence-associated genes. Methods Mol Biol 1533:315–333

9. Li Z, Peng J, Wen X et al (2013) Ethylene-insensitive3 is a senescence-associated gene that accelerates age-dependent leaf senescence by directly repressing miR164 transcription in Arabidopsis. Plant Cell 25:3311–3328

10. Guo Y, Gan S (2006) AtNAP, a NAC family transcription factor, has an important role in leaf senescence. Plant J 46:601–612

11. Kim HJ, Hong SH, Kim YW et al (2014) Gene regulatory cascade of senescence-associated NAC transcription factors activated by

ETHYLENE-INSENSITIVE2-mediated leaf senescence signalling in Arabidopsis. J Exp Bot 65:4023–4036

12. Qiu K, Li Z, Yang Z et al (2015) EIN3 and ORE1 accelerate Degreening during ethylene-mediated leaf senescence by directly activating chlorophyll catabolic genes in Arabidopsis. PLoS Genet 11:e1005399

13. Lim PO, Kim HJ, Nam HG (2007) Leaf senescence. Annu Rev Plant Biol 58:115–136

14. Even-Chen Z, Mattoo AK, Goren R (1982) Inhibition of ethylene biosynthesis by aminoethoxyvinylglycine and by polyamines shunts label from 3,4-[C]methionine into spermidine in aged orange peel discs. Plant Physiol 69:385–388

15. Beyer EM (1976) A potent inhibitor of ethylene action in plants. Plant Physiol 58:268–271

16. Murashige T, Skoog F (1962) A revised medium for rapid growth and bio assays with tobacco tissue cultures. Physiol Plant 15:473–497

17. Guo H, Ecker JR (2003) Plant responses to ethylene gas are mediated by SCF(EBF1/EBF2)-dependent proteolysis of EIN3 transcription factor. Cell 115:667–677

18. Stepanova AN, Yun J, Likhacheva AV et al (2007) Multilevel interactions between ethylene and auxin in Arabidopsis roots. Plant Cell 19:2169–2185

19. Jefferson RA, Kavanagh TA, Bevan MW (1987) GUS fusions: beta-glucuronidase as a sensitive and versatile gene fusion marker in higher plants. EMBO J 6:3901–3907

20. Jing HC, Schippers JH, Hille J et al (2005) Ethylene-induced leaf senescence depends on age-related changes and OLD genes in Arabidopsis. J Exp Bot 56:2915–2923

21. Shi YH, Zhu SW, Mao XZ et al (2006) Transcriptome profiling, molecular biological, and physiological studies reveal a major role for ethylene in cotton fiber cell elongation. Plant Cell 18:651–664

22. Li Z, Peng J, Wen X et al (2012) Gene network analysis and functional studies of senescence-associated genes reveal novel regulators of Arabidopsis leaf senescence. J Integr Plant Biol 54:526–539

23. He Y, Fukushige H, Hildebrand DF et al (2002) Evidence supporting a role of jasmonic acid in Arabidopsis leaf senescence. Plant Physiol 128:876–884

24. Burg SP (1968) Ethylene, plant senescence and abscission. Plant Physiol 43:1503–1511

25. Aharoni N, Lieberman M (1979) Ethylene as a regulator of senescence in tobacco leaf discs. Plant Physiol 64:801–804

26. Gepstein S, Thimann KV (1981) The role of ethylene in the senescence of oat leaves. Plant Physiol 68:349–354

Chapter 9

The Assay of Abscisic Acid-Induced Stomatal Movement in Leaf Senescence

Yanyan Zhang and Kewei Zhang

Abstract

Abscisic acid (ABA) is a sesquiterpenoid (15-carbon) hormone that comprehensively regulates plant stress responses, development, and senescence. Stomata are epidermal pores on plant surface used for exchanging gases such as carbon dioxide, water vapor, and oxygen. One of the mechanisms that ABA regulates leaf senescence is to control stomatal movement and thus water loss during leaf senescence. Here we describe the procedure of measuring stomatal movement in response to ABA treatments, which will provide a useful protocol to investigate ABA signaling in leaf senescence.

Key words Abscisic acid, Stomatal movement, Epidermal peels, Microscopy, Leaf senescence

1 Introduction

In addition to its roles in biotic and abiotic stresses and seed germination [1–3], ABA is also involved in leaf senescence [4, 5]. The genes associated with ABA synthesis and signaling are upregulated, and the endogenous ABA level increases in the process of leaf senescence [6–9]. The application of exogenous ABA also induces leaf senescence [6]. Furthermore, the ABA levels are significantly elevated under the environmental stresses such as drought, high salt concentration, and extreme temperature conditions which often induce leaf senescence [10].

Stomata, consisting of a microscopic pore surrounded by a pair of guard cells, regulate biotic and abiotic stresses including drought stresses and plant pathogen responses [2, 11]. The stomatal movement may be regulated by ABA through MAPK cascades [11]. In terrestrial plants, >95% of water loss occurs through transpiration from the stomata [12]. During leaf senescence, the ABA content is increased, water loss is accelerated, and the stomatal movement is less sensitive [5]. On the other side, stomatal movement and water loss affect the leaf senescence process. The water loss caused by

Yongfeng Guo (ed.), *Plant Senescence: Methods and Protocols*, Methods in Molecular Biology, vol. 1744, https://doi.org/10.1007/978-1-4939-7672-0_9, © Springer Science+Business Media, LLC 2018

retarded/slow stomatal movement accelerates leaf senescence, while the sensitive stomatal movement inhibits leaf senescence [5].

Here we describe the procedure of measuring stomatal movement in response to ABA treatment during leaf senescence. The procedure includes plant growing, ABA treatment, and assay of stomatal movement in leaf senescence. The efficient measurement of stomatal movement will help in understanding the mechanisms of ABA signaling, stomatal movement, and water loss in leaf senescence.

2   Materials

All solutions are prepared using distilled water and analytical or higher-grade chemicals from local suppliers. The plant materials (*Arabidopsis* Columbia ecotype) are grown in a growth chamber with 60% relative humidity under continuous light. The light intensity is 120 µmol m^{-2} s^{-1}, and the temperature is 22 °C.

2.1   Growing Arabidopsis *Plants*

1. 70% ethanol sterilization solution: Prepare 70% ethanol with 0.05% Triton X-100.

2. 95% ethanol.

3. 1/2× Murashige and Skoog (MS) medium: For 1 L medium, weigh 2.165 g MS salts with vitamins and macro- and micronutrients, and dissolve in 900 mL water. Add 30 g sucrose and 0.25 g 2-(*N*-morpholino)-ethanesulfonic acid (MES). Adjust pH to 5.6–5.8 with 1 N KOH, and add water to a volume of 1 L. Add 2.5 g phytogel and autoclave the medium at 121 °C for 15 min. Distribute the medium in 9 cm plastic or glass Petri dishes for about 30 mL/plate. Store at 4 °C.

2.2   ABA Treatment

1. 100 mM ABA stock: Weigh 26.43 mg ABA in an Eppendorf tube. Add 1 mL ethanol, and dissolve it well by gently flicking. Store at −20 °C (*see* **Note 1**).

2. 10 mM ABA stock: Dilute 100 µL 100 mM ABA stock solution by adding 900 µL water. Store at −20 °C.

3. MES-KOH buffer: Weigh 1.066 g MES and transfer to the cylinder. Add water to a volume of 50 mL. Store at room temperature.

4. EGTA buffer: Weigh 0.019 g EGTA and transfer to 40 mL water. Adjust pH to 6.15 with 1 N KOH, and add water to a volume of 50 mL. Store at room temperature.

5. Incubation buffer: Weigh 0.373 g KCl and dissolve in 400 mL water. Add 0.011 g CaCl$_2$, 50 mL EGTA stock, and 50 mL MES. Adjust pH to 6.15 with 1 N KOH. Briefly, the incubation buffer contains 10 mM KCl, 0.2 mM CaCl$_2$, 0.1 mM

EGTA, and 10 mM MES-KOH (pH 6.15). Store at room temperature.

6. ABA working solution: Dilute 10 mM ABA stock solution to about 1000 folds with incubation buffer to be working solution for ABA treatment. The dilution folds may be varied for ABA treatments with different concentrations.

3   Methods

Carry out all procedures at room temperature unless otherwise specified.

3.1 Growing Arabidopsis Plants

1. In a rotating shaker, surface sterilize *Arabidopsis* seeds for 30 min in an Eppendorf tube containing 1 mL 70% ethanol sterilization solution. Wash the seeds for 10 min with 1 mL 95% ethanol. Leave the tube in the clean hood, and let the seeds sink to the bottom of the tube.

2. Remove 700 µL of the 95% ethanol using a pipet. Resuspend the seeds by gently flicking, and dump the seeds onto a piece of sterilized filter paper in the clean hood. Wait around 20 min until the filter paper and the seeds are completely dry (*see* **Note 2**).

3. Transfer the dry seeds to plates with 1/2 MS medium by gently flicking the filter paper (*see* **Note 3**).

4. Incubate the plates at 4 °C for 3 days to cold stratify the seeds.

5. Move the plates to a growth chamber, and grow them with the light intensity being 120 µmol m^{-2} s^{-1} and the temperature of 22 °C.

6. Transplant 10-day-old seedlings to soil (*see* **Note 4**), and grow them in a growth chamber under the same growth conditions. Grow the wild-type and mutant/transgenic plants side by side to reduce the environmental differences.

7. Choose leaves at the same position (e.g., the sixth rosette leaf from mutant/transgenic and wild-type plants) with the early senescence phenotypes (with about 25% yellowing) for further analysis (Fig. 1a) (*see* **Note 5**).

3.2 ABA Induction of Stomatal Closure

1. Water the plants 1 day before the experiment.

2. Basically follow the method described by Melotto et al. [2] to collect leaf peels (Fig. 1b).

3. Collect leaf peels (nine peels from three different leaves for one treatment) using forceps from the bottom surface of the green area of early senescing leaves described above (*see* **Note 6**).

4. Float the peels on the incubation solution in a Petri dish right after peeling (Fig. 1c) (*see* **Note 7**).

Fig. 1 Collecting peels from early senescing leaves. (**a**) Early senescent leaves of the sixth rosette leaf of *Arabidopsis*. (**b**) Hold a leaf with the bottom surface facing up, rip the bottom of the leaf, and pull a piece with forceps. (**c**) Place the leaf piece with peel floating on the solution, incubate under same light conditions as for growing plants. The scale bar in (**a**) and (**c**) is 0.5 cm

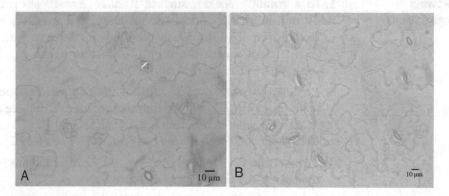

Fig. 2 Stomata observed under microscope. (**a**) The stomata were open after light induction for 2 h. (**b**) The stomata were closed 2 h after ABA treatment under light. The white line in (**a**) indicates the direction of diameter measurement

5. Place the Petri dishes under light in a growth chamber, with the light intensity of 120 μmol m^{-2} s^{-1} and the temperature of 22 °C (*see* **Note 8**). Incubate the Petri dishes for 2 h to induce stomatal opening (*see* **Note 9**).

6. Observe the peels under a microscope at 40× magnification, and take 12 photos as soon as possible (Fig. 2a). Return the peels to the incubation buffer.

7. Add ABA stock to the incubation buffer to 10 μM, and incubate the peels in light for additional 2 h.

8. Transfer the peels by forceps to a microscope slide with a drop of 10 μM ABA in incubation buffer, and cover it with a coverslip.

9. Observe the peels under a microscope, and take photos as described in **step 6** (Fig. 2b).

3.3 ABA Inhibition of Stomatal Opening

1. Water the plants 1 day before the experiment.

2. Collect and float leaf peels in incubation buffer as described in Subheading 3.2 (*see* **Note 10**).

3. Incubate the peels in darkness at 22 °C in a growth chamber for 2 h to induce stomatal closure (*see* **Note 11**).

4. Observe the peels and take photos under microscope as described above. Return the peels to the incubation buffer.

5. Add ABA stock to the incubation buffer to a final concentration of 10 μM, and incubate the peels in light for additional 2 h.

6. Transfer the peels by forceps to a microscope slide with a drop of 10 μM ABA in the incubation buffer, and cover it with a coverslip.

7. Observe the peels under a microscope and take photos.

3.4 Stomata Aperture Measurement and Data Analysis

1. Download the EZ-Rhizo software from http://www.psrg.org.uk/plant-biometrics [13].

2. Open each photo with EZ-Rhizo, and measure a scale ruler to serve as a standard.

3. Measure the diameter of around 100 stomata for each treatment (Fig. 2a) (*see* **Note 12**).

4. Record the actual aperture of each stoma.

5. Calculate the average aperture of stomata, and the statistical analyses on the obtained datasets are carried out with a student's paired t-test, using two-tailed distribution and two-sample unequal variances.

4 Notes

1. ABA stock lasts about 6 months when stored at −20 °C freezer.

2. The seeds left in the tube are rinsed by adequate 95% ethanol and can be dumped onto the filter paper completely.

3. Make sure the seeds are completely dry because residual ethanol significantly reduces germination rate.

4. Most commercial soilless mixes which include peat, vermiculite, and perlite can be used to grow *Arabidopsis*.

5. Avoid stressing the plants and the early senescing leaf should be juicy. The growth chamber needs to be kept moisturized.

6. Peel the epidermis from the green area in which the senescence has been activated. To induce stomatal opening, peel the *Arabidopsis* epidermis pieces at 6–8 am.

7. Gently peel the epidermis with your hands very close to the solution in the Petri dish.

8. The small growth chamber (e.g., Percival C36L) is located nearby the microscope to reduce the environmental effects and the time for observation.

9. Make sure 80% of the stomata are open.

10. To induce enough stomatal closure, peel the *Arabidopsis* epidermis pieces at 10–12 am.

11. Make sure 80% of the stomata are closed.

12. Measure the width of the stomata at the position of the white line in Fig. 2a, and regard it as the diameter of the stomata.

Acknowledgments

We thank Dr. Yongfeng Guo (Tobacco Research Institute, Chinese Academy of Agricultural Sciences) for his useful discussion. This work was supported by the National Science Foundation of China (31470370 and 31670277), Zhejiang Provincial Outstanding Young Scientist Award Fund (LR15C020001), and 1000 Talents Plan for young researchers of China to K.W.Z.

References

1. Finkelstein RR, Rock CD (2002) Abscisic acid biosynthesis and response. Arabidopsis Book 1:1e0058

2. Melotto M, Underwood W, Koczan J et al (2006) Plant stomata function in innate immunity against bacterial invasion. Cell 1265:969–980

3. Adie BA, Perez-Perez J, Perez-Perez MM et al (2007) ABA is an essential signal for plant resistance to pathogens affecting JA biosynthesis and the activation of defenses in *Arabidopsis*. Plant Cell 19:1665–1681

4. Zhang K, Gan S (2012) An abscisic acid-AtNAP transcription factor-SAG113 protein phosphatase 2C regulatory chain for controlling dehydration in senescing *Arabidopsis* leaves. Plant Physiol 158:961–969

5. Zhang K, Xia X, Zhang Y et al (2012) An ABA-regulated and Golgi-localized protein phosphatase controls water loss during leaf senescence in *Arabidopsis*. Plant J 69:667–678

6. Gepstein S, Thimann KV (1980) Changes in the abscisic acid content of oat leaves during senescence. Proc Natl Acad Sci U S A 77:2050–2053

7. Weaver LM, Gan S, Quirino B et al (1998) A comparison of the expression patterns of several senescence-associated genes in response to stress and hormone treatment. Plant Mol Biol 37:455–469

8. He Y, Tang W, Swain JD et al (2001) Networking senescence-regulating pathways by using *Arabidopsis* enhancer trap lines. Plant Physiol 126:707–716

9. van der Graaff E, Schwacke R, Schneider A et al (2006) Transcription analysis of *Arabidopsis* membrane transporters and hormone pathways during developmental and induced leaf senescence. Plant Physiol 141:776–792

10. Guo Y, Gan SS (2005) Leaf senescence: signals, execution, and regulation. Curr Top Dev Biol 71:83–112

11. Liu YK, Liu YB, Zhang MY et al (2010) Stomatal development and movement: the roles of MAPK signaling. Plant Signal Behav 5:1176–1180

12. Schroeder JI, Allen GJ, Hugouvieux V et al (2001) Guard cell signal transduction. Annu Rev Plant Physiol Plant Mol Biol 52:627–658

13. Armengaud P, Zambaux K, Hills A et al (2009) EZ-Rhizo: integrated software for the fast and accurate measurement of root system architecture. Plant J 57:945–956

Chapter 10

The EPR Method for Detecting Nitric Oxide in Plant Senescence

Aizhen Sun

Abstract

Nitric oxide (NO) is gaining increasing attention as a central molecule with diverse signaling functions. It has been shown that NO acts as a negative regulator of leaf senescence. In this chapter, we describe a highly selective method, electron paramagnetic resonance ([EPR], also known as electron spin resonance [ESR]), for NO determination in leaf senescence. An iron complex of ferrous and mononitrosyl dithiocarbamate ($Fe^{2+}(DETC)_2$) is used as a chelating agent for NO. Using ethyl acetate as extracting solvent, the $NOFe^{2+}(DETC)_2$ complex is extracted and determined by EPR spectrometer.

Key words Nitric oxide (NO), Leaf senescence, Electron paramagnetic resonance (EPR), Fe^{2+} $(DETC)_2$, Ethyl acetate

1 Introduction

Nitric oxide (NO) is emerging as an important multifunctional signaling molecule and plays important roles in key physiological processes such as growth and development, starting from germination to flowering, ripening of fruit, and senescence of organs [1–3]. NO acts as a negative regulator during leaf senescence [4, 5]. Similarly, endogenous NO is mainly accumulated in vascular tissues of pea (*Pisum sativum*) leaves, while this accumulation is significantly reduced in senescing leaves [6]. The NO-deficient mutants and transgenic *Arabidopsis* (*Arabidopsis thaliana*) plants expressing NO-degrading dioxygenase (NOD) show an early senescence phenotype [4, 5]. In addition, exogenous NO counteracts the promotion of leaf senescence induced by ABA and methyl jasmonate in rice [7, 8]. NO application delays yellowing and retards the onset of chlorophyll degradation during storage of broccoli (*Brassica oleracea*) florets [9, 10]. NO treatments can protect against senescence-dependent chlorophyll degradation in soybean (*Glycine max*) cotyledons [11]. Recently, NO has been proved to act as a novel negative regulator of chlorophyll catabolic

Yongfeng Guo (ed.), *Plant Senescence: Methods and Protocols*, Methods in Molecular Biology, vol. 1744,
https://doi.org/10.1007/978-1-4939-7672-0_10, © Springer Science+Business Media, LLC 2018

pathway and positively functions in maintaining the stability of thylakoid membranes during leaf senescence [12].

NO is a highly unstable free radical and is rapidly oxidized to form nitrogen dioxide (NO_2), which degrades to nitrite and nitrate in aqueous solutions. Due to the short lifetime and concentration of these radicals in tissues, its detection and quantification involve methodological difficulties, particularly in biological tissues. Several methodological approaches have been reported to assay NO in plants, including gas chromatography and mass spectrometry [1, 13], hemoglobin method by spectrophotometric measurement of the conversion of oxyhemoglobin to methemoglobin [14–16], laser photo-acoustic spectroscopy [17], spin trapping of nitric oxide with electron paramagnetic resonance (EPR) [18–20], NO electrode [21], laser photoacoustics [17, 22], and a chemiluminescent reaction involving ozone [23]. A group of fluorescent dye indicators are also available in acetylated form for intracellular measurements [6, 24, 25].

A highly specific method for NO detection is EPR (also known as electron spin resonance [ESR]), in which there is stabilization of the labile free radical with ferrous and mononitrosyl dithiocarbamate $Fe^{2+}(DETC)_2$ or other dithiocarbamate derivatives for spin trapping [26]. Additionally, EPR spectroscopy can provide some key clues relevant to the physiological NO targets. Hydrophobic $Fe^{2+}(DETC)_2$ has been used as a chelating agent for NO both in vivo and in vitro. However, it was limited by inadequate sensitivity to detect low-level endothelial NO production and insolubility in water. Although the addition of iron salt to increase levels of endogenous free iron has been recommended [27, 28], the interaction between DETC and endogenous copper gives rise to a strong $Cu(DETC)_2$ EPR signal and is associated with an inhibition of the CuZn superoxide dismutase (SOD), which in turn may affect the endogenous NO levels [29, 30]. This method was then modified by adding albumin to solubilize $Fe^{2+}(DETC)_2$ and $Na_2S_2O_4$ as a strong reductant to increase the sensitivity and stability of the EPR spectrum of the $NOFe^{2+}(DETC)_2$ complex [31]. The $NOFe^{2+}(DETC)_2$ complex is frozen at liquid nitrogen temperature and then detected by EPR at 77 K. With the modified method, the $NOFe^{2+}(DETC)_2$ complex could be extracted from water phase to organic phase using ethyl acetate as extracting solvent. The $NOFe^{2+}(DETC)_2$ complex thus can be measured on an EPR spectrometer at room temperature [32]. The hydrophilic Fe/R,R9-dithiocarbamates such as Fe/N-methyl-D-glucamine dithiocarbamate (Fe/MGD) complex are also used for EPR-based method to detect NO [33, 34]. Also, the addition of reducing agents and/or SOD is needed because its NO adduct undergoes rapid oxidation and becomes EPR silent. However, $Fe^{2+}(MGD)_2$ complex has been proved to react not only with NO but also with nitrite to produce the characteristic triplet EPR signal of $(MGD)_2Fe^{2+} \cdot NO$ [35].

In this chapter, we describe the highly specific method, EPR spectroscopy, for NO detection in plant systems. In the experiment, the spin-trapping agent $Fe^{2+}(DETC)_2$ is needed. Also, $Na_2S_2O_4$ is added as a strong reductant to increase the sensitivity and stability of the EPR spectrum of the $NOFe^{2+}(DETC)_2$ complex. The complex of $NOFe^{2+}(DETC)_2$ is extracted from water phase to organic phase with ethyl acetate and then is determined with an EPR spectrometer at room temperature.

2   Materials

1. Buffered solution: 50 mM HEPES, 1 mM dithiothreitol (DTT), 1 mM $MgCl_2$, pH 7.6. Store at 4 °C.

2. Spin-trapping agent: Dissolve diethyldithiocarbamate (DETC) in 10% bovine serum albumin (BSA) before preparation of reaction solution.

3. $FeSO_4.6H_2O$ solution: Dissolve $FeSO_4.6H_2O$ in dilute HCl solution.

4. $Fe^{2+}(DETC)_2$ reaction solution: 2 M $Na_2S_2O_4$, 3.3 mM DETC, 3.3 mM $FeSO_4$, and 33 g L^{-1} BSA (*see* **Note 1**).

5. Ethyl acetate (*see* **Note 2**).

6. 2.5-mm-internal diameter quartz tube.

7. EPR spectrometer.

3   Methods

1. Preparation of leaf extract: About 0.5 g leaves are harvested and crushed with a mortar and pestle under liquid nitrogen and then incubated in 1 mL of buffered solution for 2 min (*see* **Note 3**).

2. The homogenates are centrifuged at 13,200 × *g*, 4 °C for 2 min.

3. Recover the supernatant to a new microcentrifuge tube (*see* **Note 4**).

4. The supernatant is added to 300 μL of freshly made $Fe^{2+}(DETC)_2$ reaction solution. The mixtures are incubated for 1 h at 30 °C (*see* **Note 5**).

5. Add 300 μL ethyl acetate to the mixture (*see* **Note 6**).

6. Shake the mixture for 3 min and centrifuge at 6000 × *g*, 4 °C for 6 min (*see* **Note 7**).

7. Recover the separated ethyl acetate organic solvent phase (*see* **Note 8**).

8. The recovered samples are transferred to a 2.5-mm-internal diameter quartz tube and introduced in the EPR resonator cavity.

9. The EPR spectra were recorded at room temperature using an EPR spectrometer.

4   Notes

1. BSA is added to prevent the precipitation of the mixture of Fe^{2+} and DETC, as $Fe^{2+}(DETC)_2$ is not soluble in water. The $Fe^{2+}(DETC)_2$ reaction solution should be freshly made.

2. Organic solvent ethyl acetate is used to extracted the $NOFe^{2+}(DETC)_2$ complex which is lipid soluble.

3. Leaf cells such as protoplasts are alternative samples for NO detection. About 500 μL of cells are harvested and then are incubated in 0.6 mL of buffered solution at 37 °C for 2 min.

4. Aliquots of supernatants should be immediately used for the following assays.

5. The reduced (ferrous) form of iron traps NO in aqueous solution. However, the $Fe^{2+}(DETC)_2$ is apt to be oxidized to $Fe^{3+}(DETC)_2$, thus prevents sensitive quantitation of NO. Furthermore, NO could be rapidly oxidized to NO_2 by oxygen molecules as it has a short half-life in aqueous solution. To overcome these problems, excess $Na_2S_2O_4$ are used to reduce $Fe^{3+}(DETC)_2$ to $Fe^{2+}(DETC)_2$ and $Fe^{2+}NO(DETC)_2NO_2$ to $Fe^{2+}(DETC)_2NO$ [31].

6. The adduct $NOFe^{2+}(DETC)_2$ signal in water solution is difficult to be detected by ESR spectroscopy at room temperature because of the low concentration and water-absorbing microwaves. However, it can be extracted by organic solvent and thus measured at room temperature.

7. The $NOFe^{2+}(DETC)_2$ complex binds tightly with protein; thus after addition of ethyl acetate, the mixture should be shaken violently at least 3 min to extract $NOFe^{2+}(DETC)_2$ complex completely [32].

8. NO complex is apt to decompose at high temperature, and it is sensitive to light. Therefore, the extraction of $NOFe^{2+}(DETC)_2$ should be better kept at 0–4 °C in the dark and immediately used for detection.

References

1. Neill SJ, Desikan R, Hancock JT (2003) Nitric oxide signaling in plants. New Phytol 159:11–35

2. Crawford NM, Guo FQ (2005) New insights into nitric oxide metabolism and regulatory functions. Trends Plant Sci 10:195–200

3. Mur LA, Carver TL, Prats E (2006) NO way to live; the various roles of nitric oxide in plant-pathogen interactions. J Exp Bot 57:489–505

4. Guo FQ, Crawford NM (2005) *Arabidopsis* nitric oxide synthase1 is targeted to mitochondria and protects against oxidative damage and dark-induced senescence. Plant Cell 17:3436–3450

5. Mishina TE, Lamb C, Zeier J (2007) Expression of a NO degrading enzyme induces a senescence programme in Arabidopsis. Plant Cell Environ 30:39–52

6. Corpas FJ, Barroso JB, Carreras A et al (2004) Cellular and subcellular localization of endogenous nitric oxide in young and senescent pea plants. Plant Physiol 136:2722–2733

7. Hung KT, Kao CH (2003) Nitric oxide counteracts the senescence of rice leaves induced by abscisic acid. J Plant Physiol 160:871–879

8. Hung KT, Kao CH (2004) Nitric oxide acts as an antioxidant and delays methyl jasmonate-induced senescence of rice leaves. J Plant Physiol 161:43–52

9. Eum HL, Hwang DK, Lee SK (2009) Nitric oxide reduced chlorophyll degradation in broccoli (*Brassica oleracea* L. var. italica) florets during senescence. Food Sci Technol Int 15:223–228

10. Eum HL, Lee SK (2007) Nitric oxide treatment reduced chlorophyll degradation of broccoli florets during senescence. HortSci 42:927–927

11. Jasid S, Galatro A, Javier Villordo J et al (2009) Role of nitric oxide in soybean cotyledon senescence. Plant Sci 176:662–668

12. Liu F, Guo FQ (2013) Nitric oxide deficiency accelerates chlorophyll breakdown and stability loss of thylakoid membranes during dark-induced leaf senescence in Arabidopsis. PLoS One 8:e56345

13. Conrath U, Amoroso G, Köhle H et al (2004) Noninvasive online detection of nitric oxide from plants and some others organisms by mass spectrometry. Plant J 38:1015–1022

14. Delledonne M, Xia YJ, Dixon RA et al (1998) Nitric oxide functions as a signal in plant disease resistance. Nature 394:585–588

15. Clarke A, Desikan R, Hurst RD et al (2000) NO way back: nitric oxide and programmed cell death in *Arabidopsis thaliana* suspension cultures. Plant J 24:667–677

16. Orozco-Cárdenas ML, Ryan CA (2002) Nitric oxide negatively modulates wound signaling in tomato plants. Plant Physiol 130:487–493

17. Leshem YY, Pinchasov Y (2000) Non-invasive photoacoustic spectroscopic determination of relative endogenous nitric oxide and ethylene content stoichiometry during the ripening of strawberries *Fragaria anannasa* (Duch.) and avocados *Persea americana* (Mill.) J Exp Bot 51:1471–1473

18. Pagnussat GC, Simontacchi M, Puntarulo S et al (2002) Nitric oxide is required for root organogenesis. Plant Physiol 129:954–956

19. Huang X, Stettmaier K, Michel C et al (2004) Nitric oxide is induced by wounding and influences jasmonic acid signaling in *Arabidopsis thaliana*. Planta 218:938–946

20. Modolo LV, Augusto O, Almeida IMG et al (2005) Nitrite as the major source of nitric oxide production by *Arabidopsis thaliana* in response to *Pseudomonas syringae*. FEBS Lett 579:3814–3820

21. Yamasaki H, Shimoji H, Ohshiro Y et al (2001) Inhibitory effects of nitric oxide on oxidative phosphorylation in plant mitochondria. Nitric Oxide 5:261–270

22. Mur LA, Santosa IE, Laarhoven LJ et al (2005) Laser photoacoustic detection allows in planta detection of nitric oxide in tobacco following challenge with avirulent and virulent Pseudomonas syringae pathovars. Plant Physiol 138:1247–1258

23. Morot-Gaudry-Talarmain Y, Rockel P, Moureaux T et al (2002) Nitrite accumulation and nitric oxide emission in relation to cellular signaling in nitrite reductase antisense tobacco. Planta 215:708–715

24. Foissner I, Wendehenne D, Langebartels C et al (2000) In vivo imaging of an elicitor-induced nitric oxide burst in tobacco. Plant J 23:817–824

25. Ma W, Smigel A, Walker RK et al (2010) Leaf senescence signaling: the Ca²⁺-conducting Arabidopsis cyclic nucleotide gated channel2 acts through nitric oxide to repress senescence programming. Plant Physiol 154:733–743

26. Kleschyov AL, Wenzel P, Munzel T (2007) Electron paramagnetic resonance (EPR) spin trapping of biological nitric oxide. J Chromatogr B Analyt Technol Biomed Life Sci 851:12–20

27. Mülsch A, Mordvintcev P, Vanin A (1992) Quantification of nitric oxide in biological sam-

ples by electron spin resonance spectroscopy. Neuroprotocols 1:165–173

28. Kleschyov AL, Mollnau H, Oelze M et al (2000) Spin trapping of vascular nitric oxide using colloid Fe(II)-diethyldithiocarbamate. Biochem Biophys Res Commun 275: 672–677

29. Mügge A, Elwell JH, Peterson TE et al (1991) Release of intact endothelium-derived relaxing factor depends on endothelial superoxide dismutase activity. Am J Phys 260:C219–C225

30. Munzel T, Hink U, Yigit H et al (1999) Role of superoxide dismutase in in vivo and in vitro nitrate tolerance. Br J Pharmacol 127: 1224–1230

31. Tsuchiya K, Takasugi M, Minakuchi K et al (1996) Sensitive quantitation of nitric oxide by EPR spectroscopy. Free Radic Biol Med 21:733–737

32. Xu Y, Cao Y, Tao Y et al (2005) The ESR method to determine nitric oxide in plants. Methods Enzymol 396:84–92

33. Kotake Y, Tanigawa T, Tanigawa M et al (1995) Spin trapping isotopically-labelled nitric oxide produced from [^{15}N]L-arginine and [^{17}O]dioxygen by activated macrophages using a water soluble Fe(11)-dithiocarbamate spin trap. Free Radic Res 23:287–295

34. Zweier JL, Wang P, Kuppusamy P (1995) Direct measurement of nitric oxide generation in the ischemic heart using electron paramagnetic resonance spectroscopy. J Biol Chem 270:304–307

35. Tsuchiya K, Yoshizumi M, Houchi H et al (2000) Nitric oxide-forming reaction between the iron-N-methyl-Dglucamine dithiocarbamate complex and nitrite. J Biol Chem 275:1551–1556

Chapter 11

Hormone Treatments in Studying Leaf Senescence

Zenglin Zhang and Yongfeng Guo

Abstract

As the last stage of plant development, senescence can be regulated by a large number of signals such as aging, reproductive growth, nutrient availability, and stresses. Various plant hormones have been shown to be involved in regulating plant senescence. For example, ethylene, abscisic acid (ABA), jasmonic acid (JA), salicylic acid (SA), and strigolactones (SLs) promote senescence, whereas cytokinins (CKs) inhibit senescence. Different hormones regulate senescence via distinct pathways, while cross talks between signaling pathways exist. In senescence-related studies, treating plants with various hormones to alter senescence is a common practice. In this chapter, we summarize experimental procedures of treating detached *Arabidopsis* leaves with a number of senescence-regulating hormones including ABA, SLs, MeJA, SA peptide hormones.

   Key words *Arabidopsis*, Phytohormones, Leaf senescence, Gene expression

1   Introduction

Senescence is the last stage of plant development that involves both degenerative processes and remobilization of nutrients like nitrogen and carbon from senescing tissues to sink organs [1]. Plant senescence can be regulated by various environmental factors (such as photoperiod, stresses, nutrient starvation) and endogenous factors (including plant age, phytohormones, reproductive growth). Hormones play a crucial role in plant developmental processes including leaf senescence. Ethylene, JA, ABA, and SA could promote senescence; auxin, gibberellic acid (GA), and CKs could delay the senescence process [2–4].

The role of ABA in leaf senescence regulation is well established. Exogenous application of ABA could promote leaf senescence [5, 6]. ABA content increases in senescing leaves and exogenous application of ABA induced expression of *SENESCENCE-ASSOCIATED GENEs SAGs*) [5]. The expression of *RPK1*, a gene encoding ABA-inducible receptor kinase, increases during senescence. Mutation of *RPK1* caused reduced sensitivity to ABA treatments and delay in leaf senescence [7]. *VNI2*

Yongfeng Guo (ed.), *Plant Senescence: Methods and Protocols*, Methods in Molecular Biology, vol. 1744,
https://doi.org/10.1007/978-1-4939-7672-0_11, © Springer Science+Business Media, LLC 2018

(*VND-INTER-ACTING2*), a NAC transcription factor gene, was shown to be upregulated during leaf senescence and induced by ABA treatments or salt stress. Through regulating the expression of *COR* (*COLD-REGULATED*) and *RD* (*RESPONSIVE TO DEHYDRATION*) genes, such as *COR15A*, *COR15B*, *RD29A*, and *RD29*, VNI2 integrates ABA-mediated abiotic stress signals into the senescence process [8, 9]. *SAG113*, whose expression is regulated by ABA and senescence, functions to inhibit stomatal closure in senescing leaves (*see* Chapter 9). Loss of function of *SAG113* caused delayed leaf senescence, whereas overexpression lines displayed accelerated senescence. Further study showed that AtNAP, a positive regulator of leaf senescence, regulates the expression of *SAG113*, illustrating a strong link between ABA signaling and leaf senescence [10].

Like ABA, exogenously supplied methyl jasmonate accelerated leaf senescence [11]. Moreover, jasmonate content increases in senescing leaves. Expression of genes involved in JA synthesis and signaling such as *LOX3*, *AOC1*, *AOC4*, *OPR3*, *MYC2*, *JAZ1*, *JAZ6* and *JAZ8* increased during leaf senescence [12–14].

In addition to its roles in defense responses against pathogens, salicylic acid (SA) is also a positive regulator of leaf senescence. SA content and expression of genes involved in SA biosynthesis and signaling increase in senescing leaves. Transgenic plants overexpressing the *NahG* gene, which encodes the SA-degrading enzyme NAPHTHALENE OXYGENASE, displayed delayed senescence. Transcript profiling analysis revealed that about 20% of the detected *SAGs* showed differential expression in *NahG*-ox transgenic plants compared with wild type [15]. In addition, SA treatments promoted expression of a number of *SAGs* including *WRKY6*, *WRKY53*, *WRKY54*, *WRKY70* and *SEN1* [16, 17].

There is evidence which shows that strigolactones also play an important role in regulating leaf senescence. Grafting experiment revealed that mutant plants deficient in strigolactone biosynthesis showed delayed leaf senescence under darkness [18]. In addition, loss of function of *MAX2*, a strigolactone-signaling component, caused stay-green phenotypes [19, 20].

Besides traditional hormones, our recent work showed that a small peptide is also involved in senescence regulation. Like in animals, peptide hormones have been shown to play crucial roles in cell-cell communications [21]. In *Arabidopsis thaliana*, more than 1000 genes encoding putatively secreted peptides have been identified [22]. A number of peptide signals have been shown to be key regulators of plant development and responses to environment [23].

Various signal factors may regulate leaf senescence through different mechanisms; therefore, various strategies or systems are needed to identify senescence regulatory factors. In this chapter, we provide a series of protocols of treating detached leaves to induce senescence with different plant hormones including ABA, SLs, MeJA, SA and peptide hormones.

2 Materials

1. Plant materials: *Arabidopsis thaliana* ecotype Columbia (Col).

2. 70% ethanol: Mix 15 mL water with 35 mL 100% ethanol into a 50 mL sterilized graduated corning tube supplemented with 0.05% (v/v) Triton X-100.

3. 100% ethanol.

4. Incubation buffer (1/2 MS, 3 mM MES, pH to 5.8): Weigh 0.22 g MS powder, 0.064 g MES powder, add water to 100 mL, adjust pH to 5.8 with 1 N KOH, and store at room temperature.

5. ABA stock solution (100 mM): Dissolve 26.4 mg ABA in 1 mL ethanol, and store at −20 °C.

6. ABA working solution (10 μM): Add 10 μL ABA stock to 100 mL incubation buffer.

7. SL stock solution (25 mM): Dissolve 5 mg GR24 (a synthetic SL analogue) in 671 μL methanol, and store at −20 °C.

8. SL working solution (5 μM): Add 10 μL SL stock to 50 mL incubation buffer.

9. MeJA stock solution (100 mM): Add 114.6 μL 95% MeJA to 5 mL ethanol, and store at −20 °C.

10. MeJA working solution (100 μM): Add 20 μL MeJA stock solution to 20 mL incubation buffer.

11. SA stock solution (1 M): Dissolve 1.38 g SA in 10 mL ethanol, and store at −20 °C.

12. SA working solution (5 mM): Add 100 μL SA stock solution to 20 mL incubation buffer.

13. Peptide stock (10 mM): Dissolve synthetic peptide with DMSO, and store at −70 °C.

14. Peptide working solution (10 μM): Add 10 μL CLE peptide stock to 10 mL incubation buffer (*see* **Note 1**).

15. 1/2 MS solid medium: Weigh 0.22 g MS powder and 0.8 g agar, add water to 100 mL, adjust pH to 5.8 with 1 N KOH, autoclaved at 121 °C for 20 min, pour Petri dishes, and store at 4 °C.

16. Plastic Petri dishes and filter papers.

17. Chlorophyll extraction solution: 95% ethanol.

18. Chlorophyll fluorometer: Opti-Sciences Inc., USA.

19. RNA extraction reagents: TRIzol Reagent, chloroform, isopropanol, 75% ethanol (DEPC water preparation), and DEPC water.

20. Real-PCR system: SYBR Green Master Mix for qPCR, qPCR plates and optical foils, primers, real-time PCR instrument.

3 Methods

3.1 Prepare Arabidopsis *Plants*

1. Put *Arabidopsis* seeds in a 1.5 mL centrifuge tube.
2. Add 1 mL 70% ethanol to sterilize the seeds for 5 min.
3. Briefly centrifuge to settle the seeds, and remove the 70% ethanol by pipetting, and then wash the seeds using 100% ethanol for 1 min.
4. Remove the ethanol by pipetting (*see* **Note 2**), and wash the seeds three times with sterilized water.
5. Put the seeds on ½ MS solid medium plate using a pipette.
6. Incubate the medium plate at 4 °C for 3 days.
7. Keep the plates in a growth chamber (22 °C, continuous light) for 8 days.
8. Transplant the seedlings to soil, and grow in the growth chamber (22 °C, continuous light) (*see* **Note 3**).

3.2 Harvest Leaves from Adult Plants

1. Grow *Arabidopsis* on soil in a growth chamber for 4 weeks (*see* **Note 4**).
2. Select mature green leaves at the same position (we usually use the fifth rosette leaf), and detach the leaves for further analysis (*see* **Note 5**).

3.3 ABA Treatment of Detached Leaves for Inducing Senescence

1. Put two layers of sterilized filter paper in Petri dishes, and soak the filter paper completely with ABA working solution or with incubation buffer as a negative control (*see* **Note 6**).
2. Arrange the detached leaves from Subheading 3.2 adaxial side-up on the surface of filter paper (*see* **Note 7**).
3. Wrap the Petri dishes with Parafilm.
4. The detached leaves are incubated on the filter papers under continuous light or dark condition at 22 °C for an adequate amount of time (*see* **Note 8**).
5. Observe the progression of senescence via color change of treated leaves, and take photos every day.

3.4 SL Treatment of Detached Leaves for Inducing Senescence

1. Put two layers of sterilized filter paper in Petri dishes, and soak the filter paper completely with SL working solution or with incubation buffer as a negative control (*see* **Note 9**).
2. Arrange the detached leaves on the surface of filter paper as described in Subheading 3.3.
3. Wrap the Petri dishes with Parafilm.
4. The detached leaves are incubated on the filter paper under darkness at 22 °C for 6–7 days (*see* **Note 10**).
5. Observe the progression of senescence via color change of treated leaves, and take photos every day.

3.5 MeJA Treatment of Detached Leaves for Inducing Senescence

1. The detached leaves are floated on 2 mL MeJA working solution or incubation buffer in Petri dishes (*see* **Note 11**).

2. Wrap up the dishes with silver paper, and incubate under darkness at 22 °C for 6 days.

3. Observe the progression of senescence via color change of treated leaves, and take photos every other day.

3.6 SA Treatment of Detached Leaves for Inducing Senescence

1. The detached leaves are floated on 3 mL SA working solution or incubation buffer in Petri dishes.

2. Wrap up the dishes with silver paper and incubate under darkness at 22 °C for 8 days.

3. Observe the progression of senescence via color change of treated leaves, and take photos every day.

3.7 Peptide Treatment of Detached Leaves for Inducing Senescence

1. Put two layers of sterilized filter paper in Petri dishes, and soak the filter paper with peptide working solution or incubation buffer (*see* **Note 12**).

2. Arrange the detached leaves on the surface of filter paper as described in Subheading 3.3.

3. Wrap the Petri dishes with Parafilm.

4. The detached leaves are incubated on the filter paper under continuous light at 22 °C for 6–7 days (*see* **Note 13**).

5. Observe the progression of senescence via color change of treated leaves, and take photos every other day.

6. Measure senescence parameters such as chlorophyll content, harvest some of the treated leaves, and freeze in liquid nitrogen for extracting RNA and gene expression analysis (*see* **Note 14**).

3.8 Chlorophyll Content Measurement

1. Take one to two treated leaves into a new Petri dish, and wash them two to three times with ultrapure water.

2. Put 40–80 mg detached leaves mentioned above in a 2 mL centrifuge tubes, and add 1 mL of chlorophyll extraction solution (*see* **Note 15**).

3. Keep the incubate leaves in the dark for 24 h at room temperature.

4. Measure absorbance of the solution at 665 and 649 nm, and take chlorophyll extraction solution as a negative control.
   Chlorophyll content is calculated according to the following formula [24]:

- Chlorophyll a (μg/mL) = 13.95 × A665—6.88 × A649.

- Chlorophyll b (μg/mL) = 24.96 × A649—7.32 × A665.

- Total chlorophyll (mg/g FW) = (Ca + Cb) × V/W.

- V: volume of extraction solution, 1 mL in this protocol.

- W: weight of freshly detached leaves.

3.9 Fv/Fm Measurement

1. Collect treated leaves using tweezers (*see* **Note 16**).
2. Remove hormone working solution or incubation buffer from the leaves by filter papers.
3. Keep the leaves in darkness for 1 h.
4. Measure the Fv/Fm value with a chlorophyll fluorometer.

3.10 Analysis of the Expression of Leaf Senescence Maker Genes

1. Take out the treated leaves from −70 °C with liquid nitrogen.
2. Ground the leaf tissues into fine powder in liquid nitrogen.
3. Add 1 mL TRIzol Reagent, and vortex till fully mixed.
4. Let the mixture stand at room temperature for 15 min.
5. Centrifuge for 10 min at 4 °C and 10,000 rpm ($11800 \times g$) with a microcentrifuge.
6. Transfer the supernatant to a new centrifuge tube.
7. Transfer 0.2 mL chloroform into the tube, and vortex violently; keep at room temperature for 3 min.
8. Centrifuge for 15 min at 4 °C and 10,000 rpm ($11800 \times g$) with a microcentrifuge.
9. Transfer the uppermost layer liquid (about 500 μL) into a new centrifuge tube (*see* **Note 17**).
10. Add equal volume of isopropanol (about 500 μL), mix well, and incubate at room temperature for 30 min.
11. Centrifuge for 10 min at 4 °C 10,000 rpm ($11800 \times g$) with a microcentrifuge, and discard the supernatant.
12. Add 1 mL 75% ethanol (DEPC water), and wash the pellet.
13. With a microcentrifuge, centrifuge for 3 min, at 4 °C and 5000 rpm ($5900 \times g$), and discard the liquid.
14. Air-dry the pellet, dissolve the pellet with 30 μL DEPC water, and store at −80 °C until use.
15. Assess the RNA quality in 1.0% agarose gel at 5 V/cm.
16. First-strand cDNA is synthesized according to the instructions of the manufacturer.
17. Quantitative PCR is performed according to the instructions of the manufacturer.

4 Notes

1. It is best to prepare the CLE peptide incubation solution fresh each time.
2. Ensure the ethanol volatilizes completely; usually, we let the centrifuge tube air-dry for no less than 10 min. Otherwise, the germination rate of the seeds will be reduced.

3. After transplanting to soil, keep the seedlings covered by a clean transparent plastic cap for 4–5 days to ensure good survival.

4. The growth status of the plants has a big impact on the senescence phenotypes of individual leaves. Make sure the tested plants are in a similar healthy growth status.

5. Take fully expanded green leaves. Biological replication is essential to get reliable results. We usually put eight to ten leaves in each Petri dish as one replicate and set up five to six replicates.

6. We usually soak the filter paper with 3 mL ABA working solution or incubation buffer. In this treatment, the incubation buffer should be supplied with 0.3 μL ethanol.

7. The incubation buffer should cover the filter paper but not more than that. Enough detached leaves should be used for phenotype observation and measurement of senescence parameters such as chlorophyll content, Fv/Fm, and so on.

8. We record senescence-related phenotypes of the treated leaves in the first 4–5 days in light or 3–4 days in dark condition after treatment, during which we take pictures and measure chlorophyll contents of the leaves once every day.

9. We usually soak filter paper by 3 mL SL working solution or incubation buffer. In this treatment, the incubation buffer should be supplied with 0.6 μL methanol.

10. We record senescence-related phenotypes of the treated leaves in the first 6–7 days after treatment, during which we take pictures and measure chlorophyll contents of the leaves once every day.

11. The incubation buffer should be supplied with 2 μL ethanol.

12. Usually, we soak the filter paper with 4 mL peptide working solution or incubation buffer. The incubation buffer should be supplied with 4 μL DMSO in this study.

13. We record senescence-related phenotypes of the treated leaves in the first 5–7 days after treatment, during which we take pictures and measure chlorophyll content of the leaves every other day.

14. We usually use fresh leaves for chlorophyll measurement and frozen leaves to extract RNA for gene expression analysis.

15. We usually do four to six biological replicates for chlorophyll measurement.

16. Fv/Fm values are measured when visible yellowing of the treated leaves is observed. We usually carry out 15 biological replicates for Fv/Fm measurement.

17. Try to avoid transferring the organic phase; transfer the supernatant to a new centrifuge tube very carefully.

Acknowledgments

This work was supported by the National Natural Science Foundation of China (31600991) (to Z.Z.), the Science Foundation for Young Scholars of the Tobacco Research Institute, the Chinese Academy of Agricultural Sciences (2015B02) (to Z.Z.), and the Agricultural Science and Technology Innovation Program (ASTIP-TRIC02).

References

1. Avice JC, Etienne P (2014) Leaf senescence and nitrogen remobilization efficiency in oilseed rape (Brassica napus L.) J Exp Bot 65:3813–3824

2. Sarwat M, Naqvi AR, Ahmad P et al (2013) Phytohormones and microRNAs as sensors and regulators of leaf senescence: assigning macro roles to small molecules. Biotechnol Adv 31:1153–1171

3. Khan M, Rozhon W, Poppenberger B (2014) The role of hormones in the aging of plants—a mini-review. Gerontology 60:49–55

4. Arrom L, Munne-Bosch S (2012) Hormonal regulation of leaf senescence in Lilium. J Plant Physiol 169:1542–1550

5. Finkelstein R (2013) Abscisic Acid synthesis and response. Arabidopsis Book 11:e0166

6. Zeevaart JAD, Creelman RA (2003) Metabolism and physiology of abscisic acid. Ann Rev Plant Physiol Plant Mol Biol 39:439–473

7. Lee IC, Hong SW, Whang SS et al (2011) Age-dependent action of an ABA-inducible receptor kinase, RPK1, as a positive regulator of senescence in Arabidopsis leaves. Plant Cell Physiol 52:651–662

8. Song Y, Xiang F, Zhang G et al (2016) Abscisic acid as an internal integrator of multiple physiological processes modulates leaf senescence onset in Arabidopsis thaliana. Front Plant Sci 7:181

9. Mou W, Li D, Luo Z et al (2015) Transcriptomic analysis reveals possible influences of ABA on secondary metabolism of pigments, flavonoids and antioxidants in tomato fruit during ripening. PLoS One 10:e0129598

10. Zhang K, Gan SS (2012) An abscisic acid-AtNAP transcription factor-SAG113 protein phosphatase 2C regulatory chain for controlling dehydration in senescing Arabidopsis leaves. Plant Physiol 158:961–969

11. Kim J, Chang C, Tucker ML (2015) To grow old: regulatory role of ethylene and jasmonic acid in senescence. Front Plant Sci 6:20

12. Balbi V, Devoto A (2008) Jasmonate signalling network in Arabidopsis thaliana: crucial regulatory nodes and new physiological scenarios. New Phytol 177:301–318

13. Schommer C, Palatnik JF, Aggarwal P et al (2008) Control of jasmonate biosynthesis and senescence by miR319 targets. PLoS Biol 6:e230

14. Seltmann MA, Hussels W, Berger S (2010) Jasmonates during senescence: signals or products of metabolism? Plant Signal Behav 5:1493–1496

15. Abreu ME, Munne-Bosch S (2009) Salicylic acid deficiency in NahG transgenic lines and sid2 mutants increases seed yield in the annual plant Arabidopsis thaliana. J Exp Bot 60:1261–1271

16. Morris K, Mackerness SA, Page T et al (2000) Salicylic acid has a role in regulating gene expression during leaf senescence. Plant J 23:677–685

17. Besseau S, Li J, Palva ET (2012) WRKY54 and WRKY70 co-operate as negative regulators of leaf senescence in Arabidopsis thaliana. J Exp Bot 63:2667–2679

18. Ueda H, Kusaba M (2015) Strigolactone regulates leaf senescence in concert with ethylene in Arabidopsis. Plant Physiol 169:138–147

19. Yamada Y, Umehara M (2015) Possible roles of strigolactones during leaf senescence. Plants (Basel) 4:664–677

20. Kusaba M, Tanaka A, Tanaka R (2013) Stay-green plants: what do they tell us about the molecular mechanism of leaf senescence. Photosynth Res 117:221–234

21. Chilley P (2003) Polypeptide hormones: signaling molecules in plants. Vitam Horm 66:317–344

22. Lease KA, Walker JC (2006) The Arabidopsis unannotated secreted peptide database, a resource for plant peptidomics. Plant Physiol 142:831–838

23. Czyzewicz N, Yue K, Beeckman T et al (2013) Message in a bottle: small signalling peptide outputs during growth and development. J Exp Bot 64:5281–5296

24. Kong J, Dong Y, Xu L et al (2014) Effects of foliar application of salicylic acid and nitric oxide in alleviating iron deficiency induced chlorosis of Arachis hypogaea L. Bot Stud 55:1–12

Part IV

Stress-Induced Senescence in Plants

Stress-Induced Senescence in Plants

Chapter 12

Methods to Study Darkness-Induced Leaf Senescence

Yi Song and Lin Li

Abstract

Leaf senescence is an intergral part of plant development, involving actively regulated molecular and biochemical processes, e.g. diverse transcriptome reprogramming, macromolecules degradation and nutrient remobilization. Natural and environment-induced leaf senescence directly affects crop yield and the shelf life of green vegetables. Darkness is considered as an inducer of leaf senescence and has been well used in laboratory setting. Here we described the setup of darkness-induced senescence in both detached and attached leaves and the methods to measure senescence-related parameters.

Key words Leaf senescence, Darkness

1 Introduction

Leaf senescence is a complex degenerative process associated with many biological changes such as chlorophyll and protein degradation, nutrient translocation, and cell death, which constitute the last stage of leaf development [1–4]. Multiple environmental factors could affect the plant senescence process, including high or low temperature, drought, pathogen attack, and unfavorable light environment.

Although several days of darkness treatment is unable to induce whole-plant senescence [5], darkness can induce senescence in detached leaves and intact leaves attached to the whole plant. Compared to tip-to-base development of natural senescence, darkness-induced senescence is rather rapid and synchronous, which made it an ideal model to study senescence symptoms [6].

Leaf senescence triggers a variety of biological and molecular changes. Chlorophyll degradation is the most remarkable phenotype of leaf senescence [8–10]. Protein degradation and lipid metabolism are also affected during the senescence process. So

Yongfeng Guo (ed.), *Plant Senescence: Methods and Protocols*, Methods in Molecular Biology, vol. 1744,
https://doi.org/10.1007/978-1-4939-7672-0_12, © Springer Science+Business Media, LLC 2018

total protein content, ion leakage, and the PSII photosynthesis efficiency Fv/Fm are all indicatives of leaf senescence [11].

Besides physiological parameters, molecular markers have also been applied for senescence detection. Many senescence-associated genes have been reported [12], such as *SAG12* and *SEN1* [13, 14]. *SAG12* is a natural senescence-specific marker, but its transcript level did not change rapidly during early stages of darkness-induced senescence [15]. *SEN1* is rather a general surrogate due to its rapid and robust induction during both natural and darkness-induced senescence [16]. *CAB* and *RBCS* genes are involved in photosynthesis, and their expression is negatively correlated with senescence process.

2   Materials

Prepare all solutions using ultrapure water and analytical grade reagents. Prepare and store all reagents at room temperature unless indicated otherwise.

1. Filter papers and Petri dishes.

2. Aluminum foil.

3. Chlorophyll extraction solution: 95% acetone/ethanol (v/v = 2:1) (*see* **Note 1**).

4. Protein extraction solution: 50 mM Tris–HCl, pH 8.0, 10 mM NaCl, 0.1 M PMSF, and 0.1 M DTT.

3   Methods

3.1   Darkness- or Shade-Induced Senescence in Detached Leaves

1. Wild type or mutants of *Arabidopsis* plants were grown on soil in a growth chamber or growth room at 20 °C (minimum temperature during nighttime) to 24 °C (maximum temperature during daytime) under cool-white fluorescent light (90–100 μmol m^{-2} s^{-1}) in long-day (16 h light/8 h dark) or short-day (8 h light/16 h dark) conditions.

2. Put three to four layers of filter paper in suitable Petri dishes, soak the filter paper completely with pure water or 3 mM MES (PH5.8) or 1/2 MS, and use tissue to absorb excessive water.

3. Green rosette leaves (the third and fourth, *see* **Note 2**) detached from 3- or 4-week-old plants were sampled for darkness-induced senescence. Carefully arrange your detached leaves on the surface of filter papers (*see* **Note 3**, Fig. 1).

4. Detached leaves were incubated in darkness by wrapping the whole dish with double-layer aluminum foil at 22 °C.

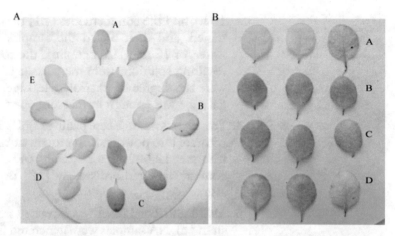

Fig. 1 Orderly arranged leaves in darkness-induced senescence experiments. Visible phenotypes were showed in detached leaves from five different genotypes which were kept in darkness for 4 days

5. After darkness incubation, rosette leaves were sampled under dim green light. Fv/Fm could be measured at this moment. Then frozen in liquid nitrogen for further analysis, including changes of the senescence markers, including chlorophyll quantification, ion leakage rate determination, immunoblot analysis, and qRT-PCR analysis (*see* **Notes 4–6**). Sequential measurement can be conducted until significant senescence phenotypes were observed (Fig. 1).

3.2 Darkness-Induced Senescence in Attached Leaves

1. Attached third and fourth rosette leaves were wrapped in aluminum foil on 4-week-old whole plants.

2. The wrapped leaves could be harvested at the indicated days, such as 2, 4, 6, and 9 days later [17].

3.3 Measurement of Leaf Senescence Parameters

1. Chlorophyll content: The detached leaves used for chlorophyll extraction were incubated in extraction solution for 2–3 h in darkness at room temperature. Then the absorbance at 645 and 663 nm was measured. Chlorophyll content was calculated using the formula $(20.23A_{645} + 8.023A_{663})$/fresh weight (mg/g F.W.).

2. Fv/Fm: Chlorophyll fluorescence (Fv/Fm) was measured with a Li-Cor 6400 photosynthesis-fluence measuring system (Li-Cor, Lincoln, Neb., USA). Leaves were dark adapted at least 20 min before measuring; minimum (Fo) and maximum fluorescence (Fm) were obtained using LCF model according to the manufacturer's protocol. The variable fluorescence, Fv/Fm, was calculated as (Fm − Fo)/Fm.

3. Ion leakage rate (electrolyte leakage): The detached leaves were incubated in deionized water for at least 2 h (less than

10 h), and the conductivities (C1) of the solutions were determined, and then boiled the samples in the same deionized water for 15 min. After cooling, the conductivities (C2) of the resulting solutions were measured again. The ratios of C1:C2 were the degree of electrolyte leakage. In fact, 5 mL centrifugal tube was enough to measure one to two leaves.

4. Protein content: Keep plant samples in liquid nitrogen and triturate into powder. Add protein extraction solution and centrifuge at $12,000 \times g$ for 30 min at 4 °C. Then quantify protein content via Bradford's assay [18] or other protein content assay.

5. qRT-PCR: Total RNA was extracted from treated leaves. First-strand cDNA samples were generated from total RNA samples by reverse transcription using an AMV reverse transcriptase. First-strand cDNA were used as templates for RT-PCR-based gene expression analysis. Bio-Rad iQ5 could be used for gene expression measuring. The oligonucleotide primer sequences used to amplify specific cDNAs are described below:

CAB-R: 5′ CCAGAGGCATTCGCTGAGTTG 3′

CAB-F: 5′ CCTTACCAGTGACGATGGCTTG 3′

RBCS-R: 5′ CCACCCGCAAGGCTAACAAC 3′

RBCS-F: 5′ TTCGGAATCGGTAAGGTCAGG 3′

SEN1-R: 5′ GTCATCGGCTATTTCTCCACCT 3′

SEN1-F: 5′ GTTGTCGTTGCTTTCCTCCATC 3′

SAG12-R: 5′ TGGATACGGCGAATCTACTAACG 3′

SAG12-F: 5′ GCTTTCATGGCAAGACCACATAG 3′

ACT2-R: 5′ CGCTCTTTCTTTCCAAGCTC 3′

ACT2-F: 5′ AACAGCCCTGGGAGCATC 3′

4 Notes

1. It is better to put chlorophyll extraction solution at −20 °C.

2. For all above mentioned experiments, relatively identical leaf stage and age are critical for reliable and reproducible results. To assure the relatively same developmental stage, only leaves from the same leaf position could be used for senescence analysis. In most conditions, the third and fourth rosette leaves were chosen because they were mature and fully expanded for 4-week-old plants. Moreover, they were relatively younger leaves and were suitable for auto-induced senescence analysis.

3. If leaves from more than one genotype were used in darkness-induced senescence experiment, it is vital to put leaves from all

genotypes in one Petri dish, because different Petri dishes may cause uncertain systematic error. Ordered arrangement of leaves is important for subsequent photo taking and measurement of senescence markers. If more biological replicates were needed for your experiment and a single Petri dish was not enough, you could conduct many Petri dishes for individual biological replicate experiment, but every dish should contain leaves from all genotypes.

4. Since senescence is a complex degenerative biological process, the more senescence markers you detected, your data will be more reliable. In fact, it is perfect to simultaneously obtain physiological senescence markers such as chlorophyll content and ion leakage and molecular senescence markers such as *SEN1* expression level.

5. For multiple genotype analysis, their original development stages may be different, so relative senescence marker changes are better than absolute values of senescence parameters. For example, you can monitor the chlorophyll content of one leaf before and after treatment and calculate the change in chlorophyll content. This method could be used to compare senescence process between one leaf and another.

6. To evaluate the senescence score between different genotypes, proper statistical analysis should be used. Generally, the more biological replicates you have, the more likely you can detect subtle changes. Based on the variation for each assay, more than seven biological replicates are better for chlorophyll content and ion leakage measurements, three biological replicates for qRT-PCR analysis, and at least seven biological replicates for Fv/Fm.

References

1. Pennell RI, Lamb C (1997) Programmed cell death in plants. Plant Cell 9:1157–1168

2. Lin JF, Wu SH (2004) Molecular events in senescing Arabidopsis leaves. Plant J 39:612–662

3. Lim PO, Kim HJ, Nam HG (2007) Leaf senescence. Annu Rev Plant Biol 58:115–136

4. Kim JH, Woo HR, Kim J et al (2009) Trifurcate feed-forward regulation of age-dependent cell death involving miR164 in Arabidopsis. Science 323:1053–1057

5. Weaver LM, Amasino RM (2001) Senescence is induced in individually darkened Arabidopsis leaves, but inhibited in whole darkened plants. Plant Physiol 127:876–886

6. Buchanan-Wollaston V, Page T, Harrison E et al (2005) Comparative transcriptome analysis reveals significant differences in gene expression and signaling pathways between developmental and dark/starvation-induced senescence in Arabidopsis. Plant J 42:567–585

7. Thomas H, Ougham H, Canteret P et al (2002) What stay-green mutants tell us about nitrogen remobilization in leaf senescence. J Exp Bot 53:801–808

8. Eckhardt U, Grimm B, Hörtensteiner S (2004) Recent advances in chlorophyll biosynthesis and breakdown in higher plants. Plant Mol Biol 56:1–14

9. Hortensteiner S, Krautler B (2011) Chlorophyll breakdown in higher plants. Biochim Biophys Acta 1807:977–988

10. Oxborough K, Baker NR (1997) Resolving chlorophyll a fluorescence images of photosynthetic efficiency into photochemical and nonphotochemical components—calculation of qP and Fv'/Fm' without measuring Fo'. Photosynth Res 54:135–142

11. Liu X, Li Z, Jiang Z et al (2010) LSD: a leaf senescence database. Nucleic Acids Res 39(Database):D1103–D1107

12. Gan S, Amasino RM (1995) Inhibition of leaf senescence by autoregulated production of cytokinin. Science 270:1986–1988

13. Quirino BF, Noh YS, Himelblau E et al (2000) Molecular aspects of leaf senescence. Trends Plant Sci 5:278–282

14. Weaver LM, Gan SS, Quirino B et al (1998) A comparison of the expression patterns of several senescence-associated genes in response to stress and hormone treatment. Plant Mol Biol 37:455–469

15. Oh SA, Lee SY, Chung IK et al (1996) A senescence-associated gene of Arabidopsis thaliana is distinctively regulated during natural and artificially induced leaf senescence. Plant Mol Biol 30:739–754

16. Van der Graaff E (2006) Transcription analysis of Arabidopsis membrane transporters and hormone pathways during developmental and induced leaf senescence. Plant Physiol 141:776–792

17. Bradford MM (1976) A rapid and sensitive method for the quantitation of microgram quantities of protein utilizing the principle of protein-dye binding. Anal Biochem 72:248–254

Chapter 13

Salt Treatments and Induction of Senescence

Yasuhito Sakuraba, Dami Kim, and Nam-Chon Paek

Abstract

High salinity, one of the most severe abiotic stresses encountered by land plants, often results from water deficit and also induces whole-plant senescence. Thus, salt treatment provides a useful technique for stress-mediated induction of senescence in plants. In this chapter, we describe the procedures to induce senescence in *Arabidopsis* (*Arabidopsis thaliana*) and rice (*Oryza sativa*), using NaCl or KCl. Furthermore, we present experimental approaches to measure salt stress-induced leaf senescence.

Key words Salt treatment, Senescence, NaCl, KCl, *Arabidopsis*, Rice, Chlorophyll, Ion leakage rate, Senescence-associated gene (SAG)

1 Introduction

Leaf senescence, the final stage of leaf development, destabilizes intercellular organelles and decomposes macromolecules in leaves, to relocate nutrients into developing tissues or storage organs [1]. Leaf senescence can be triggered by external, unfavorable factors, including extended darkness, drought, heat, cold, salt, and pathogen attacks [1–4]. Thus, these abiotic and biotic stresses often induce the leaf senescence pathway.

The effects of salt stress on plant senescence depend on the salt concentration. Weak salt stresses induce early flowering in plants [5], and strong salt stresses induce leaf senescence [2]. In salt stress-induced leaf senescence, high salt concentrations, similar to high concentrations of mannitol or polyethylene glycol (PEG), cause osmotic stress, which promotes leaf senescence. High salt also causes plant cells to accumulate ions (Na^+, K^+, and Cl^-), which causes toxic effects if their levels exceed the ability of cells to isolate these ions in the vacuole, thus inducing leaf senescence pathways [6].

To evaluate the leaf senescence phenotype (i.e., stay-green, delayed senescence, or early senescence) of different genotypes, researchers have used extended dark treatments, in addition to natural age-induced senescence [7]. More recently, researchers have adopted

Yongfeng Guo (ed.), *Plant Senescence: Methods and Protocols*, Methods in Molecular Biology, vol. 1744, https://doi.org/10.1007/978-1-4939-7672-0_13, © Springer Science+Business Media, LLC 2018

salt treatments as one of the major approaches to induce leaf senescence. The phenotypes of several *Arabidopsis* stay-green plants were confirmed by both dark-induced and salt stress-induced senescence [8, 9]. By contrast, the *Arabidopsis* autophagy mutant *atg5* showed an early leaf-yellowing phenotype under dark-induced senescence conditions [10], but it showed a stay-green phenotype under salt stress conditions [11]. This observation indicates that the mechanisms of dark- and salt stress-induced senescence largely overlap but also differ from each other. Thus, we need to evaluate leaf senescence phenotypes by combining multiple approaches for artificial induction of senescence.

Plants differ greatly in their sensitivity to salt stresses. Compared with other plant species (under similar conditions of light and humidity), *Arabidopsis* and rice can be considered salt-sensitive species [6]. Thus, assays using salt stress-induced leaf senescence provide useful tools for examining senescence in *Arabidopsis* and rice. Although NaCl is usually used for salt stress or induced senescence experiments, several salt stress studies used KCl [12]. In this chapter, we describe the procedures for salt treatments of whole plants and detached leaves in *Arabidopsis* and rice, by using NaCl or KCl to cause salt stress-induced senescence. We also introduce several experimental approaches to measure leaf senescence.

2   Materials

Prepare all solutions using distilled water and analytical grade chemicals. Chemicals used in this study were purchased from the Duchefa Biochemie (the Netherlands). Store all solutions at room temperature until use.

2.1   Salt Treatment

1. 100 mM NaCl (1 L): 5.85 g NaCl, 2.4 g Murashige-Skoog medium, 3 mM MES buffer, pH 5.8.

2. 100 mM KCl (1 L): 7.46 g KCl, 2.4 g Murashige-Skoog medium, 3 mM MES buffer, pH 5.8.

3. 150 mM NaCl: In MS medium with 3 mM MES buffer, pH 5.8.

4. 150 mM KCl: In MS medium 3 mM MES buffer, pH 5.8.

5. 200 mM NaCl: In MS medium with 3 mM MES buffer, pH 5.8.

6. 200 mM KCl: In MS medium 3 mM MES buffer, pH 5.8.

7. Mock solution (1 L): 2.4 g Murashige-Skoog medium, 3 mM MES buffer, pH 5.8.

8. Petri dishes and culture plates.

2.2 Measurement of Ion Leakage Rate	1. Incubation buffer: 400 mM mannitol.
	2. 14 mL round-bottom tubes.
	3. CON6 conductivity meter (LaMotte, USA).

3   Methods

3.1   Induction of Senescence in Detached Arabidopsis Leaves via Salt Treatments

1. Detach the fourth and fifth rosette leaves from 3-week-old plants grown on soil under long-day (LD) conditions (*see* **Note 1**).

2. Float the detached leaves, adaxial side up, on 100 mM NaCl or 100 mM KCl in plastic Petri dishes (*see* **Note 2** and Fig. 1a).

3. Incubate the leaves under continuous light in a growth chamber for an adequate time, usually 3–4 days after treatments (*see* **Note 3** and Fig. 1a).

4. Measure senescence parameters, such as chlorophyll concentration and ion leakage rate (Fig. 1b, c).

Salt treatment is one of the techniques to induce senescence in laboratories. Thus, it is useful for screening salt stress tolerance of natural variants or induced mutants (e.g., the *oresara 1* mutant [*ore1*]) that exhibit a stay-green phenotype under natural and/or dark-induced senescence conditions (*see* Fig. 1d and **Note 4**).

3.2   Induction of Senescence in Whole Arabidopsis Plants via Salt Treatments

1. Use 3-week-old plants grown on soil under LD conditions in the growth chamber (*see* **Note 5**).

2. Apply 150 mM NaCl or 150 mM KCl (*see* **Note 6**) to the tray of plants to be saturated in soil.

3. Incubate under continuous light at room temperature with moderate humidity for an adequate time, until around 5 days after treatments.

4. Cut the fourth and fifth rosette leaves and measure senescence parameters (*see* Fig. 1e and **Note 7**).

3.3   Induction of Senescence in Detached Rice Leaves via Salt Treatments

1. Prepare leaf disks (1 × 0.5 cm) using the middle part of leaf blades from 1-month-old rice plants grown on soil in the growth chamber under LD conditions (*see* Fig. 2a and **Note 8**).

2. Incubate the leaf disks, adaxial side up, in 200 mM NaCl or 200 mM KCl in plastic Petri dishes (*see* **Note 9**).

3. After an adequate incubation time, usually 3–4 days (Fig. 2b), use the leaf disks for further analysis.

3.4   Induction of Senescence in Whole Rice Plants Via Salt Treatments

1. Use 1-month-old rice plants grown on soil in the growth chamber under LD conditions.

2. Apply 200 mM NaCl or 200 mM KCl to the tray of plants.

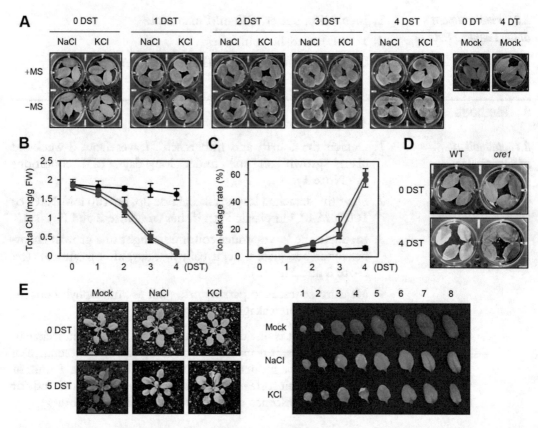

Fig. 1 Salt stress-induced leaf senescence in *Arabidopsis*. (**a–c**) Changes in leaf color (**a**), total Chl levels (**b**), and ion leakage rates (**c**) in the fourth and fifth leaves detached from 3-week-old *Arabidopsis* (Col-0) plants under 100 mM NaCl or KCl treatments. Black, red, and blue lines indicate mock, NaCl, and KCl treatments, respectively. (**d**) Detached leaf phenotype of *ore1* mutants under NaCl treatment. (**e**) The phenotype of 3-week-old whole plants under 150 mM NaCl or KCl treatments. Numbers (right panel) indicate leaf number and 9th–12th young leaves are not shown. DST, day(s) of salt treatment. DT, day(s) of mock treatment. Means and SD values were obtained from more than three biological replicates

3. Incubate at room temperature under continuous light with moderate humidity for an adequate time, until around 8 days after treatments.

4. Cut leaves and use for senescence parameter analyses (Fig. 2c).

3.5 Evaluation of Salt-Induced Leaf Senescence: Ion Leakage Rate

1. Transfer the leaf samples into 14 mL round-bottom tubes containing 6 mL of 400 mM mannitol.

2. Incubate the tubes at room temperature for 3 h with shaking (250–300 rpm).

3. Measure the conductivity of the solution in the tubes using an electro-conductivity meter (Con_{before}). Check the conductivity of 400 mM mannitol solution as a control ($Con_{control}$).

4. Boil the samples for 15 min.

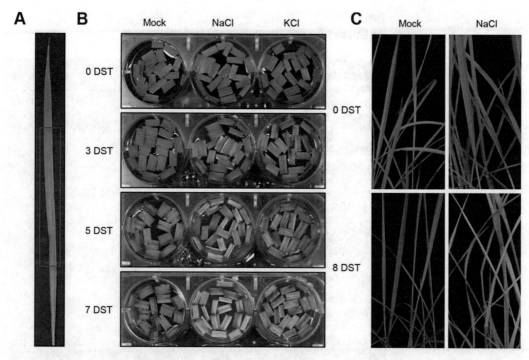

Fig. 2 Salt stress-induced leaf senescence in rice. (**a**) For the analysis of salt stress-induced senescence using detached leaves in rice, the middle parts of the leaf blade, indicated by the red rectangle, were used. (**b**) Color change in the detached leaves of 1-month-old rice plants under 200 mM NaCl or KCl treatments. (**c**) Leaf phenotypes of 1-month-old rice plants under salt stress conditions (200 mM NaCl). DST, day(s) of salt treatment

5. Cool the tubes on ice rapidly until they reach room temperature, and then remove the leaf samples (*see* **Note 10**).

6. Measure the conductivity of the solution in the tubes (Con_{after}).

7. Calculate the ion leakage rate using the formula (Con_{before} − $Con_{control}$)/(Con_{after} − $Con_{control}$) (*see* **Note 11**).

Other parameters for estimation of salt-induced senescence are listed in Table 1 (*see* **Note 12**). Among these parameters, the expression levels of senescence marker genes can be measured by qRT-PCR. In Fig. 3, we introduce the expression patterns of six well-known senescence marker genes under salt stress conditions (*see* Fig. 3 and **Note 13**).

4 Notes

1. The fourth or fifth leaves from 3-week-old *Arabidopsis* plants work well for most senescence experiments because (a) older rosettes (first to third) are very salt-sensitive, and younger rosettes (>6th) are less salt-sensitive, resulting in experimental variation, and (b) leaf size and thickness of 3-week-old plants grown under LD conditions are suitable for this experiment.

Table 1
Parameters for salt stress-induced leaf senescence

Parameters	References
Chlorophyll concentration	[2]
The levels of photosystem proteins (immunoblot analysis)	[11]
Photosynthetic efficiency (Fv/Fm)	[2]
Non-photochemical quenching	[14]
The structures of grana thylakoid and chloroplast (TEM analysis)	[11]
Malondialdehyde (MDA)	[2]
Ion leakage rate	[12]
Soluble protein leakage rate	[12]
The levels of phytohormones (ethylene, ABA, auxin, cytokinin)	[14]
Transcriptional levels of SAGs (expression analysis)	[18]
The levels of antioxidant enzymes	[12]

2. For salt treatment, Murashige-Skoog medium is often included in the NaCl solution. We have tested salt-induced senescence with and without Murashige-Skoog medium (+MS and −MS, respectively). In −MS conditions, we observed a severe leaf cell death phenotype at early stages of salt treatment and irregular retention of chlorophyll (Chl) (Fig. 1a), which resulted in experimental error. By contrast, in the presence of MS, we observed normal leaf yellowing (= Chl degradation). Thus, for the salt treatment of detached leaves, MS supplementation in the salt solution is highly recommended.

3. Incubation time (sampling time) differs depending on the type of experiment. For example, within 2 days of salt treatment (DST), the expression of *senescence-associated genes* (*SAGs*) begins to rise (Fig. 2). At 3–4 DST, we can observe significant differences in total Chl levels and ion leakage rates between mock and salt-treated leaf samples (Fig. 1b, c).

4. *ORE1* is a senescence-associated gene, and the *ore1* mutant shows a delayed senescence phenotype during both natural and dark-induced senescence [13]. Furthermore, *ore1* leaves stay green under salt stress [8], indicating that the salt stress-induced senescence pathway overlaps with dark-induced and natural senescence mediated by the transcriptional activation of ORE1.

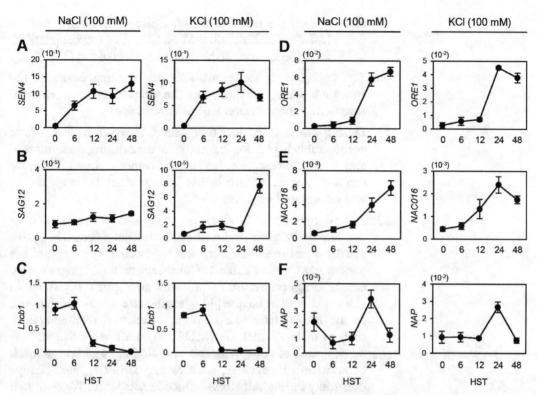

Fig. 3 Expression patterns of marker genes for salt stress-induced leaf senescence in *Arabidopsis*. Expression levels of *SEN4* (**a**), *SAG12* (**b**), *Lhcb1* (**c**), *ORE1* (**d**), *NAC016* (**e**), and *NAP* (**f**) are shown in the fourth and fifth rosette leaves of *Arabidopsis* (Col-0) plants under 100 mM NaCl or KCl treatments. The relative expression levels of the marker genes were determined by qRT-PCR and normalized to transcript levels of *GAPDH* (glyceraldehyde phosphate dehydrogenase; At1g16300). HST, hour(s) of salt treatment. Means and SD values were obtained from more than three biological replicates

5. For this experiment, 2- to 3-week-old plants are appropriate because older plants (>4 weeks old) are less sensitive to salt stress, and natural senescence begins after bolting.

6. It is necessary to increase the salt concentration to induce senescence in whole plants compared with detached leaves.

7. Leaf numbers 4 and 5 in 3-week-old *Arabidopsis* plants are appropriate for most senescence experiments, as described above (*see* Fig. 1e and **Note 1**).

8. In rice, young seedlings are the most salt-sensitive, and rice plants at the tillering stage are the least salt-sensitive [2]. Thus, young seedlings (less than 1-month-old) are appropriate for senescence experiments. We also recommend to use the middle part of the leaf blade and to avoid using the tip part, because the leaf tip senesces faster than the other parts, thus causing experimental errors.

9. Although rice is a salt stress-sensitive plant [6], it is less sensitive than *Arabidopsis*. Thus, 200 mM NaCl or KCl is appropriate for induction of senescence in the detached leaves of rice (Fig. 2b, c).

10. The conductivity values are affected by temperature of the solutions. Thus, one must cool the samples to room temperature before the measurement of conductivity.

11. During salt-induced leaf senescence, *Arabidopsis* leaves have considerably higher ion leakage rates than during dark-induced and natural senescence, because of osmotic stress. Thus, we can see clear difference in ion leakage rates between mock- and salt-treated samples.

12. Although we show the changes in total Chl levels and ion leakage rates during salt-induced leaf senescence (Fig. 1b, c), we usually measure salt-induced leaf senescence by several of the parameters listed in Table 1. Consistent with Chl degradation, Chl-binding photosystem proteins and grana thylakoid in chloroplasts also collapse [11], leading to a significant decrease of the *Fv/Fm* ratio [12]. The accumulation of metabolites, such as malondialdehyde (MDA, 2), the levels of phytohormones, such as ABA, ethylene, and auxin [14], and the levels of antioxidant enzymes, such as catalase [12], also change drastically during salt stress-induced senescence. We note that these common parameters can be used to measure all senescence-inducing conditions.

13. *SAG12* (*SENESCENCE-ASSOCIATED GENE 12*) and *SEN4* (*SENESCENCE 4*) are well-known senescence marker genes [15, 16]. As under natural and dark-induced senescence conditions, expression levels of *SEN4* and *SAG12* also increased rapidly after 100 mM NaCl or 100 mM KCl treatment (Fig. 3a, b). Photosynthesis-related genes, including *Lhcb1*, are also used as senescence markers because their expression significantly decreases at the onset of the senescence phase of leaf development [7]. *Lhcb1* mRNA levels also decreased drastically during salt stress (Fig. 3c). Three *Arabidopsis* NAC genes, *ORE1/NAC092*, *NAC016*, and *NAP/NAC029*, are involved in both leaf senescence and salt stress signaling, and thus the null mutants of these three *NAC* genes stay green during both natural senescence and under salt stress conditions [8, 9, 13, 17]. Because the expression of the three *NAC* genes significantly increased under salt stress (Fig. 3d–f), these *NAC* genes are also useful markers for salt stress-induced senescence. The expression levels of *SAG12*, *ORE1*, *NAC016*, and *NAP* increased after 24 h of salt treatment (HST). After 48 HST, however, *NAP* expression drastically decreased (Fig. 3f). Thus, we recommend 24 HST as an appropriate sampling time for the analysis of gene expression in salt stress-induced senescence. Although expression patterns of these marker genes under

NaCl and KCl treatments are considerably similar, the maximum peaks of these genes occur earlier under KCl treatments than under NaCl treatments (Fig. 3), indicating that KCl is a stronger senescence inducer than NaCl.

Acknowledgments

This work was carried out with the support of Cooperative Research Program for Agriculture Science and Technology Development (Project No. PJ011079), Rural Development Administration, Republic of Korea.

References

1. Lim PO, Kim HJ, Nam HG (2007) Leaf senescence. Annu Rev Plant Biol 58:115–136

2. Lutts S, Kinet JM, Bouharmont J (1996) NaCl-induced senescence in leaves of rice (Oryza sativa L) cultivars differing in salinity resistance. Ann Bot 78:389–398

3. Bohnert HJ, Nelson DE, Jensen RG (1995) Adaptations to environmental stresses. Plant Cell 7:1099–1111

4. Quirino BF, Normanly J, Amasino RM (1999) Diverse range of gene activity during Arabidopsis thaliana leaf senescence includes pathogen-independent induction of defense-related genes. Plant Mol Biol 40:267–278

5. Lutts S, Kinet JM, Bouharmont J (1995) Changes in plant response to NaCl during development of rice (Oryza sativa L) varieties differing in salinity resistance. J Exp Bot 46:1843–1852

6. Munns R, Tester M (2008) Mechanisms of salinity tolerance. Annu Rev Plant Biol 59:651–681

7. Weaver LM, Amasino RM (2001) Senescence is induced in individually darkened Arabidopsis leaves but inhibited in whole darkened plants. Plant Physiol 127:876–886

8. Balazadeh S, Siddiqui H, Allu AD et al (2010) A gene regulatory network controlled by the NAC transcription factor ANAC092/AtNAC2/ORE1 during salt-promoted senescence. Plant J 62:250–264

9. Kim YS, Sakuraba Y, Han SH et al (2013) Mutation of the Arabidopsis NAC016 transcription factor delays leaf senescence. Plant Cell Physiol 54:1660–1672

10. Thompson AR, Doelling JH, Suttangkakul A et al (2005) Autophagic nutrient recycling in Arabidopsis directed by the ATG8 and ATG12 conjugation pathways. Plant Physiol 138:2097–2110

11. Sakuraba Y, Lee SH, Kim YS et al (2014) Delayed degradation of chlorophylls and photosynthetic proteins in Arabidopsis autophagy mutants during stress-induced leaf yellowing. J Exp Bot 65:3915–3925. https://doi.org/10.1093/jxb/eru008

12. Santos CL, Campos A, Azevedo H et al (2001) In situ and in vitro senescence induced by KCl stress: nutritional imbalance, lipid peroxidation and antioxidant metabolism. J Exp Bot 52:351–360

13. Kim JH, Woo HR, Kim J et al (2009) Trifurcate feed-forward regulation of age-dependent cell death involving miR164 in Arabidopsis. Science 323:1053–1057

14. Ghanem ME, Albacete A, Martinez-Andujar C et al (2008) Hormonal changes during salinity-induced leaf senescence in tomato (Solanum lycopersicum L.) J Exp Bot 59:3039–3050

15. Lohman KN, Gan SS, John MC et al (1994) Molecular analysis of natural leaf senescence in Arabidopsis thaliana. Physiol Plant 92:322–328

16. Oh SA, Lee SY, Chung IK et al (1996) A senescence-associated gene of Arabidopsis thaliana is distinctively regulated during natural and artificially induced leaf senescence. Plant Mol Biol 30:739–754

17. Guo YF, Gan SS (2006) AtNAP, a NAC family transcription factor, has an important role in leaf senescence. Plant J 46:601–612

18. Allu AD, Soja AM, Wu A et al (2014) Salt stress and senescence: identification of cross-talk regulatory components. J Exp Bot 65:3993–4008. https://doi.org/10.1093/jxb/eru173

Chapter 14

Methods for Elucidation of Plant Senescence in Response to C/N-Nutrient Balance

Shoki Aoyama, Junji Yamaguchi, and Takeo Sato

Abstract

Carbon (C) and nitrogen (N) are essential elements for metabolism, and the ratio of C to N availability is called the C/N balance. C/N balance is very important for plant growth, but little is known about the detailed mechanisms of plant C/N responses. Previously a method of treating *Arabidopsis* plants with sugar-supplemented medium for studying C/N responses at early post-germinative growth stages has been developed. This method, however, cannot be used to determine physiological C/N effects in plants of mature growth stages, including senescence. Here we present two methods of analyzing responses to C/N treatments in senescing plants: transient C/N treatment with liquid medium and long-term C/N treatment with elevated atmospheric CO_2.

Key words C/N balance, Nutrient, Senescence, CO_2, Hydroponic culture system

1 Introduction

Nutrient availability, in particular the availability of carbon (C) and nitrogen (N), is very important for the regulation of plant metabolism and development. Since C and N metabolisms are coordinated, not only their independent availability, but their ratio, called the C/N balance, is important for plant growth [1, 2]. Plants utilize sugars, generated through photosynthesis, as C sources and nitrate and ammonium as N sources. In nature, C and N availability change according to environmental conditions, including atmospheric CO_2 concentration, light intensity, and rainfall [3–6].

To clarify plant responses to C/N balance, an experiment has been performed using sugar-supplemented medium. *Arabidopsis* seedlings grown in medium containing high concentrations of sugar and limited N show accumulation of purple pigments in cotyledons, along with severe inhibition of early post-germinative growth. Since this phenotype does not occur in medium containing no sugar and limited N, or in medium containing high levels of sugar and sufficient N, this phenomenon is regarded as a response

Yongfeng Guo (ed.), *Plant Senescence: Methods and Protocols*, Methods in Molecular Biology, vol. 1744, https://doi.org/10.1007/978-1-4939-7672-0_14, © Springer Science+Business Media, LLC 2018

to relatively high C/low N stress [2, 7]. Although this method can induce difference in phenotypes and is useful for first screening of C/N response mutants, C/N balance may affect plant growth throughout its life cycle, including during vegetative growth, reproductive growth, and even senescence [8–10]. In addition, sugars are not naturally found in soil but are synthesized in leaves from atmospheric CO_2 by photosynthesis.

Increased atmospheric CO_2 level has become a serious world-wide environmental problem and may disrupt ecologic processes, including plant growth [11–13]. Two methods are currently available to solve these technical problems associated with the analysis of more physiological C/N responses. The first consists of the transient exposure of mature *Arabidopsis* plants with sugar-supplemental liquid medium. This method, which requires a relatively short time, can evaluate C/N response in the vegetative growth stage, but cannot solve the C source problem. The second method is a CO_2/N response assay involving CO_2 manipulation and a hydroponic system. Instead of sugars, the C source used in this method is CO_2, the same C source as in nature. Although this method requires a relatively longer period of time, it can evaluate more physiological C/N responses throughout the plant life cycle.

We recently used these methods to show that the balance of CO_2 and N availability affects the cellular C/N balance and regulates the progression of plant senescence [14]. Plants grown in the presence of a high level of CO_2 and limited N showed a disruption of the intracellular C/N balance and accelerated leaf senescence. This progression of senescence was even observed in transient C/N response assays performed under high C/low N conditions. Here we describe in detail the experimental methods used in the transient C/N and CO_2/N response assays, as well as methods for the evaluation of senescence by measuring chlorophyll and anthocyanin.

2   Materials

Arabidopsis thaliana ecotype Columbia-0 was used for all the experiments. Sterilize all media in an autoclave at 121 °C for 20 min.

2.1   Solutions for C/N Controlled Media

1. 50× Murashige and Skoog (MS) stock solution A (1 L) (*see* **Note 1**): 70.05 g KCl, 8.5 g KH_2PO_4, 310 mg H_3BO_3, 1205 mg $MnSO_4 \cdot 5H_2O$, 430 mg $ZnSO_4 \cdot 7H_2O$.

2. 200× MS stock solution B (1 L): 166 mg KI, 50 mg $Na_2MoO_4 \cdot 2H_2O$, 5 mg $CuSO_4 \cdot 5H_2O$, 5 mg $CoCl_2 \cdot 6H_2O$.

3. 200× MS stock solution C (1 L): 44 g $CaCl_2 \cdot 2H_2O$.

4. 200× MS stock solution D (1 L):74 g MgSO$_4$·7H$_2$O.

5. 200× MS stock solution E (1 L): 5.56 g FeSO$_4$·7H$_2$O, 40 mL 0.5 M EDTA$_2$Na solution (pH 8.0).

6. 100× Vitamin mixture (1 L): 50 mg nicotinic acid, 50 mg pyridoxine HCl, 10 mg thiamine HCl, 200 mg glycine, 10 g myoinositol.

7. 1 M KNO$_3$ solution.

8. 1 M NH$_4$NO$_3$ solution.

9. 1 M KOH.

2.2 Transient C/N Response Assay with Sugar-Supplemented Liquid Medium

1. 1× MS solution: Mix appropriate amount of MS stock solutions A–E and vitamin mixture adequately.

2. Sterilized 2 M D(+)-glucose solution (*see* **Note 2**).

3. Nitrogen solution: Mix 1 M KNO$_3$ and 1 M NH$_4$NO$_3$ 1:1 to appropriate concentrations, 30 mM for high N and 0.3 mM for low N (final concentrations) (*see* **Note 3**).

4. Normal C/N medium (low C high N): 1× MS solution supplemented with 100 mM glucose and 30 mM nitrogen. Adjust pH to 5.7 (±0.05) by adding KOH.

5. High C low N medium: 1× MS solution supplemented with 200 mM glucose and 0.3 mM nitrogen. Adjust pH to 5.7 (±0.05) by adding KOH.

6. High C high N medium: 1× MS solution supplemented with 200 mM glucose and 30 mM nitrogen. Adjust pH to 5.7 (±0.05) by adding KOH.

7. Low C low N medium: 1× MS solution supplemented with 100 mM glucose and 0.3 mM nitrogen. Adjust pH to 5.7 (±0.05) by adding KOH.

8. Solid medium: Add 0.4% gellan gum to normal C/N medium before autoclaving.

9. 12-well plates.

2.3 CO$_2$/N Response Assay with a CO$_2$ Incubator and Hydroponic System

1. 1/5× MS solution: Mix appropriate amount of MS stock solutions A–E and vitamin mixture adequately.

2. Nitrogen solution: Mix 1 M KNO$_3$ and 1 M NH$_4$NO$_3$ stock solutions with 1:1 to appropriate concentrations, 3 mM for high N and 0.3 mM for low N (final concentrations) (*see* **Note 3**).

3. Normal medium (high N): 1/5× MS solution supplemented with 3 mM nitrogen. Adjust pH to 5.7 (±0.05) by adding KOH.

4. Low N Medium: 1/5× MS solution supplemented with 0.3 mM nitrogen. Adjust pH to 5.7 (±0.05) by adding KOH.

5. Soda lime.

6. Black-colored plastic container.

7. Black-colored plastic board.

8. Rock wool.

9. Plant growth chamber.

10. CO_2 measuring and controlling instrument.

11. CO_2 cylinder.

12. Air aspirator.

2.4 Chlorophyll and Anthocyanin Measurements

1. Methanol 1% HCl: 100% methanol containing 1% HCl

2. Chloroform

3. Acetone

4. Spectrophotometer

3 Methods

3.1 Transient C/N Response Assay with Sugar-Supplemented Liquid Medium (see Note 4)

1. Grow *Arabidopsis* plants on normal C/N solid medium for 2 weeks after germination with 16 h light/8 h dark cycles at 22 °C.

2. Pour normal C/N liquid medium into 12-well plates, transfer one plant into each well and grow for 2 days (*see* **Note 5**).

3. Pour each C/N-controlled medium (normal, high C low N, high C high N, low C low N) into new 12-well plates, and transfer one plant into each well.

4. Harvest the plants 24 h later for mRNA analysis or 72 h later for photography and to measure chlorophyll and anthocyanin.

3.2 CO_2/N Response Assay with CO_2 Incubator and Hydroponic System (Fig. 1a)

1. Cut the cap and the lower part of a 1.5 mL Eppendorf tube.

2. Cut a rock wool block to a thickness of 3 cm and core out with an 8 mm cork borer (*see* **Note 6**).

3. Put the cored rock wool into the cut tube to make a seedbed.

4. To make a hydroponic tank, cut a black-colored plastic board into a square and put it on top of a black-colored plastic tapper as a lid (*see* **Note 7**, Fig. 1b).

5. Make holes in the board at appropriate intervals to place the seedbeds (*see* **Note 8** and Fig. 1b).

6. Place *Arabidopsis* seeds onto the seedbeds and grow the plants under normal CO_2/N condition (low CO_2 = 280 ppm, high N = 3 mM nitrogen) for 2 weeks after germination in 12 h light/12 h dark cycles at 22 °C (*see* **Notes 9** and **10**, Fig. 1c).

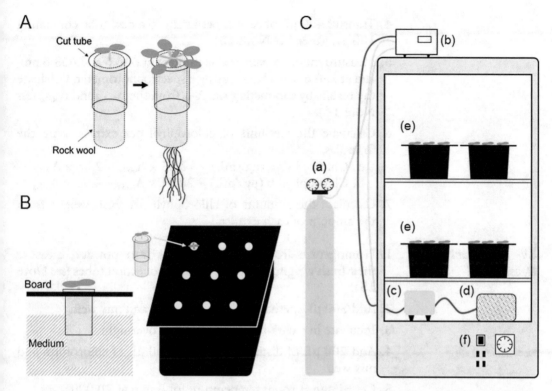

Fig. 1 Schematic model of CO$_2$/N response assay. (**A**) Model of a seedbed. The cap and under part of a 1.5 mL tube were removed, with cored rock wool placed in the cut tube. As the plant grows, its roots become longer, extending from the bottom of the rock wool into the hydroponic medium. (**B**) Model of a hydroponic system. A hydroponic tank, with a container and processed board, was prepared and filled with liquid medium. To stabilize seedbeds (left), the size of the holes in the processed board was designed to be larger than the body of a 1.5 mL tube and smaller than its top. (**C**) Model of a CO$_2$ incubator. (*a*) CO$_2$ cylinder, (*b*) CO$_2$ measuring and controlling instrument, (*c*) air aspirator, (*d*) soda lime box, (*e*) hydroponic set, (*f*) switch and light controller of the incubator. CO$_2$ concentration was manipulated by regulating the amount of CO$_2$ injected from the cylinder and the amount removed by the aspirator and soda lime

7. Transfer the plants to each CO$_2$/N condition (normal, high CO$_2$ low N, high CO$_2$ high N, low CO$_2$ low N, low CO$_2$ = 280 ppm, high CO$_2$ = 780 ppm).

8. Harvest the plants 2.5 weeks later for mRNA analysis and 4 weeks later for photography and to measure chlorophyll, anthocyanin, and carbohydrates (*see* **Note 11**).

3.3 Chlorophyll Measurement

1. Homogenize frozen plant materials into powder, measure their fresh weight, and place them in collection tubes (*see* **Note 12**).

2. Add 800 µL of 100% acetone to each tube and mix well.

3. Centrifuge at room temperature for 5 min at 20,000 × *g*.

4. Transfer 400 µL of each supernatant to a new tube containing 100 µL water (*see* **Note 13**).

5. Measure the absorbance of all samples at 646.6 and 663.6 nm, and at 750 nm as a blank, using a spectrophotometer. Calculate the results by subtracting the A_{750} from the $A_{646.6}$ and $A_{663.6}$ (*see* **Note 14**).

6. Calculate the amounts of chlorophyll per extract using the formulas.
   A: Chlorophyll a (µg/mL) = $12.25 \times A_{663.6} - 2.55 \times A_{646.6}$
   B: Chlorophyll b (µg/mL) = $20.31 \times A_{663.6} - 4.91 \times A_{646.6}$

7. Calculate the amounts of chlorophyll per fresh weight from the amount of each extract.

3.4 Anthocyanin Measurement

1. Homogenize frozen plant materials into powder, measure their fresh weight, and place them in collection tubes (*see* **Note 15**).

2. Add 300 µL methanol 1% HCl to each and mix well.

3. Incubate in a dark refrigerator (4 °C) overnight.

4. Add 200 µL of distilled water and 500 µL of chloroform and mix well.

5. Centrifuge at room temperature for 5 min at $20,000 \times g$.

6. Transfer 400 µL of each supernatant to a new tube.

7. Measure the absorbance of all samples at 532 nm (A_{532}) and at 657 nm (A_{657}), the latter as a blank, using a spectrophotometer. Calculate results by subtracting A_{657} from A_{532} (*see* **Notes 16 and 17**).

8. Calculate the amounts per fresh weight from these results (*see* **Note 18**).

4   Notes

1. Standard premixed MS basal salt mixtures cannot be used to manipulate N concentrations. To avoid the appearance of precipitates, stock solutions should be prepared separately.

2. Since glucose is easy to denature by heat, it must not be added to medium before autoclaving. Glucose should be filter sterilized. Sucrose, however, can be added before autoclave sterilization.

3. N concentration is calculated from total amounts of KNO_3 and NH_4NO_3 added to the solution. For example, 3 mM N solution contains 1 mM KNO_3 and 1 mM NH_4NO_3.

4. To avoid an outbreak of mold, preparation of media and transformation should be performed on a clean bench.

5. This operation is performed to accustom plants to liquid medium.

6. Rock wool must be cored out to an appropriate size for normal plant growth. If a larger cork borer is used and too much rock is placed in the tube, plants cannot extend their roots into the rock wool and will be unable to grow well. An 8 mm cork borer is suitable for a 1.5 mL Eppendorf tube.

7. The size of the lid should be slightly larger than the top of the tapper. To avoid an outbreak of algae, the hydroponic medium should be kept in the dark.

8. It is better for the holes to be spaced at intervals greater than 5 cm on the board.

9. Since not all plants grow well, at least 1.5 times the number of seeds should be sown compared as the number of plants required for the CO$_2$/N response analysis.

10. Cover the plants with clear wrap for the first week. Then remove the cover and grow for one additional week.

11. After 2.5 weeks, plant fresh weights differed among the plants grown under each condition, but apparent progression of senescence was not observed. After 4 weeks, senescence occurred only under conditions of high CO$_2$ and low N.

12. The amount of powder appropriate for this assay is about 5–15 mg per tube. Do not use more than that.

13. If green powder persists in this step, remove the residual supernatant and extract chlorophyll from the residue by repeating **steps 2–4**. Mix the extract together with the first extracted sample.

14. Wavelengths for absorbance and formulas for calculations have been described [15]. All samples should be diluted so that the measured absorbance falls within a range of 0.08–0.8.

15. The amount of powder appropriate for this assay is about 30–60 mg per tube.

16. The wavelength used to measure anthocyanin concentration was determined by scanning at wavelengths from 400 to 700 nm [16], with the highest value observed at 532 nm and the blank at 657 nm.

17. The supernatant after centrifugation consisted of a 6:4 mixture of methanol 1% HCl and distilled water without anthocyanin. Any sample showing an absorbance over 0.8 should be diluted with a 6:4 mixture of methanol 1% HCl and distilled water.

18. Since anthocyanin is a general term, describing several water-soluble pigments, its amount is difficult to determine accurately. Therefore, anthocyanin amounts are shown as relative levels in this method.

Acknowledgments

We thank Dr. Juntaro Negi (Kyushu University) for technical advice on the CO_2 manipulation system and Drs. Takushi Hachiya (Nagoya University) and Junpei Takano (Osaka Prefecture University) for technical advice on the hydroponic culture method. This work was supported by the Japan Society for the Promotion of Science (JSPS) Grants-in-Aid for Scientific Research (No. 24770035, 15K18819 and 17K08190) to TS, on Innovation Areas (No. 24114701 and No. 25112501) to JY, and in part by The Akiyama Foundation to TS. SA was supported by the Research Fellowship for Young Scientists from the JSPS (15J01802) and the Plant Global Education Project from the Nara Institute of Science and Technology (2013–2014).

References

1. Coruzzi GM, Zhou L (2001) Carbon and nitrogen sensing and signaling in plants: emerging 'matrix effects'. Curr Opin Plant Biol 4:247–253

2. Martin T, Oswald O, Graham IA (2002) Arabidopsis seedling growth, storage lipid mobilization, and photosynthetic gene expression are regulated by carbon:nitrogen availability. Plant Physiol 128:472–481

3. Gibon Y, Blasing OE, Palacios-Rojas N et al (2004) Adjustment of diurnal starch turnover to short days: depletion of sugar during the night leads to a temporary inhibition of carbohydrate utilization, accumulation of sugars and post-translational activation of ADP-glucose pyrophosphorylase in the following light period. Plant J 39:847–862

4. Miller AJ, Fan X, Orsel M et al (2007) Nitrate transport and signalling. J Exp Bot 58:2297–2306

5. Smith AM, Stitt M (2007) Coordination of carbon supply and plant growth. Plant Cell Environ 30:1126–1149

6. Kiba T, Kudo T, Kojima M et al (2011) Hormonal control of nitrogen acquisition: roles of auxin, abscisic acid, and cytokinin. J Exp Bot 62:1399–1409

7. Sato T, Maekawa S, Yasuda S et al (2009) CNI1/ATL31, a RING-type ubiquitin ligase that functions in the carbon/nitrogen response for growth phase transition in Arabidopsis seedlings. Plant J 60:852–864

8. Rolland F, Baena-Gonzalez E, Sheen J (2006) Sugar sensing and signaling in plants: conserved and novel mechanisms. Annu Rev Plant Biol 57:675–709

9. Wingler A, Purdy S, MacLean JA et al (2006) The role of sugars in integrating environmental signals during the regulation of leaf senescence. J Exp Bot 57:391–399

10. Watanabe M, Balazadeh S, Tohge T et al (2013) Comprehensive dissection of spatio-temporal metabolic shifts in primary, secondary, and lipid metabolism during developmental senescence in Arabidopsis. Plant Physiol 162:1290–1310

11. Long SP, Ainsworth EA, Rogers A et al (2004) Rising atmospheric carbon dioxide: plants FACE the future. Annu Rev Plant Biol 55:591–628

12. Hikosaka K, Kinugasa T, Oikawa S et al (2011) Effects of elevated CO_2 concentration on seed production in C-3 annual plants. J Exp Bot 62:1523–1530

13. Knohl A, Veldkamp E (2011) Global change: indirect feedbacks to rising CO_2. Nature 475:177–178

14. Aoyama S, Huarancca Reyes T, Guglielminetti L et al (2014) Ubiquitin ligase ATL31 functions in leaf senescence in response to the balance between atmospheric CO$_2$ and nitrogen availability in Arabidopsis. Plant Cell Physiol 55:293–305

15. Porra RJ (2002) The chequered history of the development and use of simultaneous equations for the accurate determination of chlorophylls a and b. Photosynth Res 73:149–156

16. Neff MM, Chory J (1998) Genetic interactions between phytochrome A, phytochrome B, and cryptochrome 1 during Arabidopsis development. Plant Physiol 118:27–35

Chapter 15

Study of Cotton Leaf Senescence Induced by *Alternaria alternata* Infection

Wei Liu, Wenwei Zhang, Na Zheng, Weibo Zhai, and Fangjun Qi

Abstract

Premature leaf senescence in cotton, which often happens during the mid to late growth period, has been occurring with an increasing frequency in many cotton-growing areas and causing serious reduction in yield and quality. One of the key factors causing cotton leaf senescence is the infection of Alternaria leaf spot pathogens (*Alternaria* species), which often happens when cotton plants encounter adverse environmental conditions, such as chilling stress and physiological impairment. Stressed cotton leaves are apt to be infected by *Alternaria* leaf spot pathogens (*Alternaria alternata*) because of the reduction in disease resistance, leading to the initiation of leaf senescence. Here we describe the induction of cotton leaf senescence by *Alternaria alternata* infection, including the evaluation of the disease index and measure of physiological impairment associated with cotton leaf senescence and analysis of possible molecular mechanism using microarray.

Key words Cotton, Premature leaf senescence, *Alternaria alternata* infection, Chilling stresses, Microarray

1 Introduction

Cotton (*Gossypium hirsutum* L.), the most important plant fiber production crop, is a perennial, indeterminate plant that is cultured as an annual in agronomic production systems. Cotton premature leaf senescence has been occurring with an increasing frequency in many cotton-growing countries, causing serious reduction in yield and quality [1, 2]. In China, a significant yield reduction of 20–50% was recorded in severe cotton premature leaf senescence occurred years [3]. Premature leaf senescence has developed to be one of the most serious barriers for cotton production in China in recent years.

The key factors causing and promoting cotton premature leaf senescence are still unclear [4]. In China, it was reported that the appearance of severe cotton leaf senescence in major cotton-growing areas was accompanied with unexpected short-term low

Yongfeng Guo (ed.), *Plant Senescence: Methods and Protocols*, Methods in Molecular Biology, vol. 1744, https://doi.org/10.1007/978-1-4939-7672-0_15, © Springer Science+Business Media, LLC 2018

temperature that occurred in the late cotton-growing season, especially in early to mid of August. However, it has been proved that cotton plants could recover from adverse temperature conditions, such as freezing, when temperatures returned to optimal [5]. Low temperature thus could not act as the single factor that causes cotton premature leaf senescence. Our previous field investigation showed that almost each cotton senescent leaf, usually appeared after a period of consecutive low temperature, was always accompanied with a large amount of leaf spot lesions. The main isolated pathogens from these disease lesions were identified to be *Alternaria alternata* [6].

In this work, the relationship between chilling stress and *A. alternata* infection in causing cotton leaf senescence was investigated under precisely controlled conditions with four- to five-leaf stage cotton plants, which acted as feasible experiment materials under laboratory conditions. During leaf senescence process, internal physiological and biochemical characteristics change significantly, such as the increase of membrane lipid peroxidation and membrane electrolyte osmotic potential [7], the degradation of chlorophyll [8] and proteins, the decrease of photosynthetic function [9], and so on. In order to reveal the relationship of chilling stress, *A. alternata* infection, and cotton leaf senescence, the measurements of MDA content and electrolyte leakage, irreversible decreasing in chlorophyll content, soluble protein content, and Fv/Fm ratio during Alternaria disease development promoted by chilling stress pretreatment were conducted.

In recent years, microarray analysis has been employed to identify stress responsive genes in many plant species [10–13]. Although several cotton microarray platforms exist [14, 15], there has been no significant report, to date, on the molecular mechanisms of chilling stress, pathogenic fungi infection and their combined action in cotton, or the use of array technology in identifying responsive transcripts in cotton. To address this need for a major field crop, the responsive genes were identified by a comparative microarray analysis of cotton after treatments. The identified responsive genes will provide a first glimpse into the understanding of the molecular mechanism of the process of chilling stress causing *Alternaria alternata* infection and leading to cotton leaf senescence. Real-time quantitative RT-PCR was used to verify the reliability of the microarray results.

2   Materials

Prepare all solutions using ultrapure water (prepared by purifying deionized water to attain a sensitivity of 18 M Ω cm at 25 °C) and analytical grade reagents. Prepare and store all reagents at room temperature (unless indicated otherwise).

2.1 Plant Materials and Medium

1. Cotton materials: Leaf premature senescence-resistant cultivar XLZ33 and susceptible cultivar XLZ13.

2. Potato dextrose agar (PDA): Cut 200 g peeled potatoes into about 1 cm³ pieces. Add 1000 mL water and boil for 30 min, then filter with four layers of gauze, add 2% glucose and 1.5% agar, and make up to 1000 mL with water after stirred and dissolved and autoclave at 121 °C for 20 min.

2.2 Malondialdehyde (MDA) Measurement

1. TCA buffer: 10% (w/v) trichloroacetic acid.

2. TBA buffer: 0.6% (w/v) thiobarbituric acid in 10% TCA.

2.3 RNA Isolation

1. 0.1% DEPC-treated water: Add 1000 mL water to a 1 L graduated cylinder, and then add 1 mL DEPC (see **Note 1**). Shake at 37 °C overnight, and autoclave for 20 min at 121 °C. Add to 1.5 mL centrifuge tubes (RNase-free) after cooldown. Store at 4 °C.

2. 1 M EDTA (pH 8.0): Weight 29.225 g EDTA, and transfer to a beaker containing about 70 mL water. Mix and adjust pH with about 2 g of NaOH (see **Note 2**). Make up to 100 mL with water in a graduated cylinder.

3. 1 M Tris (pH 8.0): Weight 12.11 g Tris, and transfer to a beaker containing about 70 mL 0.1% DEPC-treated water. Mix and adjust pH with about 4.2 mL of HCl (see **Note 3**). Make up to 100 mL with 0.1% DEPC-treated water in a graduated cylinder.

4. 10% SDS: Weight 1.0 g SDS and solve in 10 mL water.

5. CTAB buffer: Weight 4.0 g CTAB, 4.0 g PVP, and 23.4 g NaCl, and transfer to a 200 mL graduated cylinder containing about 40 mL water (see **Note 4**). Add 5 mL 1 M EDTA (pH 8.0). Make up to 180 mL with water, and then add 200 μL DEPC (see **Note 1**). Shake at 37 °C overnight, and autoclave for 20 min at 121 °C. Add 20 mL 1 M Tris (pH 8.0) after cooldown.

6. 8 M LiCl: Weigh 67.8240 g LiCl, and transfer to a 200 mL graduated cylinder. Make up to 200 mL with water, and then add 200 μL DEPC (see **Note 1**). Shake at 37 °C overnight and autoclave for 20 min at 121 °C.

7. 75% ethanol: Add 50 μL DEPC to a 200 mL graduated cylinder containing about 50 ml water (see **Note 1**). Shake at 37 °C overnight and autoclave for 20 min at 121 °C. Make up to 200 mL with new anhydrous ethanol after cooldown. Store at 4 °C.

8. SSTE buffer: Weigh 5.84 g NaCl and transfer to a 100 mL graduated cylinder containing about 20 mL water. Add 100 μL 1 M EDTA (PH 8.0) and 5 mL 10% SDS. Make up to 99 mL

with water, and then add 100 μL DEPC (*see* **Note 1**). Shake at 37 °C overnight and autoclave for 20 min at 121 °C. Add 1 mL 1 M Tris (PH 8.0) after cooldown.

9. 3 M/L NaAc (pH 5.0): Weigh 81.6480 g NaAc and transfer to a 200 mL graduated cylinder containing about 130 mL water. Mix and adjust pH with glacial acetic acid. Make up to 180 mL with water, and then add 200 μL DEPC (*see* **Note 1**). Shake at 37 °C overnight and autoclave for 20 min at 121 °C. Store at 4 °C.

2.4 Real-Time Quantitative RT-PCR

1. Mix I: 1 μL M-MLV enzyme, 5 μL reaction buffer, 1.25 μL dNTP mixture, and 0.625 μL RNase inhibitor and add DEPC water to 10 μL.

2. Mix II: 10.0 μL SYBR® Premix Ex Taq (Tli RNase H Plus) (2×), 0.2 μL PCR forward primer (20 μM), 0.2 μL PCR reverse primer (20 μM), 0.4 μL ROX reference dye II (50×), and 2.0 μL cDNA and add sterilized distilled water to 20.0 μL.

3 Methods

3.1 Cotton Plant Growth

1. Surface-sterilize the cotton seeds for 2 min in 75% (v/v) ethanol and 20 min in 6% (v/v) sodium hypochlorite solution.

2. Wash with sterilized water thoroughly and germinate on two sheets of moist filter paper at 28 °C in a growth chamber for 3 days.

3. Transfer to 10 cm diameter plastic pots filled with a mixture of commercial humus/commercial vermiculite (1:1, v/v) and planted in a growth chamber with a 14/10 hours photoperiod at 300 μmol photons m^{-2} s^{-1} and a 28 °C/22 °C temperature regime.

4. Irrigate and fertilize with 1/3 macro- and microelements of MS medium [16] periodically until used for experiments.

3.2 Fungus Cultivation

1. The identified *A. alternata* isolate A1 from cotton leaf spot disease lesions was kept in our lab and maintained on potato dextrose agar (PDA) slants at 4 °C [6].

2. Pick a little hypha from *A. alternata* on potato dextrose agar (PDA) slants at 4 °C, and transfer to new PDA flat in clean bench. And cultivate at 25 °C in a growth chamber.

3. Pick hyphae from the edge of a colony after 3 days, and transfer to new PDA medium. And cultivate at 25 °C in a growth chamber until used for experiments.

3.3 Cotton Plant Chilling Stress Pretreatments

1. Transfer cotton plants at four-to-five leaf stage in controlled environment chambers set with a relative humidity of 65% and a 14/10 h photoperiod at 300 μmol photons m^{-2} s^{-1}.

2. Treat the plants at low temperatures of 16/12 °C day/night for 3 days. Grow the control plants at optimal temperature of 28/20 °C day/night.

3.4 A. alternata Inoculation

1. Scrape 14-day-old cultures grown on PDA flats at 25 °C with a sterile inoculating loop, and suspend with sterile deionized water.

2. Filter through cheesecloth, and dilute to optimal concentration of 1.2×10^4 conidial/mL. Measure spore concentrations microscopically with a counting chamber.

3. Inoculate the third and fourth true leaves of seedlings from chilling stress treatments with about 1 mL freshly prepared conidial suspension per leaf by slightly brushing the leaves with a small brush, while treat mocked control leaves with sterilized water.

4. Cover inoculated leaves with a separate, loosely sealed, pre-wetted polyethylene bags.

5. After inoculation, incubate both inoculated and mocked control plants in a controlled growth chamber in the dark for 24 h at 28 ± 2 °C [17].

6. Transfer all plants at optimal temperature of 28/20 °C day/night with the photoperiod and humidity condition as described in Subheading 3.3.

3.5 Alternaria Disease Evaluation

1. Photograph diseased leaves 15 days after inoculation with a standard digital camera.

2. Analyze affected leaf area with an image software [18].

3. The degree of Alternaria disease is divided into six grades with disease scores ranging from 0 to 5 based on the percentage of affected leaf area as previously described [19] with slight modifications: 0 = no disease; 1 = minute pinhead size spots, less than 1% leaf tissue diseased; 2 = small brown to dark-brown necrotic lesions, 1–5% diseased; 3 = necrotic lesions coalescing, 5–10% diseased; 4 = necrosis lesions coalescing, 10–20% diseased; and 5 = lesions coalescing, >20% diseased and/or with desiccated and abscised leaves.

4. Calculate the disease index according to the following formula: Disease index = [($\sum$disease score × number of infected leaves of each score)/total inoculated leaves × highest score (5)] × 100.

3.6 Membrane Electrolyte Leakage Determination

1. Wash and place 10–15 leaf discs (*see* **Note 5**) excluding the main veins in 20 mL of deionized water.

2. Record water conductivity at the beginning of the incubation (initial cond.) and after incubation for 3 h with gentle shaking (cond. 3 h), at 25 °C.

3. Boil the leaf discs for 5 min, and determine maximum conductivity (max. cond.).

4. Calculate electrolyte leakage as ((cond. 3 h−initial cond.)/(max. cond.−initial cond.)) × 100 (*see* **Note 8**).

3.7 Malondialdehyde (MDA) Measurement

1. Homogenize cotton leaves (*see* **Note 5**) in 5 mL TCA buffer, and centrifuge at $13,523 \times g$ for 10 min.

2. Add 4 mL TBA buffer to a volume of 2 mL of clear supernatant.

3. Incubate the reaction mixture at 100 °C in water bath for 15 min.

4. Terminate the reaction at room temperature, and determine the absorbance of the supernatant at 450, 532, and 600 nm with a spectrometer.

5. Calculate the concentration of MDA by the formula: $C(\mu mol/L) = 6.45 \times (OD532−OD600)−0.56 \times OD450$ (*see* **Notes 6** and **7**).

3.8 Soluble Protein Measurement

1. Make standard curve with bovine serum albumin.

2. Weight 0.5 g fresh cotton leaf (*see* **Note 5**), ground into homogenates with 5 mL distilled water, and then centrifuge at $9,391 \times g$ for 10 min.

3. Add 5 mL of Coomassie Brilliant Blue G-250 to a volume of 1 mL of clear supernatant, shake the reaction mixture, and then stand for 2 min.

4. Determine the absorbance of the supernatant at 595 nm with a spectrometer.

5. Get the values of protein content from the standard curve.

6. Calculate the content of soluble protein (mg/g) as $C \times V_T/(V_S \times W_F \times 1000)$. C (μg) is the value of protein content from the standard curve; V_T (mL) is the total volume of clear supernatant; V_S (mL) is the volume of clear supernatant when determining the absorbance; W_F (g) is the weight of the fresh cotton leaves) (*see* **Notes 6** and **7**).

3.9 Measurement of Chlorophyll Content

1. Use a portable chlorophyll SPAD-502, and make sure the calibrations show that relative SPAD values depend on chlorophyll content in a linear manner over a wide range.

2. Each data point represents the mean value of ten independent measurements (*see* **Notes 5–7**).

3.10 Measurements of the Photosystem II Efficiency

1. Measure chlorophyll fluorescence of cotton leaves (*see* **Note 5**) with a chlorophyll fluorometer after dark adaption (*see* **Note 8**).

2. Get mean values of the maximum quantum efficiency of photosystem II photochemistry (Fv/Fm ratio) based on ten independent measurements (*see* **Notes 6** and **7**).

3.11 RNA Isolation

1. Add 15 μL β-mercaptoethanol to a 1.5 mL centrifuge tubes (*see* **Note 9**) with about 750 μL CTAB (*see* **Note 10**), and preheat at 65 °C.

2. Pour liquid nitrogen in a mortar (*see* **Note 11**), take about 100 mg cotton leaves (*see* **Note 12**) to the mortar, and ground to powder quickly; then transfer to the centrifuge tubes with preheated buffer, mix by vortex, and then heat for 8 min at 65 °C.

3. Add 750 μL chloroform, mix by inversion, and then centrifuge at 4 °C, 9,391 × g, for 15 min using a microfuge.

4. Transfer clear supernatant to a new 1.5 mL centrifuge tube (RNase-free), add equal volume of chloroform, and then centrifuge at 4 °C, 9,391 × g, for 15 min in a microfuge.

5. Transfer clear supernatant to a new 1.5 mL centrifuge tube (RNase-free), and add 1/3 volume of 8 M LiCl overnight at 4 °C.

6. Centrifuge at 4 °C, 13,523 × g, for 15 min in a microfuge, discard clear supernatant, and wash the precipitate with 500 μL 75% ethanol.

7. Centrifuge at 4 °C, 12,000 rpm, for 5 min in a microfuge, discard the ethanol, and dissolve the precipitate with 500 μL SSTE, add equal volume of chloroform, and then centrifuge at 4 °C, 13,523 × g, for 10 min in a microfuge.

8. Transfer clear supernatant to a new 1.5 mL centrifuge tube, add 1/10 volume of 3 M NaAc and two volumes of anhydrous ethanol, and then place at −80 °C for 2 h.

9. Centrifuge at 4 °C, 15,871 × g, for 20 min in a microfuge, wash the precipitate with 400 μL of 75% ethanol, centrifuge at 4 °C, 13,523 × g, for 5 min in a microfuge, and then discard the ethanol.

10. Dissolve RNA with 30 μL DEPC water (*see* **Note 13**), and store at −80 °C until used.

3.12 Affymetrix GeneChip Experiment and Microarray Data Analysis

1. Make biotin-labeled cRNA targets using total RNA from each cotton leaf sample.

2. Conduct all the processes for first-strand cDNA synthesis, second-strand cDNA synthesis and cRNA synthesis, cRNA purification, quality test fragmentation, and chips hybridization and washing, staining, and scanning as stipulated in the GeneChip standard protocol (eukaryotic target preparation, Affymetrix) (*see* **Note 14**).

3. Transform the image signal into the digital signal using the AGCC software (Affymetrix® GeneChip® Command Console® Software).

4. Perform data preprocessing using RMA algorithm [20].

5. Perform clustering analysis of the preprocessed data using the Cluster 3.0 software.

6. Conduct pair-wise comparisons to identify differentially expressed genes, and then analyze differentially expressed genes using SAM (significance analysis of microarray) R package, and screen differentially expressed genes as q-value ≤5% and fold change ≥2 ring ≤0.5.

7. Analyze differentially upregulated and downregulated genes using the functional categorization based on the three GO categories at p-values ≤0.05 (*see* **Note 15**).

3.13 Real-Time Quantitative RT-PCR

1. Add 2 µg of total RNA and 2 µL of random hexamers to tubes, and make up to 15 µL with DEPC water; maintain at 70 °C for 5 min to denature the RNA, then chill on ice for 1 min, and then centrifuge flashing.

2. Prepare Mix I to the chilled samples, and maintain at 37 °C for 1 h (*see* **Note 16**).

3. Design gene-specific primers for qRT-PCR with consensus sequences of *G. hirsutum* genes from GenBank with the Primer Premier 5.0 software (*see* **Note 17**).

4. Prepare Mix II in microtubes (*see* **Notes 18–22**).

5. Run the PCR program containing a pre-denaturation step of 30 s at 95 °C, followed by denaturation for 5 s at 95 °C and annealing for 34 s at 60 °C for 40 cycles.

6. Use the relative quantification method ($\Delta\Delta$CT) for quantitative evaluation of the variation between replicates (*see* **Note 23**).

4   Notes

1. DEPC is a potential carcinogen. DEPC toxicity is not very strong, but it has the strongest toxicity when inhaled.

2. Powdered NaOH can be used at first to narrow the gap from the starting pH to the required pH. From then on, it would be better to use a series of NaOH with lower ionic strengths to avoid a sudden drop in pH above the required pH.

3. Concentrated HCl can be used at first to narrow the gap from the starting pH to the required pH. From then on, it would be better to use a series of HCl with lower ionic strengths to avoid a sudden drop in pH below the required pH.

4. In order to better dissolve the solute, add PVP after CTAB is dissolved completely, and similarly, add NaCl after PVP is dissolved completely.

5. All measurements introduced below should be measured at -3 (before chilling stress treatment), 0 (just after chilling stress treatment and before inoculation), 2, 4, 6, 10, 15 days after inoculation, respectively. All measurements should be made on the third or fourth fully expanded leaves at 25 °C. And the measured leaf areas should be selected about 1–2 cm beyond from the edge of the disease spot on diseased leaves unless otherwise noted.

6. Experiments should be carried out in a completely randomized design with three replicates for each treatment. All experiments should be repeated twice, and data from each repeat experiment should be analyzed separately and in combination. Results of the two repeats analyzed separately and in combination should have similar responses. So data from the two repeats should be combined (six replicates) and analyzed together using the SAS program.

7. All data (mean 6 standard deviation) had better to be presented relatively to their values at -3 d (=100%, before chilling stress pretreatments) for better readability of the figures and to stress their changes in time. Difference should be considered significant at a probability level of $P < 0.05$.

8. Adjust the light intensity measurement before determining the Fv/Fm ratio, in order to ensure the real-time fluorescence is less than 600, preferably between 200 and 400.

9. All the tubes and tips for experiment should be pollution-free and RNase-free.

10. Transfer β-mercaptoethanol to the centrifuge tubes inside the fume hood, because β-mercaptoethanol is toxic.

11. The mortars and pestles, spoons, and tweezers for experiment should be autoclaved for 20 min at 121 °C, and then pre-cooled by pouring liquid nitrogen after the mortar cools down.

12. The third and fourth fully expanded leaves of the cotton plants should be collected and placed in liquid nitrogen immediately and stored at -80 °C until used.

13. The RNA should be purified using NucleoSpin® RNA cleanup (MN). The quantity and quality of the RNA could be determined by absorbance spectra at 260 and 280 nm (A260/280, 1.80–2.10 (protein- and DNA-free); A260/230, 2.10–2.50 (organic solvents and salt ions free)) using a ND-1000 spectrophotometer (NanoDrop) and by denaturing agarose gel electrophoresis.

14. All the reagents used in this experiment were described at www.affymetrix.com/support/technical/manuals.affx.

15. The agriGO tool http://bioinfo.cau.edu.cn/agriGO/ was used to perform the enrichment analysis using SEA (singular enrichment analysis) coupled with available background data of cotton probes [21].

16. The products of this step are cDNA samples, which should be diluted to less than 50 ng/μL for qPCR analysis.

17. The product size from each amplicon had better be ranged from 80 to 150 base pairs and also can be extended to 300 base pairs.

18. Make sure that the reagent stays on the ice when the reaction liquid was mixed. And avoid glare when the reaction liquid was mixed because SYBR® Green I is light sensitive.

19. New (pollution-free) tips and microtubes should be used to prepare the reaction liquid.

20. Perform real-time quantitative PCR assays in triplicate using a SYBR® Premix Ex Taq™ Kit (TaKaRa) with cotton poly-ubiquitin 14 (UBQ14) gene [22] as the internal standard with an ABI 7500 sequence detection system as prescribed in the manufacturer's protocol. If the kit or the detection system is changed, the component and the volume of the reaction liquid should be changed.

21. It usually helps us get better results when the final concentration of primer is 0.2 μM. But when the reaction performance is poor, the final concentration of primer can be adjusted from 0.1 to 1.0 μM.

22. In a 20 μL reaction system, the amount of cDNA template should be less than 100 ng. Because different types of cDNA template contain different copy number of the target gene, gradient dilution could be carried out to determine the optimal amount of cDNA template. But the cDNA template volume should not exceed 10% of the PCR reaction volume.

23. All qRT-PCR expression assays should be independently performed and analyzed three times under identical conditions.

Acknowledgments

This work was supported by the National Natural Science Foundation of China (31371898), State Key Laboratory of Plant Disease, and Insect Biology Open Foundation (SKLOF201615).

References

1. Dong HZ, Li WJ, Tang W et al (2005) Research progress in physiological premature senescence in cotton. Cotton Sci 17:56–60

2. Wright PR (1998) Research into early senescence syndrome in cotton. Better Crops Int 12:14–16

3. Liu L, Chen Y, Wang SZ et al (2010) Comparison and evaluation of cotton varieties resistant to premature aging. Plant Prot 37:107–111

4. Qi FJ, Jian GL, Li JS (2013) Discrimination on the relationship among cotton premature senescence, red leaf disease and Alternaria leaf spot disease. Cotton Sci 1:81–85

5. Kargiotidou A, Kappas I, Tsaftaris A et al (2010) Cold acclimation and low temperature resistance in cotton: Gossypium hirsutum phospholipase Da isoforms are differentially regulated by temperature and light. J Exp Bot 61:2991–3002

6. Li S, Zhang WW, Qi FJ et al (2011) Effect of environment conditions on conidia germination of cotton Alternaria leaf spot. Cotton Sci 23:472–475

7. Woo HR, Kim JH, Nam HG et al (2004) The delayed leaf senescence mutants of Arabidopsis, ore1, ore3 and ore9 are tolerant to oxidative stress. Plant Cell Physiol 45:923–932

8. Hörtensteiner S, Matile P (2004) How leaves turn yellow: catabolism of chlorophyll. In: Noodén LD (ed) Plant cell death processes. Elsevier Academic Press, San Diego, pp 189–202

9. Gergoff G, Chaves A, Bartoli CG (2010) Ethylene regulates ascorbic acid content during dark-induced leaf senescence. Plant Sci 178:207–212

10. Gorantla M, Babu PR, Reddy VB et al (2007) Identification of stress-responsive genes in an indica rice (Oryza sativa L.) using ESTs generated from drought-stressed seedlings. J Exp Bot 58:253–265

11. Micheletto S, Rodriguez-Uribe L, Hernandez R et al (2007) Comparative transcript profiling in roots of Phaseolus acutifolius and P. vulgaris under water deficit stress. Plant Sci 173:510–520

12. Kersey RK (2004) Microarray expression analysis to identify drought responsive genes involved in carbohydrate, and lipid metabolism in Medicago sativa leaves. Ph.D. Thesis, New Mexico State University, p 264

13. Rodriguez-Uribe L, Higbie SM, Stewart JMD et al (2011) Identification of salt responsive genes using comparative microarray analysis in Upland cotton (Gossypium hirsutum L.) Plant Sci 180(3):461–469

14. Arpat A, Waugh M, Sullivan JP et al (2004) Functional genomics of cell elongation in developing cotton fibers. Plant Mol Biol 54:911–929

15. Udall JA, Flagel LE, Cheung F et al (2007) Spotted cotton oligonucleotide microarrays for gene expression analysis. BMC Genomics 8:81

16. Murashige T, Skooge F (1962) A revised medium for rapid growth and bioassays with tobacco tissue culture. Physiol Plant 15:473–479

17. Bashan Y, Levanony H, Or R (1991) Association between Alternaria macrospora and Alternaria alternata, casual agents of cotton leaf blight. Can J Bot 69:2603–2607

18. Fourie PH, du Preez M, Brink JC et al (2009) The effect of runoff on spray deposition and control of Alternaria brown spot of mandarins. Australas Plant Pathol 38:173–182

19. Mehta YR (1998) Severe outbreak of Stemphylium leaf blight, a new disease of cotton in Brazil. Plant Dis 82:333–336

20. Irizarry RA, Hobbs B, Collin F et al (2003) Exploration, normalization, and summaries of high density oligonucleotide array probe level data. Biostatistics 4:249–264

21. Ranjan A, Nigam D, Asif MH et al (2012) Genome wide expression profiling of two accession of G. herbaceum L. in response to drought. BMC Genomics 13:94

22. Artico S, Nardeli SM, Brilhante O et al (2010) Identification and evaluation of new reference genes in Gossypium hirsutum for accurate normalization of real-time quantitative RT-PCR data. BMC Plant Biol 10:49

Chapter 16

Study of Hydrogen Peroxide as a Senescence-Inducing Signal

Stefan Bieker, Maren Potschin, and Ulrike Zentgraf

Abstract

In many plant species, leaf senescence correlates with an increase in intracellular levels of reactive oxygen species (ROS) as well as differential regulation of anti-oxidative systems. Due to their reactive nature, reactive oxygen species (ROS) were considered to have only detrimental effects for long time. However, ROS turned out to be more than just toxic by-products of aerobic metabolism but rather major components in different signaling pathways. Considering its relatively long half-life, comparably low reactivity, and its ability to cross membranes, especially hydrogen peroxide, has gained attention as a signaling molecule. In this article, a set of tools to study hydrogen peroxide contents and the activity of its scavenging enzymes in correlation with leaf senescence parameters is presented.

Key words Hydrogen peroxide, Leaf senescence, *Arabidopsis*, Oilseed rape, Plants, Catalase, Ascorbate peroxidase, Superoxide dismutase, Lipid peroxidation, SAGs, SDGs, Guaiacol, Chlorophyll contents, Anti-oxidative systems

1 Introduction

Senescence is an age- and development-dependent process which can take place on tissue, organ as well as on whole organism level. In the following, we will focus on leaf senescence, which is controlled by two underlying key processes: (1) sequential leaf senescence which reallocates nutrients from old to newly developing leaves. This is achieved by a metabolic shift from anabolic to catabolic processes. Sequential leaf senescence is mainly under the control of the growing apex and is arrested when no more new leaves develop and the plant starts to flower and later sets fruits and seeds. (2) During flower induction and anthesis, the second process takes over, and monocarpic leaf senescence governs the remobilization of nutrients from the leaves to the now developing flowers and fruits. This process is crucial for fruit and seed development as it has a major impact on yield quantity and quality. In many cases, reproductive development has control over leaf senescence, since

Yongfeng Guo (ed.), *Plant Senescence: Methods and Protocols*, Methods in Molecular Biology, vol. 1744,
https://doi.org/10.1007/978-1-4939-7672-0_16, © Springer Science+Business Media, LLC 2018

removal of reproductive organs can lead to regreening of already senescent leaves. This so called correlative control can be observed particularly in soybean and pea, while *Arabidopsis* does not show such a behavior [1].

In general, leaf senescence can be divided into three phases. During the *initiation phase,* the interplay between hormones and environmental and developmental cues determines the time point to trigger senescence. The correct timing is crucial, as too early induction would reduce the plant's capability to assimilate CO_2, while too late onset of leaf senescence would narrow the time frame for remobilization of nutrients to the developing fruits. During the *reorganization phase,* a shift from anabolic to catabolic processes occurs, and massive transcriptomic changes take place. Almost 6500 differentially regulated genes have been identified via reverse-genetic approaches and large-scale transcriptome profiling [2]. The progression of reorganization is easily visible to the naked eye as besides other macromolecules chlorophyll is degraded, thus converting the leaf's color from green to yellow. This phase entails detoxification of degradation intermediates and by-products and the remobilization of salvaged nutrients, and it is also accompanied by a loss of anti-oxidative capacity. The *terminal phase* completes the process of leaf senescence. During this phase, the remaining cellular components which have formerly been necessary to maintain control of the whole process are degraded. The nuclear DNA is fragmented; membranes are deteriorated; the remaining compartments like, e.g., mitochondria and nuclei, are disintegrated; and thus cell integrity and viability are irreversibly lost [3].

Reactive oxygen species (ROS), especially hydrogen peroxide, play a pivotal role throughout all three mentioned phases. During initiation and reorganization phase, hydrogen peroxide has been shown to be necessary for a successful induction and progression of the senescence program. Living cells balance production and scavenging of ROS to keep ROS levels in all cellular compartments under tight control. For *Arabidopsis* and oilseed rape, an increase in intracellular hydrogen peroxide concentrations correlates with the onset of senescence. A temporal loss of anti-oxidative capacity during senescence initiation can be observed and appears to be mainly achieved by a loss of CATALASE2 (CAT2) and ASCORBATE PEROXIDASE 1 (APX1) activities. In the case of *Arabidopsis*, this activity loss is accomplished by the transcriptional repression of the *CAT2* gene by the transcription factor G-box binding factor 1 (GBF1) [4]. As a consequence, the increased levels of hydrogen peroxide lead most likely to an inhibition of APX1 activity on the protein level. APX appears to be rendered sensitive against its own substrate exactly at this time point [5]. This loss of anti-oxidative capacity culminates in an intracellular accumulation of hydrogen peroxide which coincides with leaf senescence induction. Several senescence-associated transcription factors have been shown to be highly responsive to hydrogen peroxide (see, e.g., [6–8]), and, additionally, scavenging of hydrogen peroxide has been shown to have severe senescence-delaying effects [9]. During the terminal phase of the senescence program, an even more substantial second increase of hydrogen peroxide contents occurs in some plants. These ROS are thought to be mainly originating from macromolecule degradation processes like, e.g., lipid degradation via β-oxidation and membrane deterioration or the disruption of the electron transport chains. In addition, this higher production of ROS is reinforced by a decreasing anti-oxidative capacity in senescent tissues.

A fairly new method for the estimation of relative hydrogen peroxide contents is the fluorescent sensor HyPer [10]. To create this sensor, the regulatory domain from the bacterial hydrogen peroxide-sensing transcription factor OxyR was implemented in a circularly permutated YFP (cpYFP). This regulatory domain undergoes conformational changes upon oxidation via H_2O_2 or reduction via the glutaredoxin (GRX) and thioredoxin (TRX) systems, thus also inducing a conformational change in the cpYFP. When reduced, the HyPer protein has its excitation maximum at 420 nm, and when oxidized at 500 nm, emission is in both cases at 516 nm. Sequential excitation at both wavelengths followed by emission ratio calculation allows the determination of relative hydrogen peroxide contents.

The system has been shown to be able to detect short-term ROS bursts in plants as well as in animal systems [11–13]. Unfortunately, this sensor cannot be used to assess senescence-specific long-term changes in hydrogen peroxide   contents since oxidation of the HyPer protein scavenges hydrogen peroxide molecules which are essential to trigger senescence. Although the kinetics of reduction have been shown to be very slow [10, 14], it seems to be sufficient for effective scavenging in the case of a slow increase in hydrogen peroxide contents as given during senescence, and thus inducing a delay in senescence induction [9]. Nevertheless, when used under inducible promoter systems and as localized variants, this tool might shed further light on the senescence-specific H_2O_2-signaling cascade in the future.

Considering the impact of ROS during senescence induction and progression, the study of these molecules has become more and more important over the past. However, due to ROS inherent reactivity and instability, specific measurement of ROS is still considered to be problematic. Therefore, a set of tools to measure and estimate ROS contents in correlation with other senescence parameters is presented here to describe this complex process in detail.

2 Materials

2.1 H₂DCFDA

1. Stock solution: 0.4 mg carboxy-H$_2$DCFDA solved in 400 μL DMSO, dilute 1:1 with distilled water.

2. Working solution: 400 μL of stock solution are added to 39.6 mL MS medium.

3. MS medium: 4.3 g MS medium without vitamins, 30 g sucrose, solve in 600 mL distilled water, adjust pH to 5.7–5.8 with KOH, and add water to 1 L.

4. 40 mM Tris-HCl, pH 7.0.

5. Hydrogen peroxide (30%).

6. 0.5 M NaOH.

7. Fluorescence reader: excitation filter ~480 nm, emission filter ~520 nm.

2.2 Guaiacol-Based H₂O₂ Measurement Buffer (Modified After Tiedemann [15] and Maehly and Chance [16])

Reaction buffer: 50 mM potassium phosphate buffer pH 7.0, 0.05% guaiacol (v/v), add shortly before use, 2.5 μmL^{-1} horseradish peroxidase.

2.3 Catalase Zymograms (After Chandlee and Scandalios [17])

1. Protein extraction buffer: 100 mM Tris, 20% glycerol (w/v), pH 8.0; add 30 mM DTT final concentration shortly before use.

2. Stacking gel: 3.5% acrylamide, 0.5 M Tris, pH 6.8.

3. Separating gel: 7.5% acrylamide, 1.5 M Tris, pH 8.8.

4. PAGE running buffer: 25 mM Tris, 250 mM glycine, pH 8.3.

5. Staining solution 1: 0.01% hydrogen peroxide solution, set up just before use.

6. Staining solution 2: 1% FeCl$_3$ and 1% K$_3$[Fe(CN)$_6$] (w/v), stir at least 1 h before use.

2.4 APX Zymograms (After Mittler and Zilinskas [18])

All buffers are best prepared shortly before use.

1. Stacking gel: 5% acrylamide, 0.5 M Tris-HCl, pH 6.8, 10% glycerol (v/v).

2. Separating gel: 10% acrylamide, 1.5 M Tris-HCl, pH 8.8, 10% glycerol (v/v).

3. Electrophoresis buffer: 25 mM Tris, 250 mM glycine, 2 mM ascorbic acid, pH 8.3.

4. Protein extraction buffer: 50 mM potassium phosphate, pH 7.8, 2% Triton-X 100 (v/v), 5 mM ascorbic acid, 35 mM β-mercapto-ethanol, 2% polyvinylpyrrolidone (w/v).

5. Staining solution I: 50 mM potassium phosphate, pH 7.0, 2 mM ascorbic acid.

6. Staining solution II: 50 mM potassium phosphate, pH 7.0, 20 mM ascorbic acid, 0.5 μM hydrogen peroxide.

7. Developer solution: 50 mM potassium phosphate, pH 7.8, 14 mM TEMED, 2.45 mM nitroblue tetrazolium (NBT).

2.5 SOD Zymograms (After Baum and Scandalios [19])

1. Protein extraction buffer: 100 mM Tris, 20% glycerol (w/v), pH 8.0; add 30 mM DTT final concentration before use.

2. Stacking gel: 5% acrylamide, 0.5 M Tris-HCl, pH 6.8.

3. Separating gel: 13% acrylamide, 1.5 M Tris-HCl, pH 8.8.

4. Electrophoresis buffer: 250 mM glycine, 25 mM Tris, pH 8.3.

5. Staining buffer: 50 mM potassium phosphate, pH 7.8, 1 mM EDTA, 0.2% N,N,N,N-Tetramethylethylenediamine (TEMED), 2.6 μM riboflavin, 1.2 mM NBT.

2.6 Catalase Assay Buffers (After Cakmak and Marschner [20])

1. Phosphate buffer: 25 mM potassium phosphate, pH 7.0, add 2 mM EDTA after titration.

2. 10 mM hydrogen peroxide solution.

2.7 APX Assay Buffers (After Nakano and Asada [21])

1. Extraction buffer: 100 mM potassium phosphate, pH 7.0, 5 mM ascorbic acid, 1 mM EDTA.

2. Reaction buffer: 25 mM potassium phosphate, pH 7.0, 0.1 mM EDTA, 1 mM hydrogen peroxide, 0.25 mM ascorbic acid.

2.8 Lipid Peroxidation Buffers (After Hodges, DeLong [22] and Janero [23])

1. Extraction buffer: 0.1% trichloroacetic acid.

2. Reaction buffer 1: 20% trichloroacetic acid, 0.01% butylated hydroxytoluene.

3. Reaction buffer 2: 20% trichloroacetic acid, 0.5% thiobarbituric acid.

3 Methods

3.1 Plant Preparation and General Considerations for Senescence Phenotyping

Basically, two approaches can be followed: (i) all plants are sown at once and harvested periodically or (ii) the plants are sown in the desired intervals and harvested at once. In addition, stratification of imbibed seeds for 1–2 days at 4 °C should always be carried out to ensure synchronous plant development. What has to be considered concerning the planting regime are the planned experiments. Especially the here described hydrogen peroxide measurement with a fluorescent dye is very prone to high variations when carried out with different dye solutions; furthermore long-term storage of one batch of dissolved dye is not recommended.

For all of the methods described in the following, it is necessary to sample defined leaf positions. Although grouping of leaves into, e.g., old (position 1–4), middle aged (position 5–8), and young (position 9–12 and above) is possible, this will result in higher variance of the results. Therefore, the leaves should be color-coded according to their sequence of emergence during early developmental stages so that, even in later stages, positions within the rosette can clearly be assigned to each individual leaf. Figure 1 shows the typical arrangement and shape of the leaves. It goes without saying that the attached markings should in no way hinder growth and nutrient supply of or light incidence onto the leaf to ensure proper development.

When studying hydrogen peroxide as a senescence-inducing signal, growth conditions have to be considerably controlled. Too high light irradiation induces excess photon energy, giving rise to plastid ROS production. Suitable light conditions strongly depend on the used plant species and ecotype; in our case working with intensities below 110–120 $\mu E\ m^{-2}\ s^{-1}$ has proven to be viable for *Arabidopsis* ecotype Col-0. Almost all biotic stresses (e.g., gnats feeding on the plants, fungal infestation, etc.) will elicit ROS signaling cascades as well as premature senescence induction. Abiotic stresses (e.g., drought, heat, or cold) also will induce signaling cascades including ROS production; thus continuous monitoring and logging of growth conditions are highly recommended.

Furthermore developmental indicators as, e.g., the time point of bolting and flowering as well as the development of first pods should be monitored. By that, plant lines with a general alteration in development can easily be distinguished from plant lines with altered senescence (*see* **Note 1**).

3.2 Chlorophyll Contents

As mentioned above, the loss of chlorophyll is the first indicator for the progression of senescence visible to the naked eye. Speaking in general terms, two different approaches are available to determine chlorophyll contents. Non-invasive measurements are relying on light transmittance through the leaf. The two most common devices for this are the atLEAF+ (FT Green LLC, Wilmington, DE, USA) and the SPAD-502 chlorophyll meter (Soil Plant Analysis Development, Minolta Camera Co., Ltd., Japan). Both devices operate in a similar manner, varying only slightly in the utilized wavelengths. The SPAD-502 measures light transmittance at 650 and 940 nm, while the atLEAF+ uses 660 and 940 nm. For a measurement, the leaf is clamped into the device; hence it is not necessary to remove it. Considering reliability and reproducibility, both devices have been shown to deliver comparable results [24]. Additionally, FT Green implemented the conversion from atLEAF+ values to SPAD values already in their software.

For reproducible monitoring of chlorophyll contents with one of these devices, a few simple points have to be kept in mind: (1)

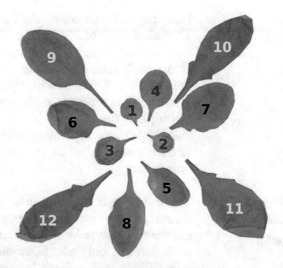

Fig. 1 *Arabidopsis* rosette. Numbers indicate sequence of leaf emergence. Suggested grouping indicated as follows: blue numbers = leaf 1–4/group old, black numbers = leaf 5–8/group middle aged, yellow numbers = leaf 9, and above/group young. Spiral growth direction is clockwise

Orientation of the measured leaves should always be the same. (2) To avoid positional effects, several measurements distributed over the whole leaf should be made. (3) The device's measuring area should always be completely covered by the leaf.

Chlorophyll extraction and the following photometric measurement has the advantage to deliver results formatted as µg chlorophyll per mg fresh weight as well as it gives the opportunity to determine Chl A to Chl B ratios. Furthermore, positional effects as they may occur with noninvasive measurements are avoided by homogenization of whole leaves. However, a major drawback of this approach is the necessity to remove the measured leaf, eliminating it for further analysis, and, in comparison with the noninvasive approach, it is more time-consuming.

Another quick and easy method to estimate senescence progression via chlorophyll contents is to assign each leaf on a plant to four categories according to its color (e.g., green, green-yellow, yellow, and brown) and calculate the proportion of each category in relation to the total number of leaves. The upside of this method is the minimal requirements in time and equipment and that the difference in total number of rosette leaves of individual plants is taken into account. The downside is that the categorization relies on the individual judgment of the experimenter.

The following results were gained using the atLEAF+ chlorophyll meter. Fifteen biological replicates were measured with at least three measurements per leaf. Despite the high number of replicates, standard deviation increases as soon as chlorophyll contents start to decrease. Another typical behavior can be observed in the later stages. While chlorophyll degradation has not started in leaf 11 during week 6, 1 week later, its chlorophyll levels are almost as low as in leafs 5 and 7. This markedly faster and sudden decrease in chlorophyll contents usually marks the point, when sequential senescence is shifted to whole plant senescence.

A delayed senescence phenotype is accompanied with a lag in chlorophyll degradation. As an example, Fig. 2 also shows an *Arabidopsis* line expressing HyPer. These plants show a severe senescence delay, while the development of the plants remains untouched.

3.3 Hydrogen Peroxide Measurement

3.3.1 H₂DCFDA

The used dye is in a non-polar state enabling it to cross cellular membranes. An intracellular esterase deacetylates the dye, rendering the molecule polar and thus trapped inside the cell. Additionally, oxidation can only take place on the deacetylated dye; thus only intracellular oxidants will be measured, as extracellular oxidized dye is not able to enter the cell and will be rinsed off before homogenization. However, carboxy-H₂DCFDA is a general ROS indicator; although H₂O₂-specific dyes like di-amino-benzidine do exist, these dyes are often highly toxic. This is why using a non-toxic dye in combination with other analytic tools is preferred.

As mentioned above, sampling the same leaf position is crucial for reliable data. Also the measurement needs to be normalized to either leaf area or leaf weight. A combination of both (e.g., leaf discs with the same diameter which are weighed) usually gives the best results.

Weigh sample and/or measure sample area. Incubate sample for 45 minutes in working solution. Rinse sample twice in distilled water and dab off remaining fluid. Freeze sample in a new tube on liquid nitrogen. Homogenize sample in 500 μL 40 mM Tris pH 7. Centrifuge for 15 min at >12,000 × g and 4 °C. Measure empty plate and Tris to cover variation in plate and Tris buffer. Measure sample in plate reader (excitation, 480 nm; emission, 520 nm).

When working with several batches of dye solutions or to make independent experiments better comparable, calibration of the dye is needed. Two options are given; both include chemical deacetylation followed by oxidation of the dye via hydrogen peroxide. Here one has to keep an eye on bubble formation due to the generation of oxygen during the oxidation reaction, as these will strongly influence the measurement in a plate reader. Dilutions indicated here are adapted to our fluorescence plate reader (Tristar LB 941, Berthold Technologies); if too low values or values exceeding dynamic range of your reader occur, change dilutions after the dye oxidation.

Maximal oxidation: Mix 0.755 mL 0.5 M NaOH solution with 1 mL working solution (molar ratio 40:1) → dye deacetylation. Incubate 30 min at room temperature. Add 500 μL 30% hydrogen peroxide solution to 156 μL deacetylated dye. Shake at

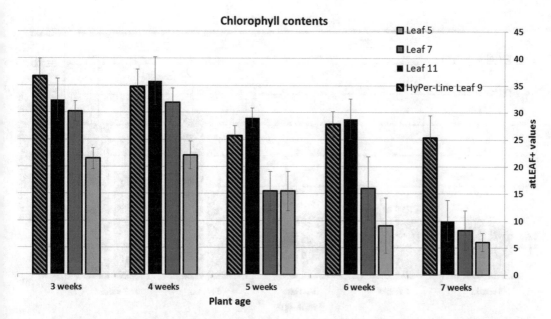

Fig. 2 atLEAF+ measurements over plant development. Leaf positions No. 5, 7, and 11 of *Arabidopsis* Col-0 plants and position 9 of *Arabidopsis* plants expressing HyPer were sampled in weekly rhythm as indicated. Values are mean of 15 or 10 biological replicates of wild-type and HyPer plants, respectively. Error bars indicate standard deviation. atLEAF+ values start decreasing at week 5 in wild-type plants, while chlorophyll contents of HyPer plants remain constant

room temperature for 1 h in the dark. Meanwhile measure deacetylated dye and all other used buffers (MS medium, 40 mM Tris working solution, etc.) to be able to correct variations in the used buffers. Measure fully oxidized, deacetylated dye. Values can be used to correct for variation in dye batches.

Calibration curve: Perform dye deacetylation as described above. Prepare tubes with different hydrogen peroxide concentrations in a volume of 24 μL (e.g., 5, 50, 75 and 100 μM final concentration hydrogen peroxide after the addition of 156 μL deacetylated dye to each). Add 156 μL deacetylated dye to each tube. Incubate at room temperature for 10 min in the dark. Add 700 μL 40 mM Tris pH 7 to each tube. Measure each hydrogen peroxide-dye mix and all other used solutions. Offset and slope can be used to correct for variation in dye batches.

The senescence-specific increase in hydrogen peroxide levels coincides with the decrease in chlorophyll contents. Figure 3 shows the hydrogen peroxide contents of leaf 5 of the same plants as depicted in Fig. 2. In wild-type plants, hydrogen peroxide contents start to increase at the same time point when atLEAF+ values for leaf 5 starts to decrease. In consistence, the HyPer lines show a much less pronounced increase in hydrogen peroxide contents, and the decrease in chlorophyll contents is markedly delayed and also less pronounced. The abovementioned second increase in

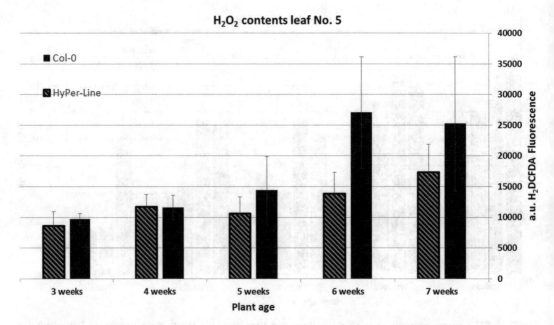

Fig. 3 Hydrogen peroxide contents over plant development. Leaf position No. 5 of *Arabidopsis* Col-0 and HyPer-expressing plants were sampled in a weekly rhythm as indicated. Values are mean of 15 or 10 biological replicates of wild-type and HyPer plants, respectively. Error bars indicate standard deviation. H_2O_2 contents start increasing in wild-type plants in week 5, while concentrations in HyPer start increasing in week 6. Additionally, increase in wild-type plants is much more pronounced than in HyPer plants

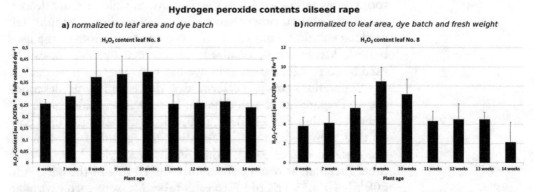

Fig. 4 Hydrogen peroxide contents during development of *Brassica napus* (cv. Mozart). Data represents mean values of at least three biological replicates. Error bars indicate standard deviation. For measurements, leaf discs with constant diameter were used. (**a**) Hydrogen peroxide contents normalized to leaf area. (**b**) Same data as in (**a**) but additionally normalized to mg fresh weight (fw) [9]

H_2O_2 contents is not observed here due to too short sampling time frame.

Transferring this knowledge onto other plant species, Fig. 4 shows a comparable pattern observed in *Brassica napus* (cv. Mozart). Also here, peak values of hydrogen peroxide contents coincide with the beginning of chlorophyll degradation (data not shown). While in *Arabidopsis* hydrogen peroxide contents increase again in the latest stages of development, this could not be observed for oilseed rape.

3.3.2 Guaiacol-Based H$_2$O$_2$ Measurement (Modified After [15] and Maehly and Chance [16])

Punch 8–10 leaf discs with same diameter and determine fresh weight. Add 2 mL of reaction buffer to the leaf material. Add peroxidase. Incubate sample for 2 h in the dark. Take off supernatant. Determine OD$_{470}$. Hydrogen peroxide concentration can be calculated with the following formula:

$$C(H_2O_2)(\mu M) = \left(\frac{4OD_{470\,nm}}{26.6}\right) \times 1000$$

This assay gives the opportunity to estimate absolute concentrations. A major downside of this method is the necessity to injure the measured leaves. While using H$_2$DCFDA allows whole leaves to be incubated in the reaction buffer, thus restricting the area of inflicted injury only to the small diameter of the petiole, this method necessitates punching of leaf discs and, furthermore, long incubation times of the injured material. In that way, the injured area is extended as well as the time frame for possible stress responses, hence increasing the possibility of artifacts. Nevertheless, this method has been successfully used in our lab to determine senescence-specific changes in intracellular hydrogen peroxide contents (*see* Fig. 5).

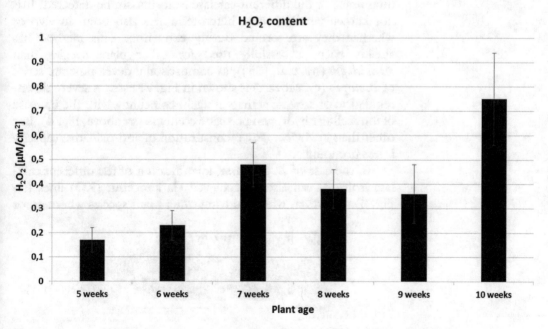

H$_2$O$_2$ content

Fig. 5 Hydrogen peroxide measurements during plant development using guaiacol. Hydrogen peroxide content was measured in leaf discs and is indicated in μM/cm^2. Error bars indicate standard deviation of three replicates [5]

3.4 Anti-Oxidative Systems

An additional indicator for changing ROS contents is the differential regulation of the anti-oxidative enzymatic systems. A downregulation of these systems in most cases results in an increase of reactive oxygen species. An upregulation is a reliable indicator for already elevated ROS production most likely due to stress conditions. To monitor the activity of these enzymes, several possibilities are given; native PAGE and following in-gel activity staining gives the opportunity to visualize changes on isoform level, but quantification is not easily feasible in a reproducible manner. In contrast to in-gel activity assays, photometric assays can quantify enzyme activities, while an isoform-specific measurement is not possible.

3.4.1 Qualitative In-Gel Activity Assays: Catalases (After Chandlee and Scandalios [17])

Freeze ~100 mg leaf material in liquid nitrogen (samples can be kept at −80 °C for several months). Homogenize on ice in 400 μL extraction buffer. Centrifuge 30 min at >12,000 × g at 4 °C. Take off supernatant, and determine protein concentration. Load 10 μg protein on gel, and let gel run. Remove stacking gel, and incubate separating gel for exactly 2 min in staining solution 1. Rinse gel 2–3 times with distilled water. Incubate gel up to 10 min in staining solution 2; incubation should be stopped 1–2 min after bands become visible. Rinse gel in distilled water. Gels can be preserved in 7% solution of acetic acid.

As illustrated in Fig. 6, senescence-specific downregulation of CAT2 activity occurs, coinciding with the increase in hydrogen peroxide content (*see* Fig. 5). Furthermore, in addition to the CAT2 and CAT3 homotetramers, heterotetramers consisting of monomers of the different catalase isoforms can be detected. It is not known whether these heterotetramers also occur in vivo or whether they are formed during extraction. This method has already been successfully used for other plant species than *Arabidopsis* (*see*, e.g., [25]); as comparison a developmental series of *B. napus* (cv. Mozart) is shown in Fig. 7. Here, a severe downregulation of catalase activity is again coinciding with the increase of intracellular hydrogen peroxide contents (*see* above, Fig. 4). But other than in *Arabidopsis*, reconstitution of anti-oxidative capacity is less pronounced.

In the case of *A. thaliana*, identification of the different catalase isoforms can also be achieved via knockout (KO) lines (*see* Fig. 8). However, when studying other plant species where no or

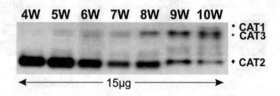

Fig. 6 Catalase activity zymogram overdevelopment of *A. thaliana*. CAT activity staining after native PAGE as described from *A. thaliana* Col-0 [5]

CAT activity

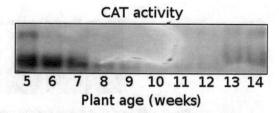

5   6   7   8   9   10 11 12 13 14
Plant age (weeks)

Fig. 7 Catalase activity zymogram overdevelopment of *B. napus*. CAT activity staining after native PAGE as described from *B. napus* (modified after [9])

Fig. 8 Catalase zymogram of *Arabidopsis* knockout plants. Catalase extracts prepared from wild-type and catalase knockout plants. Activity staining after native PAGE was conducted as described. 10 μg of the raw protein extract were loaded. Isoforms and heterotetramer composition as denoted [29]

only a limited number of knockout lines are available, identification of CAT isoforms is most easily accomplished via zymograms combined with 3-Amino-1,2,4-triazole (3-AT) treatment. 3-AT is a catalase inhibitor acting more efficiently on CAT2 and CAT1 homologues than on CAT3 homologues [26]. Thus, when implemented in a catalase in-gel activity assay, first CAT2 activity will vanish, while CAT3 activity will decrease more slowly. Figure 9 shows a 3-AT-treated catalase zymogram from oilseed rape. Here, catalase extract was analogously prepared as described above, and 15 μg were loaded on each lane. The gel was then cut into four pieces. After incubation in a 10 mM 3-AT bath for the denoted periods, activity staining was carried out as mentioned above. Each band could be assigned according to 3-AT's known effectiveness on different catalase homologues (*see* Fig. 9).

3.4.2 Qualitative In-Gel Activity Assays: Ascorbate-Peroxidases (After Mittler and Zilinskas [18])

Homogenize 100–200 mg leaf material in liquid nitrogen. Add 300 μL protein extraction buffer and resuspend homogenate. Centrifuge for 30 min with >12,000 × *g* at 4 °C. Start gel pre-electrophoresis (thus the ascorbic acid from the running buffer can run into the gel) for 20–30 min. Take off supernatant and determine protein concentration. Load 25 μg per sample on gel and let gel run. Incubate gel thrice for 10 min each in staining solution I. Incubate gel for 20 min in staining solution II. Put gel for 1 min in 50 mM potassium phosphate buffer pH 7.0. Develop gel for 5–10 min in developer solution. Rinse gel with distilled water. Gels can be preserved in 7% solution of acetic acid.

During senescence induction, APX1 activity declines in parallel to CAT2 activity loss. Despite the redundancy of the APX system in *Arabidopsis* (eight isozymes), APX zymograms of leaf material

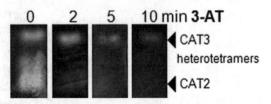

Fig. 9 3-AT-treated catalase activity zymogram of *B. napus*. Identification of isoforms in *B. napus* extracts. 3-AT incubation time prior activity staining is indicated above the lanes. Activity bands could be assigned to denoted catalases [9]

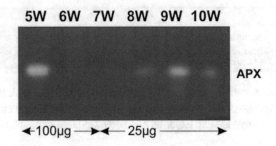

Fig. 10 Ascorbate peroxidase activity zymogram over development of *A. thaliana*. APX activity staining after native PAGE as described. Plant age is indicated in weeks. Amount of loaded raw protein extract is indicated below the lanes [5]

usually result in one prominent band representing the cytosolic APX1 (*see* Fig. 10).

For other plant species like, e.g., oilseed rape or tobacco, several activity bands occur, and a less clear senescence-associated regulation is observed. As example, zymograms prepared from oilseed rape (*B. napus* cv. Mozart) extracts are presented (*see* Fig. 11). Isoform identification without available KO plants is not easily feasible. Especially, as senescence-associated negative regulation of APX activity seems to take place on posttranslational level, thus rendering isoform-specific transcript quantification pointless.

3.4.3 Qualitative In-Gel Activity Assays: Superoxide Dismutases (After Baum and Scandalios [19])

Freeze ~100 mg leaf material in liquid nitrogen (samples can be kept at −80 °C for several months). Homogenize on ice in 400 μL extraction buffer. Centrifuge 30 min at >12,000 × *g* at 4 °C. Take off supernatant; determine protein concentration. Load 30–40 μg per lane. Incubate gel for 30 min in staining buffer. Rinse gel twice with water. Illuminate gel until staining is sufficient. Gels can be preserved in 7% solution of acetic acid.

Superoxide dismutases convert superoxide anions to hydrogen peroxide. Thus enzymatic activity is a direct indicator for hydrogen peroxide production. Isoform identification can be achieved by incubation in 5 mM hydrogen peroxide or 2 mM KCN. Manganese SOD is insensitive to either treatment, CuZn-SOD is sensitive to CN^-, and the Fe-SOD is inhibited by H_2O_2 [27]. During *Arabidopsis* plant development, the activity of a plastid-localized

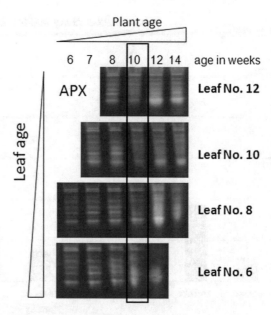

Fig. 11 Ascorbate peroxidase activity zymogram overdevelopment of *B. napus*. APX activity staining after native PAGE as described. Plant age is indicated in weeks [9]

isoform (Fe-SOD) is upregulated, while Mn-SOD isoform activity remains unchanged (for data *see* [28]).

3.4.4 Quantitative Assays: Catalase Assay (After Cakmak and Marschner [20])

Homogenize 2–4 leaves (ca. 50–150 mg; weigh before!) in 500–1000 µL phosphate buffer on ice. Centrifuge for 30 min with >12,000 × *g* at 4 °C. Take off supernatant and determine volume and protein concentration. Mix 50 µL protein extract with 850 µL phosphate buffer in a quartz cuvette. Add 100 µL 10 mM hydrogen peroxide solution (best when cuvette placed in photometer already). Measure OD$_{240}$ for up to 5 min in the desired interval. Molar extinction coefficient ε of hydrogen peroxide is 43.6 mM cm^{-1} at 240 nm. Units per mg can be calculated via Lambert-Beer'sche formula (*see* below).

3.4.5 Quantitative Assays: Ascorbate-Peroxidases (After Nakano and Asada [21])

Homogenize 100 mg leaf material on liquid nitrogen. Resuspend in 200 µL protein extraction buffer. Centrifuge for 30 min at >12,000 × *g* and 4 °C. Take off supernatant and determine protein concentration. Mix 900 µL reaction buffer with 100 µL protein extract (best when cuvette placed in photometer already). Measure OD$_{290}$ for up to 5 min in desired interval. Molar extinction coefficient ε for ascorbic acid is 2.8 mM cm^{-1} at 290 nm.

$$A = \frac{\Delta E \ / \min \times V}{\in \times d}(\times \text{Dilution})$$

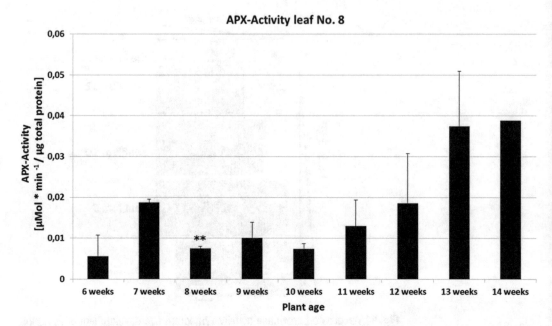

Fig. 12 Quantitative ascorbate peroxidase activity measurements during development of *B. napus*. Samples were taken from leaf No. 8. Data represents mean of at least three biological replicates. Error bars indicate standard deviation. Comparison of means and the determination of statistically differences rest on *T*-tests with **P < 0.01. Tests were made pair wise (7w–8w, 8w–9w, 9w–10w, and 10w–11w) [9]

A = activity (U = μmol min^{-1}), V = total volume, d = Layer thickness, $\Delta E/$min = change of absorbance per minute, ε = extinction coefficient.

In contrast to activity zymograms, with this assay downregulation in weeks 8 and 9 is very prominent and even upregulation in later stages is unmistakably visible (*see* Fig. 12).

3.5 Lipid Peroxidation (After Hodges, DeLong [22] and Janero [23])

Lipid peroxidation can be used as an indicator for oxidative stress in plants as well as membrane deterioration and senescence progression. The here utilized method is an improved thiobarbituric acid-reactive-substances assay (TBARS) developed by Prange et al. in 1999. Malondialdehyde is formed upon auto-oxidation and enzymatic degradation of polyunsaturated fatty acids. MDA in turn reacts with two molecules of TBA, yielding a pinkish-red substance with an absorbance maximum at 532 nm [22, 23]. To correct for non-TBA complexes also absorbing at 532 nm in the sample, besides the TBA reaction, another reaction without TBA is used. Additionally, besides lipid peroxides, sugars can complex with TBA. To take this into account, a further measurement at 440 nm is implemented. The molar absorbance of sucrose at 532 nm and 440 nm is 8.4 and 147, respectively, resulting in a ratio of 0.0571.

Weigh plant material (25–150 mg are recommendable). Homogenize plant material on liquid nitrogen. Add 500 μL 0.1% TCA solution. Centrifuge for 5–10 min at 10,000 × g. Add 200 μL supernatant to (1) 800 μL 20% TCA/0.01% butylated hydroxytoluene and (2) 800 μL 20% TCA/0.5% TBA. Vortex and incubate for 30 min at 95 °C (*attention*, pronounced formation of gases, open tubes every few minutes to prevent bursting or use screw-cap cups). Cool samples to room temperature on ice. Centrifuge again if needed. Measure absorption at 440, 532, and 600 nm. Calculate MDA concentration via following formulas:

1. $A = [(\mathrm{Abs}_{532\,nm\,+\,TBA} - \mathrm{Abs}_{600\,nm\,+\,TBA}) - (\mathrm{Abs}_{532\,-TBA} - \mathrm{Abs}_{600\,-TBA})]$.

2. $B = [(\mathrm{Abs}_{440\,nm\,+\,TBA} - \mathrm{Abs}_{600\,nm\,+\,TBA})\,0.0571]$.

3. MDA equivalents (nmol * mL^{-1}) = (A-B/157000) 10^6

For example, in a comparison of the *Arabidopsis wrky53* KO line with wild-type plants, a decelerated senescence progression as well as a delayed induction of SAG expression (data not shown) can be observed. Chlorophyll contents initially decrease at the same rate from 31 to 38 DAS but during the later stages, it dramatically decrease in WT plants which is not the case for *wrky53* plants, thus hinting to a slower senescence progression rate. In consistence, lipid peroxidation starts increasing at the same time point in both plant lines, but again the slope is higher in WT plants (*see* Fig. 13a, b).

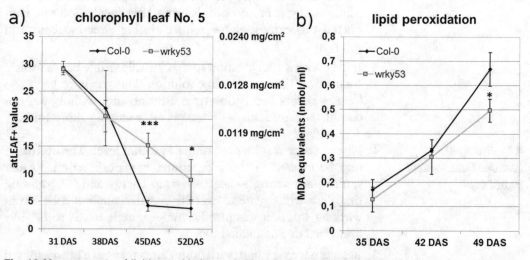

Fig. 13 Measurements of lipid peroxidation and chlorophyll contents during development of *A. thaliana*. (**a**) Chlorophyll contents of *Arabidopsis* Col-0 and *wrky53* plants. Left axis indicates atLEAF+ values, right axis the calculated chlorophyll contents in mg/cm². (**b**) Lipid peroxidation in Col-0 and *wrky53* plants. Plant age is indicated in days after seeding (DAS). Values represent mean of at least three biological replicates; error bars indicate standard deviation. Comparison of means and the determination of statistically differences rest on *T*-tests with *P < 0.05 and ***P < 0.001

4 Notes

The methods described above are fairly robust and easy in terms of conduction. However, as all other methods, also these have some pitfalls ready at their fingertips. In the following, we want to point out the most common mistakes that can occur and give some advice how to circumvent these.

4.1 Catalase Zymograms

1. Band separation is insufficient/smearing: Used buffers might be at wrong pH/wrong glycine concentration → check pH of each buffer used; prepare new gel and running buffers. DTT in extraction buffer was not added/too old → add DTT shortly before usage. Gel might be overloaded → reduce amount of protein.

2. If background is dark green but no bands are visible: Proteins might be damaged → extraction should take place on ice at all times; extracts cannot be stored, always use freshly prepared extracts; vortexing of extracts is not recommendable. DTT in extraction buffer was not added/too old. Hydrogen peroxide concentration in the staining solution was too high, so that not all molecules were degraded by the amount of catalase loaded on the gel.

3. If background of gel is too bright to distinguish bands: Hydrogen peroxide concentration in the staining solution was too low → hydrogen peroxide solutions have to be prepared shortly before usage; prepare new staining solution 1; stock solution might be too old. Rinsing after incubation was conducted too extensively → reduce rinsing after incubation in solution 1.

4. If gel darkens after staining so that initially visible bands vanish: Incubation time in staining solution 2 might have been too long → reduce incubation time. Rinsing after staining was not conducted sufficiently → rinse gel after staining thoroughly.

4.2 Hydrogen Peroxide Measurement (H_2DCFDA)

1. Fluorescence readings are on background level: The dye batch may be expired → check functionality via calibration. Reader settings are wrong → increase lamp energy and/or counting time; use white plate to amplify signal. Sample incubation in working solution was insufficient → sample needs to be fully immersed during incubation.

2. Results have very high variation: Plants may have been exposed to different conditions → check light intensities in different positions in growth chamber; plants were standing too close to each other, dark-induced senescence may have occurred; check for pests; see log files for temperature, etc. Samples vary in weight and area/leaf position → ensure reproducible sampling.

The used dye batches may differ → correct for different batches via calibration. Working solution oxidized during sampling → keep working solution dark and cool until incubation with sample.

3. Calibration does not work properly: Deacetylation was insufficient → set up new sodium-hydroxide solution. Bubbles disturb measurement → ensure there was no bubble formation during measurement by tapping your plate until all bubbles are out. Oxidation was insufficient → use new hydrogen peroxide solution. Measured values exceed reader range → use different dilutions of oxidized dye to determine proper range for your reader.

4.3 Lipid Peroxidation

1. Measured values exceed dynamic range of photometer (usually results in 9.9999 values): Blank sample not set up properly → prepare new blank. Reader not set to zero for all wavelengths → re-blank. Bubble formation in cuvette disturbs measurement → tap cuvette several times before measuring to degas solution. Precipitate might have formed during incubation at 95 °C → centrifuge again before measuring.

2. Variation between replicates is very high: The used scale might be not accurate enough → switch scale or weigh out higher amounts of material. Homogenization of material might be insufficient → sample needs to be thoroughly homogenized for reproducible measurements. Bubbles or precipitate might have disturbed measurement → *see* above.

3. Measured values level at one maximal value: Too much plant material for one reaction was used → reduce used plant material or scale up reaction volume. Values might exceed dynamic range of reader → *see* above.

4.4 Gene Expression Analysis

1. When studying hydrogen peroxide as senescence-inducing signal, besides chlorophyll degradation, correlation to senescence-associated gene expression is also crucial. Therefore, analysis of SAG and SDG expression should be carried out with the same material used for phenotypic analysis. Many of these genes have proven their value, e.g., *SAG13* (At2G29350) encodes a short-chain alcohol dehydrogenase and serves as a reliable marker for early senescence stages. As further marker genes for early senescence, stages *CAT3* (At1G20620) and *WRKY53* (At4G23810) can be used. In contrast, expression of the cysteine protease *SAG12* (At5G45890) is upregulated in later stages of leaf senescence. There are many more genes with specific senescence-associated expression patterns which can be used for detailed classification of specific alteration of the senescence process like, e.g., delayed senescence induction and decelerated senescence progression. Combined with analysis of SDG expression, valu-

able datasets can be obtained, making interpretation of results gained with the methods mentioned above more reliable. Valuable SDGs are, for example, *CAB* (At1g29930), *CAT2* (At4G35090), and *RuBisCO* (At1G67090).

References

1. Lim PO, Kim HJ, Nam HG (2007) Leaf senescence. Annu Rev Plant Biol 58:115–136

2. Breeze E, Harrison E, Mchattie S et al (2011) High-resolution temporal profiling of transcripts during Arabidopsis leaf senescence reveals a distinct chronology of processes and regulation. Plant Cell 23:873–894

3. Zimmermann P, Zentgraf U (2005) The correlation between oxidative stress and leaf senescence during plant development. Cell Mol Biol Lett 10:515–534

4. Smykowski A, Zimmermann P, Zentgraf U (2010) G-Box binding factor1 reduces CATALASE2 expression and regulates the onset of leaf senescence in Arabidopsis. Plant Physiol 153:1321–1331

5. Zimmermann P, Heinlein C, Orendi G et al (2006) Senescence-specific regulation of catalases in Arabidopsis thaliana (L.) Heynh. Plant Cell Environ 29:1049–1060

6. Miao Y, Laun T, Zimmermann P et al (2004) Targets of the WRKY53 transcription factor and its role during leaf senescence in Arabidopsis. Plant Mol Biol 55:853–867

7. Balazadeh S, Kwasniewski M, Caldana C et al (2011) ORS1, an H(2)O(2)-responsive NAC transcription factor, controls senescence in Arabidopsis thaliana. Mol Plant 4:346–360

8. Wu A, Allu AD, Garapati P et al (2012) JUNGBRUNNEN1, a reactive oxygen species-responsive NAC transcription factor, regulates longevity in Arabidopsis. Plant Cell 24:482–506

9. Bieker S, Riester L, Stahl M et al (2012) Senescence-specific alteration of hydrogen peroxide levels in Arabidopsis thaliana and oilseed rape spring variety Brassica napus L. cv. Mozart. J Integr Plant Biol 54:540–554

10. Belousov VV, Fradkov AF, Lukyanov KA et al (2006) Genetically encoded fluorescent indicator for intracellular hydrogen peroxide. Nat Methods 3:281–286

11. Costa A, Drago I, Behera S et al (2010) H_2O_2 in plant peroxisomes: an in vivo analysis uncovers a Ca(2+)-dependent scavenging system. Plant J 62:760–772

12. Niethammer P, Grabher C, Look AT et al (2009) A tissue-scale gradient of hydrogen peroxide mediates rapid wound detection in zebrafish. Nature 459:996–999

13. Pase L, Layton JE, Wittmann C et al (2012) Neutrophil-delivered myeloperoxidase dampens the hydrogen peroxide burst after tissue wounding in zebrafish. Curr Biol 22:1818–1824

14. Markvicheva KN, Bilan DS, Mishina NM et al (2011) A genetically encoded sensor for H_2O_2 with expanded dynamic range. Bioorg Med Chem 19:1079–1084

15. Tiedemann AV (1997) Evidence for a primary role of active oxygen species in induction of host cell death during infection of bean leaves with Botrytis cinerea. Physiol Mol Plant Pathol 50:151–166

16. Maehly AC, Chance B (1954) The assay of catalases and peroxidases. Methods Biochem Anal 1:357–424

17. Chandlee JM, Scandalios JG (1984) Analysis of variants affecting the catalase developmental program in maize scutellum. Theor Appl Genet 69:71–77

18. Mittler R, Zilinskas BA (1993) Detection of ascorbate peroxidase activity in native gels by inhibition of the ascorbate-dependent reduction of nitroblue tetrazolium. Anal Biochem 212:540–546

19. Baum JA, Scandalios JG (1981) Isolation and characterization of the cytosolic and mitochondrial superoxide dismutases of maize. Arch Biochem Biophys 206:249–264

20. Cakmak I, Marschner H (1992) Magnesium deficiency and high light intensity enhance activities of superoxide dismutase, ascorbate peroxidase, and glutathione reductase in bean leaves. Plant Physiol 98:1222–1227

21. Nakano Y, Asada K (1981) Hydrogen peroxide is scavenged by ascorbate-specific peroxidase in spinach chloroplasts. Plant Cell Physiol 22:867–880

22. Hodges DM, Delong JM, Forney CF et al (1999) Improving the thiobarbituric acid-reactive-substances assay for estimating lipid

peroxidation in plant tissues containing antho-cyanin and other interfering compounds. Planta 207:604–611

23. Janero DR (1990) Malondialdehyde and thio-barbituric acid-reactivity as diagnostic indices of lipid peroxidation and peroxidative tissue injury. Free Radic Biol Med 9:515–540

24. Zhu J, Tremblay N, Liang Y (2012) Comparing SPAD and atLEAF values for chlorophyll assessment in crop species. Can J Soil Sci 92:645–648

25. Skadsen RW, Scandalios JG (1987) Translational control of photo-induced expression of the Cat2 catalase gene during leaf development in maize. Proc Natl Acad Sci 84:2785–2789

26. Havir EA (1992) The in vivo and in vitro inhibition of catalase from leaves of Nicotiana sylvestris by 3-amino-1,2,4-triazole. Plant Physiol 99:533–537

27. Asada K, Yoshikawa K, Takahashi M et al (1975) Superoxide dismutases from a blue-green alga, Plectonema boryanum. J Biol Chem 250:2801–2807

28. Ye Z, Rodriguez R, Tran A et al (2000) The developmental transition to flowering represses ascorbate peroxidase activity and induces enzymatic lipid peroxidation in leaf tissue in Arabidopsis thaliana. Plant Sci 158:115–127

29. Zentgraf U, Zimmermann P, Smykowski A (2012) Role of intracellular hydrogen peroxide as signalling molecule for plant senescence. In: Nagata T (ed) Senescence. InTech , ISBN: 978-953-51-0144-4, DOI: 10.5772/34576

Chapter 17

Identification of Postharvest Senescence Regulators Through Map-Based Cloning Using Detached *Arabidopsis* Inflorescences as a Model Tissue

Donald A. Hunter, Rubina Jibran, Paul Dijkwel, David Chagné, Kerry Sullivan, Aakansha Kanojia, and Ross Crowhurst

Abstract

Postharvest deterioration of fruits and vegetables can be accelerated by biological, environmental, and physiological stresses. Fully understanding tissue response to harvest will provide new opportunities for limiting postharvest losses during handling and storage. The model plant *Arabidopsis thaliana* (*Arabidopsis*) has many attributes that make it excellent for studying the underlying control of postharvest responses. It is also one of the best resourced plants with numerous web-based bioinformatic programs and large numbers of mutant collections. Here we introduce a novel assay system called AIDA (the Arabidopsis Inflorescence Degreening Assay) that we developed for understanding postharvest response of immature tissues. We also demonstrate how the high-throughput screening capability of AIDA can be used with mapping technologies (high-resolution melting [HRM] and needle in the *k*-stack [NIKS]) to identify regulators of postharvest senescence in ethyl methanesulfonate (EMS) mutagenized plant populations. Whether it is best to use HRM or NIKS or both technologies will depend on your laboratory facilities and computing capabilities.

Key words Postharvest, Senescence, Whole genome sequencing, WGS, High-resolution melting, HRM, Ethyl methanesulfonate, EMS

1 Introduction

Many vegetable and fruit tissues deteriorate rapidly when harvested because of disruption in their energy, hormone, nutrient, or water supply. Understanding the molecular changes that control timing of this deterioration will enable strategies to be developed for maintaining quality of the produce postharvest. To achieve this we developed a system, the Arabidopsis Inflorescence Degreening Assay (AIDA), which is based on the degreening of detached dark-incubated immature inflorescences of *Arabidopsis* held in water or other chemical treatments. The results from AIDA may be particularly relevant for broccoli because *Arabidopsis* is in the same family (Brassicaceae) as

Yongfeng Guo (ed.), *Plant Senescence: Methods and Protocols*, Methods in Molecular Biology, vol. 1744,
https://doi.org/10.1007/978-1-4939-7672-0_17, © Springer Science+Business Media, LLC 2018

broccoli, its immature inflorescences resemble a miniature broccoli head, and transcription changes and degreening occur in both of their inflorescences with similar timing. One of the major advantages of the immature *Arabidopsis* inflorescence over the broccoli head is simply its small size enabling replicate treatments to be easily conducted within the space of a microtiter plate. This makes the system very convenient for rapidly investigating an array of chemical, controlled atmosphere and other treatments on postharvest longevity of the harvested tissue [1].

The reliability of AIDA makes it easy to identify degreening mutants within T-DNA, chemical, and radiation-induced mutant collections of the Arabidopsis Biological Resource Center (ABRC) and the Nottingham Arabidopsis Stock Centre (NASC). In this chapter we describe how the screening capability of AIDA in conjunction with recent mapping (HRM)- and sequencing-based technologies (NIKS) can be used for identifying EMS-induced genetic changes that control senescence.

HRM is a post-polymerase chain reaction (PCR) method for genotyping presence/absence of single-nucleotide polymorphism (SNP) markers in mapping populations [2–4]. In HRM, SNP markers are detected as changes in the melting profile of PCR amplicons. PCR is first performed with primers to amplify a genomic DNA (gDNA) region containing a known validated SNP between the two ecotypes of interest (e.g., Landsberg *erecta* [Ler-0] and Columbia [Col-0]). After amplification, the PCR products are denatured at 95 °C and then quickly reannealed at 40 °C. This rapid denaturing and reannealing are the critical steps of the technique. If the plant being analyzed contains only Ler-0 or only Col-0 DNA in the amplified region, then during reannealing at 40 °C the DNA will form perfect complementary double-stranded Ler-0 or Col-0 duplexes (homoduplexes). However, if recombination has occurred at the SNP locus, both Ler-0 and Col-0 sequences will have been amplified, which during reannealing will produce a mixture of Ler-0 and Col-0 homo- and heteroduplexes. These homo- and heteroduplexes can be distinguished because they have different stabilities and therefore melting profiles when they are reheated from 65 °C to 95 °C. This enables the researcher to genotype a plant in an outcrossed population segregating for a trait of interest as either homozygous for one of the parents or heterozygous. The HRM technique can therefore be used for mapping a mutation of interest but ultimately still requires an independent sequencing step to identify the causal mutation.

NIKS was recently developed by Nordstrom et al. [5] as a method for reference-free genome comparison. It is based solely on the frequencies of short subsequences within WGS data that target identification of mutagen-induced, small-scale, and homozygous differences between two highly related genomes, independent of their inbred or outbred background. Nordstrom et al. [5] used it to find small sets of mutations including causal changes in non-reference rice cultivars and in the nonmodel species *Arabis alpina*. The NIKS pipeline incorporates the bioinformatic tools of Jellyfish [6] for analysis of k-mers, Velvet [7] for localized de novo assembly and BLAST+ [8]. Here, we provide an alternate "run" script for NIKS that provides additional features to the original NIKS script. AIDA draws on the power of NIKS but also subsequently leverages the available annotation information for ecotypes of *Arabidopsis* (http://1001genomes.org/index.html) for rapidly identifying putative EMS-induced lesions resulting in delayed degreening/senescence.

Both technologies can be used independently or in conjunction with each other for mutation identification. Whether you use both or choose one will depend on your laboratory capabilities. HRM is a rapid, simple, non-computationally intensive method that involves individual genotyping of many individuals and eventual sequencing. NIKS by contrast is computationally intense requiring a significant investment in central processing unit time. It can give you causal SNP candidates in as little as a few days.

2 Materials

2.1 Plant Materials

1. Seeds can be obtained from the ABRC (https://abrc.osu.edu/) at Ohio State University or NASC (http://arabidopsis.info). *Ler*-0 EMS M2 seeds can be purchased from LEHLE Seed Company (Round Rock TX, USA; www.lehleseeds.com). Alternatively EMS material can be generated following the protocol of Kim et al. [9]. Designation of generations mentioned: M3 *Ler*-0 plants are progeny of original M2 recessive trait mutants that were allowed to self-fertilize; M4 *Ler*-0 plants are progeny of self-fertilized M3 plants; F1 *Ler*-0 plants are progeny from backcross or outcross of M4 plant with *Ler*-0 or Col-0, respectively (i.e., they will have one chromosome from *Ler*-0 mutant and one from wild type); and F2 plants derive from self-fertilized F1 plants. These plants segregate for trait of interest (mapping population).

2.2 AIDA

1. Black plastic 2 L container with lid (e.g., ice cream container).

2. Blotting paper.

3. Microtiter plate: 96 well.

4. Scalpel.

2.3 Chlorophyll Analysis

1. 96% ethanol.
2. Microfuge tube.
3. NanoDrop 1000 spectrophotometer.

2.4 Tissue Collection for HRM

1. Deep well plate.
2. Plastic spatula.
3. Rubber cover seals for plate.
4. Microcentrifuge tube.

2.5 Genomic DNA Isolation for WGS

1. Extraction Buffer 1: 400 mM sucrose; 10 mM Tris–HCl, pH 8.0; 10 mM $MgCl_2$; 5 mM β-mercaptoethanol added before use.
2. Extraction Buffer 2: 250 mM sucrose; 10 mM Tris–HCl, pH 8.0; 10 mM $MgCl_2$; 1% Triton™ X-100; 5 mM β-mercaptoethanol added before use.
3. Extraction Buffer 3: 1.7 M sucrose; 10 mM Tris–HCl, pH 8.0; 2 mM $MgCl_2$; 0.15% Triton™ X-100; 5 mM β-mercaptoethanol added before use.
4. Lysis buffer: 160 mM Tris–HCl, pH 7.5; 32 mM EDTA, pH 8.0; 1.2 M NaCl; 25.5% (w/v) sorbitol; 12% (w/v) CTAB. CTAB does not solubilize well, so mix overnight on a magnetic stirrer. Just before use, add 15 mg of sodium bisulfite ($Na_2S_2O_5$) to 10 mL of lysis buffer.
5. TE pH 8.0: 10 mM Tris–HCl, pH 8.0; 0.1 mM EDTA, pH 8.0.
6. PCI: Phenol/chloroform/isoamyl alcohol (25/24/1).
7. Falcon™ 50 mL conical centrifuge tube.
8. Miracloth.
9. Chloroform.
10. Isopropanol.
11. 3 M sodium acetate (NaOAc): pH 5.2.
12. 70% ethanol.
13. Agarose gel electrophoresis setup including gel image capture system.

2.6 HRM Analysis

1. Real-time PCR instrument capable of performing HRM. We use a LightCycler 480® instrument (Roche Diagnostics).
2. LightCycler® 480 High Resolution Melting Master kit (Roche Diagnostics).
3. Primer stock solution: 10 μM.
4. Genomic DNA: 1–50 ng per reaction.

2.7 NIKS-Based WGS

1. Genomic DNA: From at least 50 individuals of each F2 EMS mutant line of *Arabidopsis* trait positive for a phenotype of interest.

2. Illumina HiSeq2000 sequence: 100 bp paired end sequence of pooled individual DNA (equimolar) yielding preferably four-fold or more genome coverage per individual.

3. Command line access to a computer running a distribution of Linux (*see* **Note 1**).

4. BASH shell as the default shell.

5. Software for NIKS pipeline.

6. Prerequisite software: svn (for obtaining NIKS code), gcc/g++ (for compiling components with NIKS pipeline code or other required tools as necessary), Jellyfish version 1 [6], BLAST+ (blastn and makeblastdb), Velvet (shuffleSequences_fastq.pl, velveth, and velvetg), Java 1.6, Bash (*see* **Note 2**), wget (commonly available in most Linux distributions).

3 Methods

3.1 AIDA

1. Grow plants under long days. When the florets of the immature inflorescences are just starting to open, use tweezers to pull off the primary immature inflorescences and place on paper (*see* **Note 3**).

2. Using a scalpel carefully excise the immature inflorescence at the junction between the peduncle and pedicel of the lowermost unopened floret of the inflorescence.

3. Place green immature inflorescence in 200 μL sterile water (or chemical treatment) within a well of a microtiter plate (*see* **Note 4**).

4. Place the microtiter plate on blotting paper that has been moistened with 50 mL of water inside the black plastic 2 L container.

5. Cover the container and place in a 21 °C incubator.

6. Inspect inflorescences at intervals and remove at day of interest.

3.2 Chlorophyll Analysis

1. Weigh inflorescences into microfuge tubes and store at −80 °C until all inflorescences from all treatment times are collected.

2. Remove inflorescences from −80 °C and add 30 × volume of 96% ethanol (i.e., 5 mg/150 μL). We do not grind the tissue (*see* **Note 5**).

3. Place in the dark at 4 °C for 48 h.

4. Remove 2 μL of the ethanol extractant from each sample, and measure absorbance with NanoDrop 1000 spectrophotometer

at wavelengths 649 nm and 665 nm. Calculate the chlorophyll concentration according to equations of Wintermans and De Mots [10] (*see* **Note 6**).

3.3 Mapping Population Creation for HRM and NIKS (See Note 7)

1. Grow EMS-treated M2 L*er*-0 seeds (*see* **Note 8**) and screen inflorescences using AIDA. Identify lines with altered timing of degreening.

2. Harvest selfed seed (M3 seeds) from recessive trait M2 plants and sow and rescreen inflorescences using AIDA to confirm stability of trait in M3 generation plants.

3. Harvest selfed seed (M4 seeds) from recessive trait M3 plants and sow and rescreen using AIDA to confirm stability of trait in M4 generation plants (*see* **Note 9**).

4. To create mapping populations, grow M4 mutants and out-cross and backcross (*see* **Note 10**) with Col-0 (for HRM analysis) and L*er*-0 (for NIKS analysis) wild-type ecotypes, respectively, to produce F1 seeds.

5. Sow F1 seeds and check for absence of phenotype with AIDA (*see* **Note 10**).

6. Sow F2 seeds from each cross into separate trays, and check segregation of the phenotype of F2 progeny with AIDA (Fig. 1). A 1:3 ratio of mutant to wild-type phenotype indicates the degreening trait is monogenic and recessive.

7. Select a positive plant within each of the backcrossed and out-crossed populations, and repeat the backcrossing and out-crossing to wild types to reduce further the number of EMS-induced mutations in their backgrounds.

8. Harvest leaf disk tissue from F1 plants from the second back- and outcrosses into microfuge tubes and store at −80 °C for eventual DNA isolation. The isolated DNA will be used as a control in HRM analysis. Collect F2 seeds and keep F2 seeds obtained from each F1 plant separate (*see* **Note 11**).

9. For low-resolution mapping using HRM, sow F2 seeds of out-cross and harvest leaves from a minimum of 30 recessive trait positive F2 progeny for DNA isolation (*see* Subheading 3.4). This number will enable mapping to a chromosome arm.

10. For NIKS analysis, sow F2 seeds of backcross and harvest leaves (three leaves per plant) from >50 trait positive progeny and pool for DNA isolation and WGS. Crush pooled tissue under liquid N_2, and preferably proceed immediately to isolating genomic DNA for NIKS analysis (*see* Subheading 3.6). Alternatively, freeze in ~2 g amounts at −80 °C in Falcon tubes until needed.

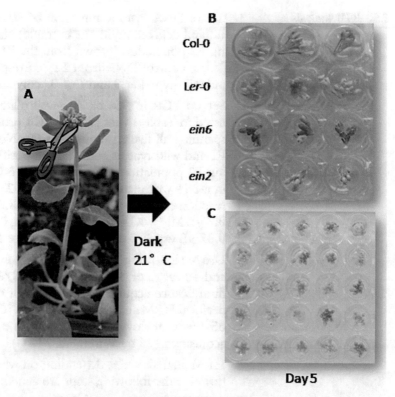

Fig. 1 AIDA can be used to uncover key regulators of postharvest degreening in mutant collections. (**a**) Immature inflorescence ready for detachment. Inflorescences are detached and placed in water within microtiter plates, which are then transferred into a covered darkened container and held at 21 °C. (**b**) Delayed degreening of *ethylene insensitive6* (*ein6*, ABRC #CS8055 in Ler-0 background) and *ethylene insensitive2-1* (*ein2-1*, ABRC #CS3071 in Col-0 background) at day 5 of dark incubation compared with wild types. (**c**) An ethyl methanesulfonate generated mutant that is segregating for the delayed degreening trait in the F2 generation at day 5 of dark incubation

3.4   Genomic DNA Isolation for HRM

3.4.1   Commercial DNA Isolation

1. With gloves on, detach cauline leaf (size of small fingernail) and place into well of deep well plate on dry ice. Use a plastic spatula to push the leaf to the bottom of well. Wipe spatula with 75% ethanol between samples.

2. Once samples are all collected, cover the plate with a rubber seal, freeze in liquid N_2, and store at −80 °C.

3. Freeze-dry the tissue and courier to commercial company for DNA isolation (*see* **Note 12**).

3.4.2   CTAB Protocol

1. With gloves on detach ~100 mg leaf tissue into microfuge tube.

2. Follow the CTAB method of Dellaporta et al. [11] as described by Stepanova and Alonso [12].

3.5 HRM Analysis

1. Isolate DNA from a minimum of 30 individual trait positive (altered degreening in our example) plants from the segregating F2 population (grown from the F1 Ler-0 X Col-0 seeds). Isolate control DNA from F1 (corresponds to heteroduplex) and wild-type Ler-0 and Col-0 ecotypes (homoduplexes).

2. Set up PCR in 96- or 384-well plate format using primers listed in Table 1 that amplify PCR products containing SNPs spanning all five chromosomes (*see* **Note 13**). Use gDNA of F1 and wild type as controls and positives from the segregating F2 population. An example of 96 DNA samples screened in a 384-well plate is shown in Fig. 2. Each individual HRM reaction consists of 7 μL containing 3.5 μL HRM master mix, 0.7 μL Mg^{2+}, 0.14 μL F primer stock, 0.14 μL R primer stock, 0.52 μL water, 2 μL gDNA (*see* **Note 14**).

3. Use the following PCR conditions: denaturation 95 °C/5 min and 40 cycles of (95 °C/10 s, 55 °C/30 s, 72 °C/15 s [with fluorescence acquisition]). At end of the PCR amplification, perform HRM analysis as follows: 95 °C/1 min, 40 °C/1 min, 65–95 °C at slow ramping rate with continuous fluorescence acquisition (25 measurements/1 °C).

4. HRM analysis varies depending on which instrument is used; however the following steps are generic.

5. Remove the suboptimal PCR amplifications from the HRM analysis.

6. Normalize the melting profile by selecting 1 °C windows before and after the melting point drop in fluorescence (*see* **Note 15**). The real-time instrument software for HRM will use these windows to calculate normalized melting curves.

7. Group the melting curves to three types, matching each of the Ler-0, Col-0, and heterozygous controls.

8. Use the SNP-based amplicons whose melting peaks clearly differentiate Ler-0, Col-0, and F1 heterozygous genotypes for testing linkage between the SNPs and the degreening mutation.

9. Identify SNPs that are linked to the degreening trait by analyzing the SNP profile in the population of trait positive plants. SNPs that are predominantly homoduplex for Ler-0 in the plant population but exhibit a heteroduplex profile in a very small subset of plants are informative. They indicate that a recombinant event has occurred in these plants between the marker SNP and the causal mutation. The frequency that this occurs in the plant population can be used to determine linkage of SNP to mutation.

Table 1

HRM primers for low-resolution mapping of Ler-0 and Col-0 crosses

Chr.	Col-0 Pos. (bp) (TAIR10)	ID	Forward primer	Reverse primer	SNP (C/L)	Profile identifies
1	592,939	1-1	CGAATTGAGAAATGGATGGAG	CCAGAAGAAACGGAGGAAGA	A/G	L/C/H (MP)
1	3,229,810	1-2	AAAACTCCCAGAACCAATC	TCCTCTCGTTTCTCAACACTTC	A/C	L/C/H (MP)
1	8,385,017	1-3	TTCGTTTTTGCTTTCCCAAC	TGATTGTCGCCTTACGTGTC	G/A	[L/C]H (MP
1	11,466,023	1-4	ACACTGGTCCTCCTCCTGTT	CCGAAGAGCTTGTGACTCCT	A/T	L/C/H (MP)
1	18,037,462	1-5	GCTTCTGTTTCGTTGTCTTGG	TGTGGAAGTGCAGGAGAGAG	A/G	L/C/H (MP)
1	20,080,144	1-6	CAGACCTCACATCAAAGCACA	CAATGCAAACCCATTATCC	C/A	L/C/H (TS)
1	26,783,875	1-7	ACGATATTGGGGACTCTGCT	TCGAAGAGGTTCATTGCTGTT	A/G	L/C/H (MP)
2	592,782	2-1	GCCACGAATCTCAAACCAAC	TCACATGGGAAGTAGTCCTG	T/A	L/C/H
2	839,806	2-2	AATTGTGCAGTGATGGGTTG	TCTGGAGGGTCTTCAAATGG	G/T	L/C/H (MP)
2	2,365,073	2-3	GGCACTGAAACCACAACTCC	TGCCTCCAAGTCCACTGATT	T/C	L/C/H (MP)
2	9,422,080	2-4	CTGACGATGAATTGCCCTTT	GAATCTCTCTTCTCTCCATAACA	A/C	L/C/H (MP)
2	12,428,124	2-5	CACACATAACAACAGACCCACTTC	CAACTTTGAGCCAGTTTGGT	G/C	L/C/H (MP)
2	19,594,990	2-6	GAGCCGACAGGGGATTAGTT	GCAGCGTTTGAGCTGTTACTT	T/C	L/C/H (MP)
3	580,130	3-1	TTAATTCCGGCTGGTTTGTT	TGATTCCTTCAGGCAAAGGT	C/G	[L/C]H (MP)
3	3,895,917	3-2	ATCCAGTTCACCTTCCATCC	GAACAATTACGGATATGACTTCCAG	T/G	L/C/H (MP)
3	9,340,724	3-3	CTGAGACAAAGGGAAAAGTGAA	ACGGTGTTGATTCTGGAGTG	T/A	L/C/H (TS)
3	21,298,252	3-4	GGCCCAAGAAGAAGGAAACA	TACTACATGGCTCCCGAGGT	T/G	L/C/H (MP)
4	5,077,557	4-1	TTGCGGCTAACTTTTTCA	CGGAAACACAATTCGGAGA	A/G	L/C/H (MP)

(continued)

Table 1
(continued)

Chr.	Col-0 Pos. (bp) (TAIR10)	ID	Forward primer	Reverse primer	SNP (C/L)	Profile identifies
4	8,297,805	4-2	GAATTACAGAGAGGTGGACACTGG	TACCACGAGAAGAGAGCGTTGA	A/T	L/C/H (MP)
4	10,482,224	4-3	GACGAGCTTCCTCCAACAGA	GCCGGTCGAAGAGATGGTAAG	C/A	L/C/H (MP)
4	12,195,803	4-4	GTGAAACCAATCTTCTCAAACTTCC	ATTGTGGGTGAGGGAAAACA	G/A	L/C/H (MP)
4	14,075,769	4-5	ATACCTTGCCGTTTGGGTCT	GTCTCCTTATGCGCCATCTC	G/A	[L/C]H (MP)
5	342,423	5-1	CCCTTTTATCTCCACCACATT	CAAAACGAGTCTTCAGGATGG	A/G	L/C/H (MP)
5	6,708,012	5-2	TTCGTCTTCTCATCTTTGCTTG	TCAGACAGGTGCCTTAAATCAGT	G/T	L/C/H (MP)
5	15,854,978	5-3	CAACATCATACATCGTAACTTCCAG	CCTTTACTTCCAAGAACGAGTCA	T/G	L/C/H (MP)
5	19,397,570	5-4	CACCTTTTGAGAGCGGTTATTT	TCGTTACTTGAATTGCCTCGT	G/C	[L/C]H (MP)

SNP positions were obtained by interrogating the TAIR10 Col-0 genomic sequence downloaded from arabidopsis.org/home/tair/Sequences/whole_chromosomes/. Please note that the position of the SNP is slightly different according to the Multiple SNP Query Tool. Pos., SNP position from start of chromosome genomic sequence. (C/L), Col-0/*Ler*-0. MP, Melting Peak Difference Plot. TS, Normalized and Temp-Shifted Difference Plot. L/C/H means distinguishes *Ler*-0/Col-0/Heterozygous profiles. [L/C] means difficult to distinguish between homozygote parents in mapping population.

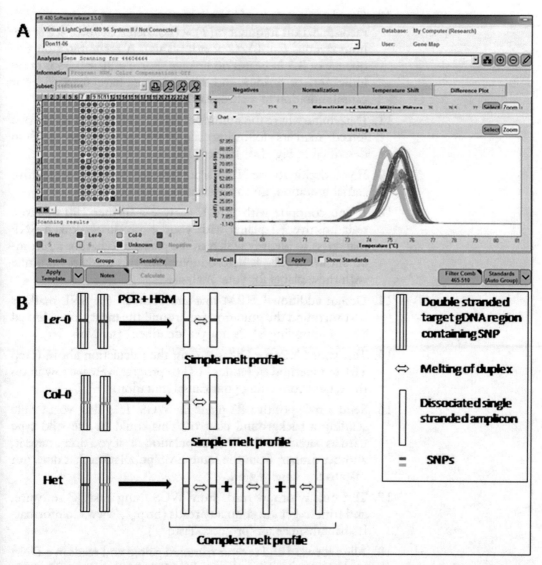

Fig. 2 Mapping SNPs using HRM analysis. HRM analysis can indicate SNP presence or absence in an individual of an F2 mapping population. (**a**) Screenshot of HRM analysis being done on 96 individuals obtained from an F2 mapping population whose inflorescences showed delayed dark-induced degreening. Mutation for degreening is in L*er*-0 background and is recessive. If SNP is close to mutation, then it is less likely that recombination during gamete formation of selfed F1 individuals will separate the SNP from the causal mutation. Therefore almost all individuals showing delayed degreening would be L*er*-0 (red profile) sequence for the SNP. The amount of linkage can be determined by calculating recombination frequency (*F*). An F value of ~0.3 indicates linkage, whereas 0.5 indicates no linkage. Using the equation in Subheading 3.5, **step 10** ([45 blue het +2 × 22 green Col-0]/2 X96 plants analyzed) = 0.46, it can be seen that this SNP marker is unlinked to the causal mutation. (**b**) Schematic illustrating the homo- and heteroduplexes of a SNP marker that give rise to the HRM profiles seen in panel A above. Note the mixture of homo- and heteroduplexes, which gives rise to the complex melting profile of the individual that is heterozygous for the SNP

10. Calculate the extent of SNP linkage to mutation by determining recombination frequency: (F) = (no. heterozygotes + 2 × no. homozygous Col-0)/(2 × no. plants). A recombination frequency of 0.5 indicates no linkage. For $n = 30 < 0.35$ is good indication of linkage and suggests you have mapped your causal mutation to a chromosomal arm.

11. Determine where the mutation resides with respect to the two linked markers following the three point cross method described in Fig. 4 of Jander [13].

12. If you decide to use NIKS from here to identify your putative causal mutation, go to Subheadings 3.6 and 3.7.

13. If you continue with HRM, analyze a further 100 or more trait positive F2 plants with your most closely linked SNP marker(s) to identify which of the plants also harbor a recombinant event (i.e., is heteroduplex) for this marker. Continue with these plants for your analysis.

14. Design additional HRM markers between the SNP markers that surround the mutation (or around the most closely linked SNP if bounding SNPs not yet identified) (*see* **Note 16**).

15. Test your SNPs for linkage using the calculation above (**step 10**) and method of Jander [13] to progressively narrow in on the region containing your causal mutation.

16. Send a trait positive F2 plant for WGS. It is also worthwhile sending a background control. This could be the wild type used as parent for the EMS population or, if you don't have it, another mutant from the same EMS population that does not map to the same region.

17. Test quality of raw reads from WGS using FastQC software, and trim to get good-quality reads (http://www.bioinformatics.babraham.ac.uk/projects/fastqc/).

18. Align forward and reverse trimmed paired end reads in a DNA analysis program such as BioMatters Geneious Desktop Software (http://www.geneious.com/) [14]. Extract region of TAIR10 L*er*-0 reference that the mutation has been mapped to, and assemble reads of both mutant and background control to this region. Scan for SNPs that are present in the mutant and absent in the control background.

3.6 Genomic DNA Isolation for NIKS (See Note 17)

1. Crush ~2 g of freshly collected pooled leaf tissue of your recessive trait positive F2 L*er*-0 X L*er*-0 plants (from Subheading 3.3) under liquid N_2, or alternatively remove crushed frozen pooled leaf tissue from −80 °C (*see* **Note 18**).

2. Add 20 mL of chilled Extraction Buffer 1 and vortex to resuspend the leaf material.

3. Filter through two layers of Miracloth into a sterile 50 mL Falcon tube.

4. Centrifuge crude extract at 1940 × g at 4 °C for 20 min and discard the supernatant.

5. Resuspend the nuclei-containing pellet in 1 mL of ice-chilled Extraction Buffer 2, centrifuge at 12,000 × g for 10 min at 4 °C, and discard the supernatant.

6. Resuspend the nuclei-containing pellet in 300 μL ice-chilled Extraction Buffer 3. Use a plastic pestle to gently homogenize solution, and make sure all cell clumps are broken up.

7. Overlay the 300 μL of resuspended nuclei gently onto a further 300 μL of chilled Extraction Buffer 3 in a sterile microfuge tube.

8. Pellet the nuclei by centrifuging at 14,000 × g for 1 h at 4 °C and discard supernatant.

9. Add 500 μL lysis buffer and 100 μL 5% Sarkosyl and 5 μL 100 mg/mL RNase A to the pellet and incubate at 65 °C for 1 h to lyse the nuclei, and remove RNA.

10. Add 600 μL PCI to the nuclei lysate and mix for ~2 min by gently tilting tube up and down until an emulsion is seen (*see* **Note 19**).

11. Centrifuge at 19,350 × g (top speed in microcentrifuge) for 3 min at 4 °C and transfer the supernatant to a new tube containing 600 μL PCI and repeat the cleaning step.

12. Centrifuge at 19,350 × g for 3 min at 4 °C, and transfer the supernatant to a fresh tube containing 1 volume of chloroform and mix by tilting.

13. Centrifuge at 19,350 × g for 3 min at 4 °C and transfer supernatant to a new tube and precipitate the DNA with 1/10 volume 3 M NaOAc and an equal volume of chilled isopropanol and mix by gently tilting tube up and down to fully mix and incubate on ice for 30 min.

14. Pellet DNA by centrifugation at 19,350 × g for 30 min at 4 °C, and wash pellet twice with 700 μL 70% ethanol.

15. After removing residual ethanol with pipette, air dry for 10 min at room temperature and resuspend in 50 μL TE pH 8.0.

16. Check quality by resolving in 0.8% agarose gel. Image capture the resolved DNA to include with samples sending to next-generation sequencing service provider.

17. Check quantity of DNA on a NanoDrop 1000 spectrophotometer. Typically 2–4 μg gDNA is obtained from 2 g of leaf tissue.

18. Typically 2–5 μg DNA is submitted at a concentration of 100 ng/μL in 1× TE buffer for 100 bp paired end run on an Illumina HiSeq2000 sequencer.

19. We send the samples by courier samples in screw-top parafilm-sealed tubes at room temperature.

3.7 NIKS Analysis
(See Note 20)

1. Obtain your sequence reads in standard FASTQ format from your NGS sequencing service provider, and transfer them to your Linux server.

2. Log in to your Linux server and download the NIKS pipeline code from the NIKS svn repository as follows (*see* **Note 21**):

   $ cd ${HOME}

   $ svn checkout svn://svn.code.sf.net/p/niks/code-0/trunk niks-code-0 #or use svn checkout https://svn.code.sf.net/p/niks/code-0/trunk niks-code-0. Otherwise obtain the code from https://sourceforge.net/p/niks/code-0/

3. If the *k*-mer counting program jellyfish is in your path, then determine which version you have as follows:

   $ which jellyfish # to find out where your jellyfish binary is installed if you did not install it yourself.

   $ jellyfish --version # Note there are 2 '-'before version.

   If jellyfish is not on your system or if your version is not version 1, then download the jellyfish distribution code from http://www.cbcb.umd.edu/software/jellyfish/ and compile it as follows (*see* **Note 22**):

   $ cd /home/<your_username>

   $ wget http://www.cbcb.umd.edu/software/jellyfish/jelly-fish-1.1.11.tar.gz

   $ tar zxvf jellyfish-1.1.11.tar.gz

   $ cd /home/<your_username>/jellyfish-1.1.11/

   $ ./configure --prefix=/path/to/install/jellyfish-1.1.11

   $ make && make install

   $ export /path/to/install/jellyfish-1.1.11/bin:$PATH

4. Change directory to the niks-code-0/kmerPipeline/src subdirectory, and build the jellyfish_sorting_key binary. The build instructions are supplied in the README in the src directory of the checkout code base. Use the compile command appropriate to your version of jellyfish. For example, to compile jellyfish_sorting_key for jellyfish 1.1.11, do the following:

   $ cd path/to/directory/niks-code-0/kmerPipeline/src

   $ more README.txt

   $ cd jellyfish_sorting_key

   $ JELLYFISH_INSTALL_PREFIX=/path/to/jellyfish-1.1.11

```
$ g++ -pthread -ljellyfish-1.1 -L${JELLYFISH_INSTALL_
    PREFIX}/lib    -I${JELLYFISH_INSTALL_PREFIX}/
    include/jellyfish-`VERSION`

-o ../scripts/jellyfish_sorting_key jellyfish_sorting_key.cc
```

5. If BLAST+ is not installed on your system, download and install the appropriate version from ftp://ftp.ncbi.nlm.nih.gov/blast/executables/blast+/LATEST/. For example,

```
$  wget    ftp://ftp.ncbi.nlm.nih.gov/blast/executables/
    blast+/LATEST/ncbi-blast-2.2.29+-1.x86_64.rpm

$ rpm -ivh ncbi-blast-2.2.29+-1.x86_64.rpm # (as root user
    if installing system wide)
```

6. Download the reference genome for *Arabidopsis* as follows (change the paths to one that is appropriate to your Linux filesystem) (*see* **Notes 23** and **Note 24**):

```
$ mkdir –p /input/ComparativeDataSources/Arabidopsis/
    thaliana/TAIR10/

$  cd  /input/ComparativeDataSources/Arabidopsis/thali-
    ana/TAIR10/

$  wget  ftp://ftp.arabidopsis.org/home/tair/Sequences/
    whole_chromosomes/TAIR10_chr*.fas
```

7. Download the associated protein sequences from TAIR10. AIDA will use both the descriptions as well as the sequences. The descriptions are extracted to a separate file later in the protocol:

```
$  wget  ftp://ftp.arabidopsis.org/home/tair/Sequences/
    blast_datasets/TAIR10_blastsets/TAIR10_pep_
    20101214_updated
```

8. Download the genome for the appropriate ecotype (L*er*-0 in this example) from the 19 genomes project (*see* **Note 23**):

```
$ mkdir –p /input/ComparativeDataSources/Arabidopsis/
    thaliana/Landsberg/erecta/Ler-0/

$  cd  /input/ComparativeDataSources/Arabidopsis/thali-
    ana/Landsberg/erecta/Ler-0/

$ wget

http://mtweb.cs.ucl.ac.uk/mus/www/19genomes/fasta/
    UNMASKED/ler_0.v7.fas
```

9. Download the annotations (in gff3 format) associated with the specific ecotype (*Ler*-0 in this example) from the 19 genomes project, and uncompress them as follows:

```
$ wget

http://mtweb.cs.ucl.ac.uk/mus/www/19genomes//
    annotations/consolidated_annotation_9.4.2011/gene_
    models/consolidated_annotation.Ler_0.Col_0.gff3.bz2
```

$ bzip2 –d consolidated_annotation.Ler_0.Col_0.gff3.bz2

10. Download our NIKS wrapper script from https://github.com/rosscrowhurst/NIKS-Wrapper and place them in your path.

11. Ensure all the required software tools are in your path (*see* **Note 25**).

12. Ensure that your FASTQ sequence files are uncompressed (*see* **Note 26**).

13. Change directory to your working directory.
    $ cd /workspace/NIKS

14. Run our NIKS wrapper (*see* **Note 27**). For example, if

    (a) FASTQ read files for your mutants are located in "/input/genomic/plant/Arabidopsis/thaliana/Ler/mutants/1302KHS-0060",

    (b) your working directory is "/workspace/NIKS160",

    (c) your first mutant is named "ddi1",

    (d) the mutant to which you wish to compare it is called "ddi2",

    (e) your read files are called "ddi1-160i_1.fastq", "ddi1-160i_2.fastq", "ddi2-160i_1.fastq", and "ddi2-160i_2.fastq", respectively, for the paired end reads for both mutants.

    (f) you installed the NIKS code in "/software/NIKS/kmerPipeline",

    (g) you wish to use 16 cores,

    (h) your ecotype reference genome is located in "/input/ComparativeDataSources/Arabidopsis/thaliana/Landsberg/erecta/Ler-0",

    (i) your ecotype reference genome is called "ler_0.v7.fasta",

    (j) the name you wish to apply to the BLAST+ database for the ecotype genome reference is "ler_0.v7",

    (k) you wish to use a version of jellyfish other than that in your path, then issue the following command:

    $ /path/to/NIKS-kmerPipeline.sh   -DATADIR   /input/genomic/plant/Arabidopsis/thaliana/Ler/mutants/1302KHS-0060   -WORKINGBASEDIR   /workspace/NIKS160 -MUTANTA_NAME ddi1 -MUTANTB_NAME ddi2 -MUTANTA_READ1_FILE ddi1-160i_1.fastq -MUTANTA_READ2_FILE ddi1-160i_2.fastq -MUTANTB_READ1_FILE ddi2–160i_1.fastq -MUTANTB_READ2_FILE ddi2–160i_2.fastq -NIKS_SCRIPTS_DIR /software/NIKS/kmerPipeline/scripts -CORES 16 -GENOME_REFERENCE_DIR /input/ComparativeDataSources/Arabidopsis/thaliana/

Landsberg/erecta/Ler-0 -REFERENCE_GENOME_FASTA
ler_0.v7.fasta   -NAMEBLASTDBGENOME   ler_0.v7
-JELLYFISH_BIN   /software/x86_64/jellyfish-1.1.10/bin/
jellyfish (*see* **Notes 28–30**).

15. To actually run the pipeline on Open Lava, just run the fol-
    lowing code (*see* **Note 31**):

    ```
    $ cd /workspace/NIKS160/NIKS/ddi1-ddi2;NAME=ddi1-
       ddi2; bsub -n 16 -R "mem > 300000 order[ut]" -J
       ${NAME}"    -o   /workspace/NIKS160/NIKS/ddi1-
       ddi2 /ol.${NAME}.niks.stdout

    -e /workspace/NIKS160/NIKS/ddi1-ddi2/ol.${NAME}.
       niks.stderr < /workspace/NIKS160/NIKS/ddi1-ddi2/
       kmerPipeline.ddi1-ddi2.71073.sh
    ```

16. Once the pipeline has completed (potentially a day or so),
    return to your working directory, and examine the log files pro-
    duced as well as (for our example above) the files "ddi1-ddi2_
    candidates_ext_genomeAnnotation.sorted.chromosome.sum.
    cvs", "ddi1_candidates_ext.fa", and "ddi2_candidates_ext.fa".
    The "sorted.chromosome.sum" file contains a four-column
    tab-delimited data set providing the NIKS contig names, their
    location (chromosome number and base position) in the
    Arabidopsis reference genome used in the analysis, and the
    alleles present. The "*candidates_ext.fa" files contain fasta
    sequence for each NIKS contig (*see* **Note 32**).

17. To determine functional descriptions for the NIKS contigs,
    either use our "niksMergeAndGetDescriptionFromTairDesc.
    pl" script or use BLASTx to compare the NIKS contigs to a
    protein database such as Swissprot or the annotated proteins
    from *Arabidopsis* (*see* **Note 33**).

18. To compare your NIKS contigs to TAIR10 peptides using
    BLASTx (if you downloaded TAIR10_pep_20101214_updated.
    fasta as in our example above and makeblastdb binary is in your
    path), then create a blast database from it as follows:

    ```
    $ makeblastdb -in TAIR10_pep_20101214_updated.fasta
      -dbtype prot
    ```

19. Run BLASTx as follows (*see* **Note 34**):

    ```
    $ blastx -db /input/ComparativeDataSources/Arabidopsis/
       thaliana/TAIR10/TAIR10_pep_20101214_updated.fasta
       -query ddi1_candidates_ext.fa -out ddi1_candidates_ext.
       TAIR10_pep_20101214.blastx -num_threads 4

    $ blastx -db /input/ComparativeDataSources/Arabidopsis/
       thaliana/TAIR10/TAIR10_pep_20101214_updated.
       fasta -query ddi2_candidates_ext.fa -out ddi2_candidates_
       ext.TAIR10_pep_20101214.blastx -num_threads 4
    ```

20. To use "niksMergeAndGetDescriptionFromTairDesc.pl" to merge NIKS results with preexisting annotations from TAIR10 and 19 genomes project, run the tool as follows (*see* **Note 35**):

    $ tairDecFile=/input/ComparativeDataSources/Arabidopsis/ thaliana/TAIR10/ TAIR10_pep_20101214_updated. desc; chrNum=4;

    ChrFasta=/input/ComparativeDataSources/Arabidopsis/ thaliana/Landsberg/erecta/Ler-0/Chr4.fa;

    ler0_consolidated_annotations=/input/Comparative DataSources/Arabidopsis/thaliana/Landsberg/erecta/ Ler-0/19genomes/annotations/consolidated_annota- tion_9.4.2011/consolidated_annotation.Ler_0.gff3;

    niks_locations/workspace/NIKS160/NIKS/ddi1-ddi2/ ddi1-ddi2_candidates_ext_genomeAnnotation.sorted. chromosome.sum.cvs;

    niksMergeAndGetDescriptionFromTairDesc.pl -c=$chrNum -tair_desc=$tairDecFile   -tair_ref=$ChrFasta   -niks_ locations=$niks_locations

    -ler0Anno=$ler0_consolidated_annotations 1>Chr4_NIKS_ ddi1_ddi2.out 2>Chr4_NIKS_ddi1_ddi2.out

21. The output is a simple text file containing six columns (position in chromosome, AGI identifier, NIKS contig name, reference allele, alternate allele, description associated with AGI identifier) with entries ordered by position on the selected chromosome. Repeat the process for each different chromosome in which you have an interest.

22. The output files need now to be manually examined, and where appropriate the mRNA for the identified AGI sequence, the reference genome sequence, and the NIKS contigs can be aligned and visualized using your favorite tool such as BioMatters Geneious Desktop Software. As the possibilities for aligning and visualization are both numerous and common, we do not describe further the mechanisms to undertake this alignment and visualization.

4   Notes

1. You will need access to a multi-core server running a Linux OS. The minimum RAM requirement is 128 GB RAM. More realistically 256 GB to 1 or more terabytes of RAM are required dependent on the volume of sequence and the number of simultaneous NIKS pairwise analyses you wish to perform. A single pairwise comparison with 1.5× genome coverage per individual and 50 individuals in a pool requires approximately 105 Gb RAM (on our server). The same single pairwise comparison but employing 4× coverage of 110 indi-

viduals per pool will utilize approximately 200 Gb RAM. A server with 1 Tb or more RAM will allow simultaneous analysis of more than one pairwise comparison.

2. The NIKS pipeline shell scripts use process substitution. Our NIKS-kmerPipeline.sh NIKS wrapper has replaced process substitution in some places as we wished to capture the intermediates. However, if you submit your NIKS analysis to compute nodes on a grid, do check whether BASH is the default shell on all processing nodes. BASH is very common, but if your run fails, examine the shell script lines pointed to in any error messages that are generated. If those lines include process substitution, e.g., "<(some code)", then the execute host likely did not run the script using the BASH interpreter. This issue is not usually evident when running the NIKS scripts from your command line interactively.

3. Timing of degreening is not greatly affected by preharvest conditions. We screened for mutants with altered timing of degreening using plants grown in standard greenhouse conditions at different times of year, which spanned short days and long days. For other types of analyses, we typically grew plants in controlled conditions in a growth chamber with a 16-h-light photoperiod, temperature of ~22 °C, and 100 µmol m^{-2} s^{-1}. Inflorescences are ready for detachment ~5 weeks after sowing (when plant is at stages 6.00–6.10 [15]) or when the first flower of the inflorescences is starting to open. After the first harvest, plants can be cut back and left for the next series of buds to emerge, typically within a week. Although there is a reasonably wide window in which buds can be collected, do not use inflorescences from plants that are at Boyes stage >6.5. The inflorescences of these plants will be smaller due to global arresting of buds, more yellow at harvest and will yellow prematurely, i.e., will be completely degreened sooner than day 5.

4. AIDA is suited for assessing effects of plant hormones, sugars, or other chemicals on timing of degreening. We have also exposed inflorescences to controlled atmosphere conditions by simply placing the plates in sealed bags with syringes punched into each end serving as the inlet/outlet for gas from compressed cylinders. The bag remaining inflated serves as a test that the low throughput of administered gas is occurring.

5. We did bead bash the inflorescences at 500 rpm for 10 min in 20× (e.g., 5 mg/100 µL), 30×, 50×, and 100× ethanol volumes with a SPEX Geno/Grinder. This method did not extract significantly more chlorophyll than extracting from whole tissue.

6. Chlorophyll concentration calculated according to equations of Wintermans and De Mots [10]: Chlorophyll a = 13.7 A_{665}

$_{m\mu}-5.76\ A_{649\ m\mu}$ μg/mL; Chlorophyll b = 25.80 $A_{649\ m\mu}-7.60$ $A_{665\ m\mu}$ μg/mL; Chlorophyll a + b = 6.10 $A_{665\ m\mu}-20.04\ A_{649\ m\mu}$ μg/mL.

7. Clearly presented background information and step-by-step protocols for mapping of EMS-induced causal traits can be found in "Arabidopsis Protocols," *Methods in Molecular Biology*, Volume 323 [16]. We reference these rather than repeating what is already clearly explained in the format of this series. Because of this we strongly encourage the reader to view the protocols referenced and their associated notes, particularly Chapters 1, 6, 8–10, and 13.

8. Rivero-Lepinckas et al. [17] describe how to grow, collect, and preserve seeds of *Arabidopsis*.

9. It is common when screening mutagenized populations for traits for the numbers of plants showing the trait to decrease in subsequent generations. This can be due to environmental conditions or interactions caused by additional lesions in the plants. Therefore it is beneficial to test for stability of the trait in further two generations before making mapping population. Whether it is needed to self again depends on the mutagenic load.

10. Detailed methodology for crossing *Arabidopsis* lines is described in Koornneef et al. [18]. Jander [13] suggests using the mutant as the female parent because you can tell if your cross for a recessive trait was not successful since the F1 progeny will show the trait.

11. Keeping crosses separate will make sure that crosses that do not work will not interfere with segregation ratio analysis.

12. We used SlipStream Automation (www.slipstream-automation. co.nz).

13. Although most HRM markers present in Table 1 can clearly differentiate between the three genotypes, some are more difficult to analyze than others because the Ler-0 and Col-0 homoduplexes may have very similar melt profiles. Therefore we would recommend using the markers distinguished by L/C/H first. The less definitive ones that only clearly distinguish the heteroduplexes can still be informative as the closer they are to the lesion the less heteroduplexes will be identified.

14. Never perform HRM on DNA isolated by different methods. Different isolation procedures can result in different melt profiles for the same marker presumably because of the different impurities left in the extracted DNA. We have observed very different melting curves between PCR products obtained from CTAB and column-based extracts.

15. The melting profiles for each PCR product must be normalized, as there is usually variability between PCR quantities (vis-

ible as variable fluorescence intensities) across samples. The raw fluorescence values must be normalized in such a way that dsDNA (i.e., PCR products premelting) and ssDNA (i.e., PCR products post-melting) are set as 100% and 0% of bound fluorescent dye, respectively. Typically normalization is performed by selecting a region of ~1 °C before and after the dramatic drop in fluorescence, with special attention paid to selecting regions where all the curves are parallel and horizontal. Following normalization, the software automatically groups melting curves based on their similarity. Options for visually assigning the melting curve groupings include using the placement of the melting curves. Heterozygous samples containing both Ler-0 and Col-0 SNP alleles usually melt at lower temperature than homozygous samples of the Ler-0 and Col-0 types.

16. On average there is one SNP every 3.3 kbp between Ler-0 and Col-0 [13], which provides an essentially unlimited number of SNPs for use in HRM analysis. Therefore fine mapping with HRM can be performed. To find additional SNPs (between Ler-1 and Col-0) and their surrounding sequence for primer design, use the Multiple SNP Query Tool (http://msqt.wei-gelworld.org/ [19]). We recommend that researchers make themselves familiar with the resources at the http://1001genomes.org website. Design HRM primers using Primer3 (http://bioinfo.ut.ee/primer3-0.4.0/primer3 [20]) with default parameters except for the following changes: Set "Max Self Complementarity" to 4, "Max 3 Self Complementarity" to 1, "Primer GC%" min to 40 and max to 55, and "Product Size" Min 70 and Max 150. Target the SNP in Primer3 with square brackets '[]'.

17. This method is a hybrid of the nuclear enrichment method of Lutz et al. [21] and the CTAB method of Dellaporta et al. [11]. The method enables increased coverage of the nuclear genome to be obtained from the WGS data by preferentially removing chloroplast gDNA during the DNA isolation. Triton™ X-100 in Extraction Buffers 2 and 3 minimizes disruption of the nuclear membrane allowing preferential elimination of chloroplasts and mitochondria. We find that the gDNA isolated by the hybrid method has ~fourfold less DNA of the chloroplast-encoded ribosomal protein S18 (ATCG00650) than gDNA isolated by the CTAB method.

18. We recommend using fresh tissue for gDNA isolation because with frozen tissue we observed more shearing of the gDNA.

19. From **step 10** on, do not be too vigorous when mixing as this can lead to shearing of gDNA.

20. The sequencing-based NIKS technology developed by Nordstrom et al. [5] can be used on pooled genomic DNA of positive F2 plants with recessive phenotype to identify the

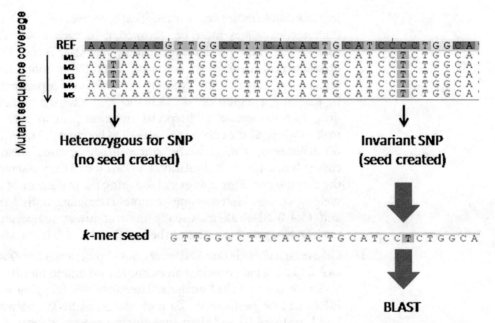

Fig. 3 NIKS draws direct comparisons between two closely related genomes by removing identical sequences through analysis of k-mers in DNASeq reads from each genome. By using pools of individuals of each genome (wild type and mutant or between two mutant phenotypes) for the comparison, NIKS is able to filter out irrelevant differences (those which vary within a pool) and identify relevant differences (those where the mutant individual are invariant at the site). The k-mers containing these relevant differences are then extended into contigs, which in turn can be annotated by comparison against preexisting sequence databases using tools such as BLAST. Although NIKS is a reference free method, the availability of annotated reference genomes speeds the annotation of contigs produced by the NIKS pipeline. REF, reference genome; M1, M2... represents the sequence of individuals in the pooled population

invariant genetic lesion responsible for the trait of interest (Fig. 3). It can be used as a stand-alone method or alternatively used to identify the genetic lesion in coarsely mapped populations. Here, we provide an alternate "run" script for NIKS that adds the following features to the original NIKS script: easy restart on failure, checks for prerequisite binaries, merges the original separate annotation script in the workflow as an optional parameter, allows passing of custom paths for binaries where multiple versions are in the path, and renames NIKS contigs to reflect names of mutants in pairwise comparisons.

21. If desired move the checked out directory created by the following steps to an appropriate location on your file system. However, it is not necessary to install it in any specific location as the path to the NIKS pipeline is passed on the command line when running our NIKS wrapper code.

22. If you encounter issues in using wget, then it is likely your firewall or your organization's firewall is blocking http traffic. Because firewall issues are beyond the scope of this protocol,

we suggest that if you encounter such problems that you either consult your system administrator in the first instance or use an alternative computer for downloading, then copy the files to your Linux server. As a last resort, you may wish to temporarily disable your firewall, re-try your wget, and then re-enable your firewall.

23. Although NIKS is designed for reference, free analysis AIDA leverages the reference genomes and annotations for *Arabidopsis* from TAIR10 [22] as well as a specific ecotype reference genome from the 19 Genomes Project [23].

24. This step may take some time as you will be downloading each chromosome separately as AIDA needs each chromosome in a separate FASTA sequence file. If you have the chromosomes already but in a multi-FASTA file, simply split them apart with a tool like fasta-explode.

25. The programs jellyfish, blastn, makeblastdb, shuffleSequences_fastq.pl, velveth, velvetg, and java must be in your path. You can check this for each by issuing the command line "for program in blastn makeblastdb shuffleSequences_fastq.pl velveth velvetg java; do which ${program}; done" on your Linux command line. Alternatively if you are working on a multiuser server where the default versions of these programs are not those you wish to use and you have installed the version required locally in your home directory (or other location), you can simply pass the full Linux path to these programs on the command line of our NIKS-kmerPipeline.sh NIKS wrapper script. The wrapper script will prompt you to supply the path if it cannot find the binary on your system.

26. Although the original NIKS pipeline code accepts either compressed or uncompressed FASTQ sequence files, our NIKS wrapper script expects uncompressed FASTQ and will fail if the sequence files are compressed. To uncompress use a command like "gzip –d filename" if the files have a .gz extension.

27. Our NIKS-kmerPipeline.sh NIKS wrapper script is an alternative way to run NIKS. Although you can run NIKS according to the pipeline's README distributed with the NIKS code, our wrapper does offer a few advantages:

(a) It checks that the requisite programs are in your path and will provide an opportunity to run them from locations on your file system of your choice if you have multiple versions of the programs as we do in our software tree.

(b) It incorporates the annotation steps that were in a separate script in the original NIKS code distribution.

(c) It includes code to be able to restart the analysis just by issuing the original command line. It will check for the presences of expected output files and if found will not repeat the step required to produce those outputs.

(d) It renames NIKS "contigs" to reflect the pairwise comparison undertaken enabling subsequent merging of results from multiple pairwise comparisons in to a single results file without issues of replacement of similarly named contigs from the different comparisons.

(e) If you wish to leverage an *A. thaliana* accession genome reference, then our wrapper script will make a blast+ database from it if you have not done this manually before running the pipeline.

(f) When passing names for mutants to our wrapper, only use an alphanumeric alphabet, e.g., ddi1, ddi2, and ddi3, and do not include spaces or control characters.

28. Our NIKS-kmerPipeline.sh NIKS wrapper will print a series of output messages to your screen. It does not run your analysis directly but instead prints out all the individual commands required to complete your NIKS analysis to a shell script that you subsequently run. This allows you to examine the commands for the full pipeline prior to running and then to run the script from your command line directly or to submit it to a job scheduling system such as Open Lava (http:// http:// www.openlava.org/).

29. The script produced by NIKS-kmerPipeline.sh will be named according to the following convention: kmerPipeline.${MUTANTA}-${MUTANTB}.$$.sh. For our example above, it will be named kmerPipeline.ddi1-ddi2.71073.sh where 71,073 represents the process ID under which NIKS-kmerPipeline.sh ran. When you run NIKS-kmerPipeline.sh, it will create a sub-directory called NIKS and within that a pairwise comparison specific sub-directory (named after the two mutants being analyzed, ddi1-ddi2 for the above example). This allows you to set up and run multiple comparisons as each will run from an independent sub-directory. Within the comparison specific sub-directory will be written in the shell script to actually run the NIKS pipeline for the specific comparison.

30. The last few lines of output from NIKS-kmerPipeline.sh are very important and should be copied/kept as they provide alternative ways by which you may run the actual NIKS analysis. At present only two options are provided: run it directly from the command line or submit to the Open Lava Job Scheduler. If you use a scheduler other than Open Lava, then modify the output to reflect your job scheduling environment. By writing out all the NIKS pipeline commands specific to your analysis to a file instead of running it immediately provides the possibility to examine the commands prior to run-

ning as well as to submit the code to your code repository for that analysis. The versions of all programs to be used are captured to this shell file as part of the process.

31. This is the output from our example above—the "bsub -n 16 -R 'mem > 300000 order[ut]'" part of the code simply submits the job to Open Lava requesting the task be run on an execute host with more than 300 Gb of RAM and to select between execute hosts, if more than one, based on current utilization states of those qualifying execute hosts.

32. Note that the NIKS contig names from our NIKS-kmerPipeline.sh NIKS wrapper are different from those of the original NIKS pipeline in that we encode the names of both mutants in the pairwise comparison undertaken into the NIKS contig names.

33. Although these steps could be included in our NIKS-kmerPipeline.sh wrapper, we have not chosen to do so as there are many different comparative databases you may wish to use.

34. Add -outfmt '6 std' to the end of the command if you want tabular output. Note BLASTx is being run interactively from the command line here instead of submitting to a job scheduler as this comparison takes only a minute or so at most.

35. This approach generally requires that you have prior knowledge about the chromosome on which your lesion is located. Using this program requires:

    (a) The descriptions from the TAIR10 proteins which and should be derived as follows:

    ```
    $ grep ">" /input/ComparativeDataSources/Arabidopsis/
    thaliana/TAIR10/TAIR10_pep_20101214_updated.
    fasta|cut-d ">" -f2 >/input/ComparativeDataSources/
    Arabidopsis/thaliana/TAIR10/TAIR10_
    pep_20101214_updated.desc
    ```

    (b) The fasta sequence of a single chromosome.

    (c) The consolidated annotations for the ecotype in gff3 format.

Acknowledgments

The authors wish to thank Plant and Food Research staff Lyn Watson and Zoe Erridge for technical assistance, Elena Hilario for comments on the gDNA isolation protocol, and David Brummell and Elena Hilario for critically reading the manuscript. We also wish to thank Kris Tham (SlipStream Automation) for technical assistance and Korbinian Schneeberger for use of the NIKS method his team developed. We thank the Massey Doctoral

Fund for providing a scholarship to RJ. The work was supported by Plant and Food Research CORE research funding program 12059 Leafy Crops.

References

1. Trivellini A, Jibran R, Watson LM et al (2012) Carbon deprivation-driven transcriptome reprogramming in detached developmentally arresting Arabidopsis inflorescences. Plant Physiol 160:1357–1372

2. Gundry CN, Vandersteen JG, Reed GH et al (2003) Amplicon melting analysis with labeled primers: a closed-tube method for differentiating homozygotes and heterozygotes. Clin Chem 49:396–406

3. Liew M, Pryor R, Palais R et al (2004) Genotyping of single-nucleotide polymorphisms by high-resolution melting of small amplicons. Clin Chem 50:1156–1164

4. Chagné D, Gasic K, Crowhurst RN et al (2008) Development of a set of SNP markers present in expressed genes of the apple. Genomics 92:353–358

5. Nordstrom KJV, Albani MC, James GV et al (2013) Mutation identification by direct comparison of whole-genome sequencing data from mutant and wild-type individuals using k-mers. Nat Biotechnol 31:325–330

6. Marçais G, Kingsford C (2011) A fast, lock-free approach for efficient parallel counting of occurrences of k-mers. Bioinformatics 27:764–770

7. Zerbino DR, Birney E (2008) Velvet: algorithms for de novo short read assembly using de Bruijn graphs. Genome Res 18:821–829

8. Camacho C, Coulouris G, Avagyan V et al (2009) BLAST+: architecture and applications. BMC Bioinformatics 10:421

9. Kim Y, Schumaker KS, Zhu JK (2006) EMS mutagenesis of Arabidopsis. In: Salinas J, Sanchez-Serrano JJ (eds) Arabidopsis protocols, 2nd edn. Humana Press Inc, Totowa, NJ, pp 101–103

10. Wintermans JF, de Mots A (1965) Spectrophotometric characteristics of chlorophylls a and b and their pheophytins in ethanol. Biochim Biophys Acta 109:448–453

11. Dellaporta SL, Wood J, Hicks JB (1983) A plant DNA minipreparation: version II. Plant Mol Biol Report 1:19–21

12. Stepanova AN, Alonso JM (2006) PCR-based screening for insertional mutants. In: Salinas J, Sanchez-Serrano JJ (eds) Arabidopsis protocols, 2nd edn. Humana Press Inc, Totowa, NJ, pp 163–172

13. Jander G (2006) Gene identification and cloning by molecular marker mapping. In: Salinas J, Sanchez-Serrano JJ (eds) Arabidopsis protocols, 2nd edn. Humana Press Inc, Totowa, NJ, pp 115–126

14. Kearse M, Moir R, Wilson A et al (2012) Geneious basic: an integrated and extendable desktop software platform for the organization and analysis of sequence data. Bioinformatics 28:1647–1649

15. Boyes DC, Zayed AM, Ascenzi R et al (2001) Growth stage-based phenotypic analysis of arabidopsis: a model for high throughput functional genomics in plants. Plant Cell 13:1499–1510

16. Salinas J, Sanchez-Serrano JJ (eds) (2006) Arabidopsis protocols. Methods in molecular biology, 2nd edn. Humana Press Inc, Totowa, NJ

17. Rivero-Lepinckas L, Crist D, Scholl R (2006) Growth of plants and preservation of seeds. In: Salinas J, Sanchez-Serrano JJ (eds) Arabidopsis protocols, 2nd edn. Humana Press Inc, Totowa, NJ, pp 3–12

18. Koornneef M, Alonso-Blanco C, Stam P (2006) Genetic analysis. In: Salinas J, Sanchez-Serrano JJ (eds) Arabidopsis protocols, 2nd edn. Humana Press Inc, Totowa, NJ, pp 65–77

19. Warthmann N, Fitz J, Weigel D (2007) MSQT for choosing SNP assays from multiple DNA alignments. Bioinformatics 23:2784–2787

20. Untergasser A, Cutcutache I, Koressaar T, Ye J et al (2012) Primer3—new capabilities and interfaces. Nucleic Acids Res e11:40

21. Lutz KA (2011) Isolation and analysis of high quality nuclear DNA with reduced organellar DNA for plant genome sequencing and resequencing. BMC Biotechnol 11:54

22. Lamesch P, Berardini TZ, Li D et al (2012) The Arabidopsis Information Resource (TAIR): improved gene annotation and new tools. Nucleic Acids Res 40:D1202–D1210

23. Gan X, Stegle O, Behr J et al (2011) Multiple reference genomes and transcriptomes for Arabidopsis thaliana. Nature 477:419–423

Part V

Molecular and Cellular Processes in Plant Senescence

Chapter 18

Chlorophyll and Chlorophyll Catabolite Analysis by HPLC

Aditi Das, Luzia Guyer, and Stefan Hörtensteiner

Abstract

The most obvious event of leaf senescence is the loss of chlorophyll. Chlorophyll degradation proceeds in a well-characterized pathway that, although being common to higher plants, yields a species-specific set of chlorophyll catabolites, termed phyllobilins. Analysis of chlorophyll degradation and phyllobilin accumulation by high-performance liquid chromatography (HPLC) is a valuable tool to investigate senescence processes in plants. In this chapter, methods for the extraction, separation, and quantification of chlorophyll and its degradation products are described. Because of their different physicochemical properties, chlorin-type pigments (chlorophylls and magnesium-free pheo-pigments) and phyllobilins (linear tetrapyrroles) are analyzed separately. Specific spectral properties and polarity differences allow the identification of the different classes of known chlorins and phyllobilins. The methods provided facilitate the analysis of chlorophyll degradation and the identification of chlorophyll catabolites in a wide range of plant species, in different tissues, and under a variety of physiological conditions that involve loss of chlorophyll.

Key words Chlorophyll, Pheophytin, Pheophorbide, Phyllobilin, Leaf senescence, Fruit ripening, HPLC, Chlorophyll catabolite

1 Introduction

Chlorophyll, the most abundant plant pigment, absorbs sunlight as a key factor for the conversion of solar energy to chemical energy during photosynthesis. At the same time, however, chlorophyll is a potentially phototoxic molecule, whose biosynthesis and degradation have to be tightly regulated. During leaf senescence, when photosynthetic activities diminish and photosystem protein components are degraded for nitrogen remobilization to other parts of the plant, chlorophyll is metabolized for its detoxification. Breakdown of chlorophyll has been used as a diagnostic tool not only for induction and progression of leaf senescence but also for monitoring other processes of plant

---

Aditi Das and Luzia Guyer contributed equally to this work.

Yongfeng Guo (ed.), *Plant Senescence: Methods and Protocols*, Methods in Molecular Biology, vol. 1744, https://doi.org/10.1007/978-1-4939-7672-0_18, © Springer Science+Business Media, LLC 2018

development, such as fruit ripening and seed maturation and as plants'
responses to biotic and abiotic stress challenges.

In recent years, a pathway of chlorophyll breakdown has been
elucidated that converts chlorophylls, green-colored tetrapyrroles
with a chlorin ring structure, to colorless, linear tetrapyrroles
termed phyllobilins [1]. This pathway consists of several enzymatic
steps that produce different short-lived catabolite intermediates
and a plant species-specific set of phyllobilins that ultimately
accumulate in the vacuole of senescing cells [1, 2]. The series of
intermediary and final catabolites are depicted in Fig. 1. The pathway is structured into two parts: (i) reactions from chlorophyll to
a "primary" fluorescent chlorophyll catabolite (pFCC) that are
localized in the chloroplast and commonly occur in all plant species
investigated so far [3] and (ii) species-specific, largely cytosolic
reactions that site-specifically modify peripheral positions of pFCC
and give rise to the more or less complex pattern of phyllobilins
detectable in a given species [4–6]. The majority of phyllobilins
that accumulate are nonfluorescent, because fluorescent precursors
rather quickly isomerize to their nonfluorescent equivalents within
the vacuole due the low pH of the vacuolar sap [5, 7]. Exceptions
are so-called "hypermodified" FCCs found in a few species, in
which modification of the propionyl side chain (Fig. 1) prevents
efficient isomerization to their respective nonfluorescent equivalents [8, 9]. Two major classes of phyllobilins are distinguished,
which are formyloxobilin-type (FCCs and NCCs) and dioxobilin-
type tetrapyrroles (DFCCs and DNCCs) (Fig. 1) [6], the latter of
which are (formally) derived from the former ones through an oxidative deformylation [5]. In some cases, NCCs were shown to be
oxidized to yellow chlorophyll catabolites [10].

Along the pathway of degradation, polarity of chlorophyll
catabolites increases together with a general shift of absorption
toward shorter wavelengths [2, 6]. These features allow (i) the
separation of chlorophyll catabolites by reversed-phase HPLC, (ii)
the identification of different groups of catabolites based on their
characteristic UV/Vis absorption spectra (Fig. 1), and (iii) the
quantification of several groups of compounds for which standards
are available or can be relatively easily produced.

In this chapter, two methods for separation, identification and
quantification of either chlorin- or phyllobilin-structured chlorophyll catabolites, are described.

2   Materials

All chemicals used should be of analytical grade.

2.1   General Solutions

1. Tris–HCl, pH 8.0.

2. 0.5 M KH_2PO_4.

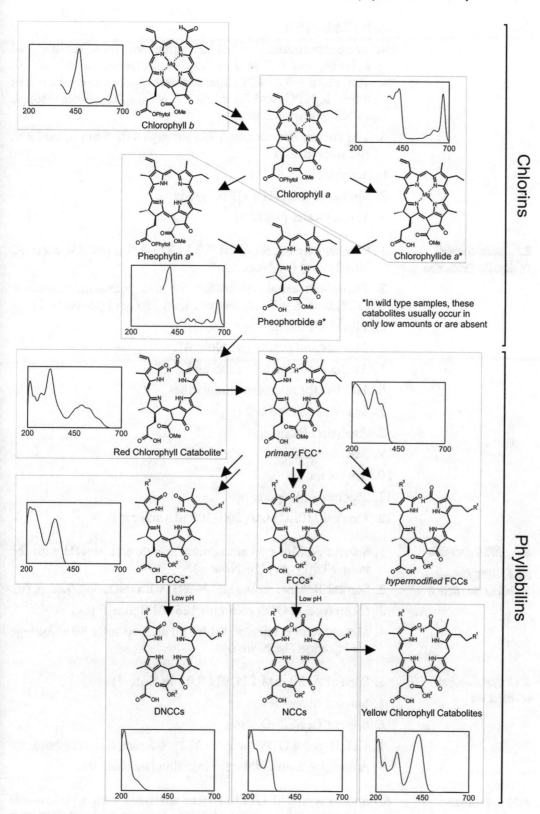

Fig. 1 Constitutional formulas and UV/Vis absorption spectra of chlorophyll-derived chlorins and phyllobilins occurring during chlorophyll breakdown

3. 0.5 M K_2HPO_4.

4. 10× concentrated (0.5 M) KPi buffer: Mix solutions of 0.5 M K_2HPO_4 and 0.5 M KH_2PO_4, until reaching a pH of 7.0. 100 mL of 0.5 M KPi buffer requires approximately 42 mL of 0.5 M K_2HPO_4 and 58 mL of 0.5 M KH_2PO_4. Filter through a 0.45 μm nylon filter.

5. 1 M ammonium acetate: Filter through a 0.45 μm nylon filter, and store at 4 ° C.

6. Acetone.

7. Methanol (MeOH): HPLC grade.

8. Milli-Q water (18 MΩ).

2.2 Chlorophyll/ Phyllobilin Extraction

1. Chlorin extraction buffer: 0.2 M Tris–HCl, pH 8.0, acetone, 10:90 (v/v) (*see* **Note 1**).

2. Phyllobilin extraction buffer: 50 mM potassium phosphate (KPi) buffer, pH 7.0, methanol, 25:75 (v/v) (*see* **Note 2**).

3. Liquid nitrogen.

4. Mortars and pestles (*see* **Note 3**).

5. Microcentrifuge tubes: 1.5 and 2 mL.

6. Holders for microcentrifuge tubes.

7. Spatula and thumb forceps.

8. Analytical balance.

9. Microcentrifuge.

10. Vortex mixer.

11. Pipettes and pipette tips.

12. Conical HPLC vials (200–300 μL) and caps.

2.3 HPLC Analysis

*2.3.1 HPLC Analysis of Chlorins (**See Note 4**)*

1. Solvent A: Mix 1 M ammonium acetate and MeOH with the ratio 20:80 (v/v) (*see* **Note 5**).

2. Solvent B: Mix acetone and MeOH with the ratio 20:80 (v/v).

3. C18 Hypersil ODS columns: 125 × 4.0 mm, 5 μm.

4. Manual syringe injector: With a 50 μL loop and a 50 μL syringe for injection (*see* **Note 6**).

2.3.2 HPLC Analysis of Phyllobilins

1. Solvent A: 50 mM KPi, pH 7.0 (*see* **Note 7**).

2. Solvent B: MeOH.

3. Solvent C: Milli-Q water.

4. C18 Hypersil ODS columns: 250 × 4.6 mm, 5 μm (*see* **Note 8**).

5. A coolable autosampler for injection (*see* **Note 9**).

2.3.3 HPLC System

A remote-controlled HPLC system interfaced with a photodiode array detector with an absorption range between 200 and 700 nm is required (*see* **Note 10**). The HPLC system should contain a pump able to produce a gradient of at least two solvents (*see* **Note 11**) and a thermostatted column compartment set to 28 ° C (*see* **Note 12**). To avoid gas formation caused by mixing of solvents, an efficient degasser should be installed with the pump system (*see* **Note 13**).

2.4 Quantification

1. Chlorin standards: Several chlorin standards can be purchased from certified suppliers or are isolated from leaf extracts and purified by analytical HPLC as described here or according to published methods (e.g., [11, 12]).

2. Calibration solutions for chlorins: Pure solutions of respective chlorins, prepared at defined concentrations (~2 mM) in acetone, are used for calibrating HPLC peak absorptions (*see* **Note 14**).

3. *Cj*-NCC-1: Phyllobilin standards are not commercially available. *Cj*-NCC-1 is the major nonfluorescent chlorophyll catabolite from senescent leaves of *Cercidiphyllum japonicum*, a deciduous tree grown in many gardens and parks. *Cj*-NCC-1 can rather easily be isolated and purified by HPLC according to published methods [13] and can be used for NCC quantifications.

4. Calibration solution for *Cj*-NCC-1 (*see* **Note 15**): *Cj*-NCC-1 dissolved in H_2O is quantified spectrophotometrically at 314 nm (log ε_{314} = 4.23, [13]) (*see* **Note 16**). Defined concentrations are used to calibrate HPLC peak absorptions and can be used for quantification of NCCs.

3 Methods

3.1 Extraction of Chlorins (See Notes 17 and 18)

1. Harvest green or senescent plant material (*see* **Note 19**) and immediately flash-freeze in liquid nitrogen. Store at −80 ° C until use.

2. Grind plant material with mortar and pestle in liquid nitrogen to a fine powder (*see* **Note 20**).

3. Weigh 50–100 mg of plant material into a microcentrifuge tube (precooled in liquid nitrogen), and keep the samples in liquid nitrogen until extraction (*see* **Note 21**).

4. For pigment extraction, work on ice or at 4 ° C (*see* **Note 22**). Add 5–10 volumes of chlorin extraction buffer (v/w) to frozen plant material, and vortex vigorously for around 10 s until the plant material is thawed (*see* **Note 23**).

5. Incubate in the dark (*see* **Note 24**) at -20 ° C for 2–16 h until all pigments are extracted into the buffer (*see* **Note 25**).

6. Pellet plant extracts by centrifugation for 2 min at 16,000 × *g* and 4 ° C with a microcentrifuge. Transfer the supernatant into a new microcentrifuge tube, and immediately flash-freeze and store in liquid nitrogen until HPLC analysis (*see* **Notes 26** and **27**).

3.2 Extraction of Phyllobilins (see Note 17)

1. Harvest green or senescent plant material (see **Note 19**), and immediately flash-freeze in liquid nitrogen. Store at −80 °C until use.

2. Grind plant material with mortar and pestle in liquid nitrogen to a fine powder (see **Note 20**).

3. Weigh 100–200 mg of plant material into a microcentrifuge tube (precooled in liquid nitrogen), and keep the samples in liquid nitrogen until extraction (see **Note 21**).

4. Add 3 volumes of phyllobilin extraction buffer (v/w), and vortex vigorously for about 10–15 s until the plant material is thawed and uniformly mixed with the buffer.

5. Sonicate for 10 min in an ice-cooled sonication bath.

6. Centrifuge the mixture at 16,000 × *g* for 5 min with a microcentrifuge, and then transfer the supernatant to a new 1.5 mL microcentrifuge tube.

7. Recentrifuge the supernatant at 16,000 × *g* for 5 min with a microcentrifuge (see **Note 28**), and transfer 100–200 μL of the supernatant to a HPLC vial (see **Note 29**).

3.3 HPLC Analysis and Quantification of Chlorins (See Note 30)

1. Remove a sample from liquid nitrogen just prior to injection, and thaw it (see **Notes 6** and **31**). Inject a defined volume (e.g., 50 μL) into the HPLC system, and run the following HPLC program.

2. HPLC program used for chlorin analysis (flow rate 1 mL min⁻¹):

Time (min)	Solvent A (%)	Solvent B (%)
0	100	0
15	0	100
25	0	100
28	100	0
32	100	0

3. Calibration: Inject standard solutions of respective chlorins, and determine the retention times and absorption spectra. Fig. 1 shows the spectra for different chlorins, and in Table 1, the retention times are listed (see **Note 32**). Integrate peak areas of standard solutions with known concentrations (see **Note 14**) at 665 nm (see **Note 33**), and determine the relation between absorption and amount/concentration.

4. Identify the chlorins present in your sample via comparing the absorption spectra and retention times of peaks with standard solutions.

5. Integrate the peak areas at 665 nm, and convert into amounts/concentrations using the factors calculated with standard solutions.

Table 1
Approximate retention times of chlorins using the HPLC system outlined in Note 32

Pigment	Retention time (min)
Chlorophyllide *b*	1.5
Chlorophyllide *a*	3
Pheophorbide *b*	7
Pheophorbide *a*	9
Chlorophyll *b*	16
Chlorophyll *a*	18
Pheophytin *b*	21
Pheophytin *a*	23

3.4 HPLC Analysis of Phyllobilins

1. Place samples in the precooled (7–10 °C) autosampler of the HPLC system (*see* **Note 34**). Run the batch of samples by injecting defined volumes (e.g., 50 μL) with the following HPLC program.

2. HPLC program used for phyllobilin analysis (flow rate 1 mL min^{-1}):

Time (min)	Solvent A (%)	Solvent B (%)	Solvent C (%)
0	80	20	0
5	80	20	0
35	40	60	0
45	40	60	0
47	40	60	0
49	0	100	0
54	0	100	0
56	0	60	40
58	80	20	0
60	80	20	0

3. Calibration: Inject defined amounts of *Cj*-NCC-1 standard solutions (*see* Subheading 2.4, **item 3**) into the HPLC. Integrate peak areas at 315 nm (*see* **Note 35**), and determine the relation between absorption and amount/concentration.

Table 2
Approximate retention times of the major *Arabidopsis* phyllobilins using the HPLC system outlined in Note 37

Phyllobilin	Retention time (min)
At-DNCC-1	21.5
At-NCC-1	23.7
At-DNCC-2	24.7
At-NCC-2	26.4
At-DNCC-3	27.3
At-NCC-3	27.8
At-NCC-4	29.1
At-DNCC-4	29.2
At-DNCC-5	30.5
At-NCC-5	36.2

4. Identify the phyllobilins that are present in your sample via screening chromatograms at 254 nm (*see* **Note 35**), and search for peaks that exhibit UV/Vis absorption spectra typical for the different known types of phyllobilins (Fig. 1) (*see* **Note 36**).

5. For quantification of NCCs, integrate the peak areas at 320 nm, and convert into amounts/concentrations using the factors calculated with the *Cj*-NCC-1 standard.

6. Retention times of the major phyllobilins from senescent leaves of *Arabidopsis* [5, 14] are listed in Table 2 (*see* **Note 37**).

4   Notes

1. Chlorophyll is sensitive to acidic pH which causes artifactual pheophytin formation [15].

2. Dilute the 10× KPi stock solution in Milli-Q H_2O. Aqueous KPi buffers are prone to bacterial contaminations. Store aliquots of the 10× stock solution at -20 ° C. Once prepared, use the extraction buffer rather quickly.

3. For small sample amounts (100 mg or less), grinding is best done in a mixer mill (e.g., Retsch MM200, Haan, Germany). For this, place the plant tissue in 1.5 mL centrifuge tubes, add 5–10 glass beads (3 mm diameter), and freeze in liquid nitrogen. Place tubes in the Teflon holders provided with the mill (precooled in liquid nitrogen), and run at a frequency of 30 Hz for 3 × 30 s with re-cooling in liquid nitrogen in between.

4. Before use, degas solvents by sonicating in a sonication bath for 5 min to prevent gas formation within the HPLC system.

5. To avoid the effect of end volume changes when mixing an aqueous solution with an organic solvent, measure MeOH and the 1 M ammonium acetate solutions in separate measuring cylinders, and subsequently combine them.

6. Autosamplers should not be employed for chlorin analysis, because of the instability of phytylated pigments even at 4 °C, which is caused by chlorophyllases potentially still active in the extracts [16]. Instead, it is crucial for chlorin extracts to be kept in liquid nitrogen until injection into the HPLC.

7. Maintaining a pH of 7.0 is critical to avoid undesired FCC/DFCC to NCC/DNCC conversion during the HPLC run that may occur at low pH [7, 17].

8. HPLC columns should be stored in MeOH/H$_2$O, 50:50 (v/v), to prevent precipitation of salts on the column.

9. Although an autosampler is not absolutely necessary, it allows running a batch of multiple samples without the need of manual injection. Consider, however, that phyllobilins are rather thermolabile and light sensitive; they should be cooled in the autosampler and be protected from light.

10. Availability of a fluorescence detector is advantageous for the detection of FCCs/DFCCs, which emit blue fluorescence at 450 nm when exited at 360 nm [6].

11. If only two solvent channels are available, reduce the concentration of the KPi buffer to 10 mM for phyllobilin analysis. This is necessary, because high MeOH concentrations used during the HPLC run could cause precipitation of phosphate salts.

12. A temperature-controlled column is recommended for phyllobilin separation to guarantee uniform retention times. This is, however, less critical for separation of chlorins.

13. Gas formation is particularly critical during phyllobilin analysis when mixing the KPi buffer (solvent A) with MeOH (solvent B). If no degasser is available, premix the KPi buffer with MeOH to the start conditions (i.e., 20% MeOH, when using the gradient provided in Subheading 3.4, step 2), degas by sonication, and adjust the HPLC gradient accordingly.

14. If the concentrations of chlorin standards are unknown, e.g., because of manual extraction and preparation of pigments (see Subheading 2.4, item 1), the concentration can be determined spectrophotometrically prior to HPLC injection. For calculating concentrations, use the following equations valid for pigments in 80% acetone [18, 19] (A = absorption at indicated wavelength):

chlorophyll *a* or chlorophyllide *a* [µg mL^{-1}] = 11.63 A$_{665}$— 2.39 A$_{649}$

chlorophyll *b* or chlorophyllide *b* [µg mL^{-1}] = 20.11 A$_{649}$— 5.18 A$_{665}$

pheophytin *a* or pheophorbide *a* [µg mL^{-1}] = 22.42 A$_{665}$— 6.81 A$_{653}$

pheophytin *a* or pheophorbide *a* [µg mL^{-1}] = 40.17 A$_{653}$— 18.58 A$_{665}$

15. Due to the commercial unavailability of phyllobilin standards and the limited spectral data available for the different classes of phyllobilins, relatively accurate quantification is so far only possible for NCCs.

16. Rather similar log ε values have been determined for NCCs from different plant species (e.g., [20–22]), allowing the approximate quantification of (unknown) NCCs from any plant sample using *Cj*-NCC-1 as standard.

17. With the method presented, chlorins or phyllobilins have been analyzed in leaves and fruits of different plant species such as *Arabidopsis* and tomato [23, 24]. If other organs or plant species are analyzed, the method might require adaptations, in particular concerning the ratio of tissue amount to extraction buffer volume and the time of extraction.

18. An extensive comparison of different methods for chlorophyll extraction, including a sub-zero temperature method using acetone similar to the one described here, has recently been published [16].

19. Senescence induction in leaves (e.g., *Arabidopsis*) can be performed by incubating detached leaves in the dark. Leaves are cut and incubated on wet filter paper in a closed container at room temperature in the dark. Induction of senescence can be detected after 3–5 days. Alternatively, leaves can be covered with aluminum foil while being attached to the plant. If induction of leaf senescence is applied to other plant species, the procedure might require adaptations.

20. Work fast and make sure that the plant material does not thaw during the procedure.

21. To avoid thawing of the tissue during transferring to the microcentrifuge tube, dip the microcentrifuge tube in liquid nitrogen, and use it as a shuffle to transfer the desired amount of tissue. Immediately, weigh the sampled tissue using a fine balance that has been tared with an empty microcentrifuge tube.

22. Chlorophyllases are ubiquitously present in plant tissues and might get activated during the process of pigment extraction. Therefore, it is very important to work quickly and in a cold environment, in order to prevent artifactual chlorophyllide and/or pheophorbide formation [16, 23].

23. The required amount of buffer depends on chlorophyll concentration. Five volumes of buffer are generally sufficient for leaves of *Arabidopsis*. For tomato or other species that contain high amounts of chlorophyll, the buffer volume may need to be increased.

24. Make sure samples are not exposed to light to prevent bleaching of the pigments.

25. Extraction time depends on the plant sample. For green or senescent *Arabidopsis* leaves, 2 h of incubation is normally sufficient. For other plant tissues that are more fibrous and/or contain high chlorophyll levels (such as tomato or barley leaves), extraction can be done overnight. Alternatively, two to three rounds of extraction each followed by a centrifugation step can be done with subsequent pooling of extracted fractions, or extracts may be sonicated for a few minutes in an ice-cooled sonication bath.

26. Transfer as much supernatant as possible, but ensure that no solids are transferred.

27. Only when pellets are colorless, all chlorins have been extracted into the acetone phase. If the pellet still appears green, multiple rounds of extraction can be performed (*see* also **Note 25**).

28. Twice centrifugation ensures that samples are free of solid particles. This may particularly be necessary with leaf samples from, e.g., Poaceae species, like barley, which are rather fibrous.

29. If HPLC analysis will be performed at a later time point, prepared HPLC vials can be frozen in liquid nitrogen.

30. The chlorin extraction procedure described in Subheading 3.1 simultaneously extracts carotenoids from plant tissues. Thus, HPLC analysis of these extracts allows the simultaneous analysis of carotenoids. Methods for carotenoid analysis and quantification have been published [25].

31. Frozen acetone develops high pressure while thawing. To avoid possible explosion of the microcentrifuge tube, release pressure by opening the microcentrifuge lid.

32. These values were obtained with the following HPLC system: column, C18 Hypersil ODS column (125 × 4.0 mm, 5 μm) (MZ-Analysentechnik); pump, Gynkotek high-precision pump model 480 (Thermo Fisher Scientific); photodiode array detector, 206 PHD (365–700 nm, linear); and software, ChromQuest version 2.51 (Thermo Fisher Scientific). For other HPLC systems, retention times may vary.

33. Although peak maxima differ between different chlorins (Fig. 1), 665 nm is an ideal wavelength for simultaneous quantification of all chlorins.

34. If sample-containing HPLC vials had been stored in liquid nitrogen, make sure no air bubbles are trapped in the conical tip of the HPLC vials.

35. 254 nm is best suited for detection of all types of phyllobilin, while, e.g., DNCCs do only weakly absorb at 315 nm [5], the wavelength suitable for quantification of NCCs.

36. Peak resolution and baseline settings are important parameters for accurately quantifying NCC peaks. Senescent leaves of different plant species tend to accumulate phyllobilin-unrelated compounds that absorb in the UV range and may co-elute with phyllobilins and, thus, may interfere with phyllobilin identification and quantification.

37. These values were obtained with the following Dionex HPLC system (Thermo Fisher Scientific): column, C18 Hypersil ODS column (250 × 4.6 mm, 5 μm); autosampler, AS-100; column compartment, TCC-100 (set to 28 °C); detectors, PA-100 photodiode array detector (200–700 nm) and RF2000 fluorescence detector (excitation at 360 nm, emission at 450 nm); and software, Chromeleon 6.8 chromatography data system. For other HPLC systems, retention times may vary.

Acknowledgments

This work was supported by grants from the Swiss National Science Foundation and by CropLife, an EU Marie-Curie Initial Training Network. Aditi Das and Luzia Guyer contributed equally to this work.

References

1. Hörtensteiner S, Kräutler B (2011) Chlorophyll breakdown in higher plants. Biochim Biophys Acta 1807:977–988

2. Christ B, Hörtensteiner S (2014) Mechanism and significance of chlorophyll breakdown. J Plant Growth Regul 33:4–20

3. Sakuraba Y, Schelbert S, Park S-Y et al (2012) STAY-GREEN and chlorophyll catabolic enzymes interact at light-harvesting complex II for chlorophyll detoxification during leaf senescence in *Arabidopsis*. Plant Cell 24:507–518

4. Christ B, Schelbert S, Aubry S et al (2012) MES16, a member of the methylesterase protein family, specifically demethylates fluorescent chlorophyll catabolites during chlorophyll breakdown in Arabidopsis. Plant Physiol 158:628–641

5. Christ B, Süssenbacher I, Moser S et al (2013) Cytochrome P450 CYP89A9 is involved in the formation of major chlorophyll catabolites during leaf senescence in *Arabidopsis*. Plant Cell 25:1868–1880

6. Kräutler B (2014) Phyllobilins—the abundant bilin-type tetrapyrrolic catabolites of the green plant pigment chlorophyll. Chem Soc Rev 43:6227–6238

7. Oberhuber M, Berghold J, Breuker K et al (2003) Breakdown of chlorophyll: a nonenzymatic reaction accounts for the formation of the colorless "nonfluorescent" chlorophyll catabolites. Proc Natl Acad Sci U S A 100:6910–6915

8. Moser S, Müller T, Ebert MO et al (2008) Blue luminescence of ripening bananas. Angew Chem Int Ed 47:8954–8957

9. Kräutler B, Banala S, Moser S et al (2010) A novel blue fluorescent chlorophyll catabolite accumulates in senescent leaves of the peace lily and indicates a divergent path of chlorophyll breakdown. FEBS Lett 584:4215–4221

10. Ulrich M, Moser S, Müller T et al (2011) How the colourless 'nonfluorescent' chlorophyll catabolites rust. Chem Eur J 17:2330–2334

11. Shioi Y, Fukae R, Sasa T (1983) Chlorophyll analysis by high-performance liquid chromatography. Biochim Biophys Acta 722:72–79

12. Perkins HJ, Roberts DWA (1962) Purification of chlorophylls, pheophytins and pheophorbides for specific activity determinations. Biochem Biophys Acta 58:486–498

13. Curty C, Engel N (1996) Detection, isolation and structure elucidation of a chlorophyll *a* catabolite from autumnal senescent leaves of *Cercidiphyllum japonicum*. Phytochemistry 42:1531–1536

14. Pružinská A, Tanner G, Aubry S et al (2005) Chlorophyll breakdown in senescent Arabidopsis leaves: characterization of chlorophyll catabolites and of chlorophyll catabolic enzymes involved in the degreening reaction. Plant Physiol 139:52–63

15. Mazaki H, Watanabe T, Takahashi T et al (1992) Pheophytination of eight chlorophyll derivatives in aqueous acetone. Bull Chem Soc Jpn 65:3212–3214

16. Hu XY, Tanaka A, Tanaka R (2013) Simple extraction methods that prevent the artifactual conversion of chlorophyll to chlorophyllide during pigment isolation from leaf samples. Plant Methods 9:19

17. Oberhuber M, Berghold J, Kräutler B (2008) Chlorophyll breakdown by a biomimetic route. Angew Chem Int Ed 47:3057–3061

18. Lichtenthaler HK (1987) Chlorophylls and carotenoids: pigments of photosynthetic biomembranes. Meth Enzymol 148:350–382

19. Strain HH, Cope BT, Svec WA (1971) Analytical procedures for the isolation, identification, estimation and investigation of the chlorophylls. Methods Enzymol 23: 452–476

20. Kräutler B, Jaun B, Bortlik K-H et al (1991) On the enigma of chlorophyll degradation: the constitution of a secoporphinoid catabolite. Angew Chem Int Ed Engl 30:1315–1318

21. Berghold J, Eichmüller C, Hörtensteiner S et al (2004) Chlorophyll breakdown in tobacco: on the structure of two nonfluorescent chlorophyll catabolites. Chem Biodivers 1:657–668

22. Scherl M, Müller T, Kräutler B (2012) Chlorophyll catabolites in senescent leaves of the lime tree (*Tilia cordata*). Chem Biodivers 9:2605–2617

23. Schenk N, Schelbert S, Kanwischer M et al (2007) The chlorophyllases AtCLH1 and AtCLH2 are not essential for senescence-related chlorophyll breakdown in *Arabidopsis thaliana*. FEBS Lett 581:5517–5525

24. Guyer L, Schelbert Hofstetter S, Christ B et al (2014) Different mechanisms are responsible for chlorophyll dephytylation during fruit ripening and leaf senescence in tomato. Plant Physiol 166:44–56

25. Guzman I, Yousef GG, Brown AF (2012) Simultaneous extraction and quantitation of carotenoids, chlorophylls, and tocopherols in Brassica vegetables. J Agric Food Chem 60:7238–7244

Chapter 19

Identification and Functional Analysis of Senescence-Associated Genes in Wheat

Geng Wang, Ke Li, and Chunjiang Zhou

Abstract

Senescence is the final stage of leaf development. During this process, different macromolecules undergo degradation, and the resulting components are transported to developing and storage tissues of the plant. *Senescence-associated genes* (*SAGs*) play important roles in this process. Identification and characterization of *SAGs* are the first steps to interpret the function of these genes and to elucidate the mechanisms of leaf senescence. One of the most effective ways to identify *SAGs* is to screen for candidate genes using massive genome-scale transcriptomic data such as microarray, RNA-seq, digital RNA expression level data, etc. The basic functional analysis of candidate genes is to observe the phenotypes of transgenic plants, in which the candidate *SAGs* are overexpressed, knocked down, or knocked out. In this chapter, we outline methods for identifying and characterizing *SAGs* by microarray analysis in wheat. Methods of gene functional analyses by screening transgenic plants are also described. The protocols described in this chapter could also be used in other plant species, especially for Poaceae plants.

Key words Leaf senescence, *Senescence-associated genes* (*SAGs*), Microarray, Overexpression, RNAi, Wheat

1 Introduction

Senescence is the final stage of leaf growth and development, which ultimately leads to death or completion of a particular leaf [1, 2]. Leaf senescence a kind of programmed cell death (PCD) process in higher plants that can be induced by endogenous factors and environmental cues [3]. This complex process involves orderly, sequential changes in cellular physiology, biochemistry, and gene expression [4]. In agriculture, leaf senescence may reduce yield in crop plants by limiting the growth phase and may also cause post-harvest spoilage such as leaf yellowing and nutrient loss in vegetable crops. Therefore, studying leaf senescence will not only contribute to our knowledge about this fundamental developmental process but also improve ways of manipulating senescence for agricultural applications [5].

Yongfeng Guo (ed.), *Plant Senescence: Methods and Protocols*, Methods in Molecular Biology, vol. 1744,
https://doi.org/10.1007/978-1-4939-7672-0_19, © Springer Science+Business Media, LLC 2018

Leaf senescence is accompanied by increased expression of senescence-associated genes (*SAGs*) and by decreased expression of senescence-downregulated genes (*SDGs*), such as those genes related to photosynthesis and protein synthesis. *SAGs* are not only involved in turnover of nucleic acid, protein, and lipid but also involved in nutrient transport. Identification and analysis of *SAGs*, especially genome-scale investigations on gene expression during leaf senescence, make it possible to decipher the molecular mechanisms of signal perception, execution, and regulation of the leaf senescence process [6].

Due to their key roles in the regulation of leaf senescence, hundreds of *SAGs* have been cloned from various plant species including *Arabidopsis*, barley, wheat, rice, maize, soybean, and cotton [7]. In the last decade, application of genomic approaches has led to the identification of thousands of potential *SAGs* [8, 9]. Microarray analyses of *Arabidopsis* cDNAs revealed that more than 8000 genes change their expression during leaf senescence [9]. According to the results of large-scale single-pass sequencing of a senescent *Arabidopsis* leaf cDNA library, approximately 2500 unique genes are expressed in senescing leaves [8]. By using suppression subtractive hybridization, 815 senescence-associated ESTs that are upregulated at the onset of rice flag leaf senescence have been isolated [10]. A group of 90 senescence-upregulated cDNAs were identified by a differential display approach with mRNA populations from young and senescing primary barley leaves [11]. In wheat, 140 senescence-upregulated genes have been identified by cDNA microarray based on a 9 K wheat uni-gene set analysis [12].

The upregulated genes obtained from transcriptomic analyses could not be named as *SAGs* directly, however. Further confirmation by experimental data, such as quantitative RT-PCR, RNA gel blot assay, etc., is needed. And detailed functional analyses are needed to test the roles of the *SAGs* in regulating senescence. Commonly, the primary analysis is to change the expression levels of *SAGs* in plants. Overexpression of a positive senescence regulator would result in premature leaf senescence, while knockout or knockdown plants of the same gene would show delayed senescence phenotypes [13, 14]. For knocking out or knocking down the expression levels of *SAGs*, RNA interference (RNAi) and CRISPR/Cas systems are well established in practice.

Once confirmation of phenotypes of transgenic plants are done, more detailed functional characterization should be designed and performed aiming at elucidating the mechanisms, such as signaling pathways the *SAGs* are involved in, interactions with other genes or proteins, specific induction by biotic or abiotic stresses, etc.

Here we describe the basic protocols for identifying and analyzing candidate *SAGs* in wheat flag leaves by microarray analysis. The methods for functional analysis and phenotyping of transgenic plants are also described.

2 Materials

2.1 Plant Materials

Common wheat (*Triticum aestivum* L.) varieties "Shiluan 02-1" and "Kenong 199" (*see* **Note 1**).

2.2 Measurement of Chlorophyll Content

1. Scissors.
2. Fine balance scale.
3. 80% acetone for chlorophyll extraction.
4. Glass tubes.
5. UV spectrophotometer.
6. Pipettes.
7. Aluminum foil.

2.3 RNA Extraction

1. RNA extraction kit (*see* **Note 2**).
2. Motor grinder.
3. Water bath.
4. RNase-free water.
5. RNase-free DNase.
6. RNeasy Mini Kit.
7. UV spectrophotometer: NanoDrop 2000.
8. Electrophoresis set.
9. Agarose, MOPS, formaldehyde, running buffer, etc., for denaturing gel.
10. Pipettes.
11. Centrifuge.

2.4 Microarray Analysis (See Note 3)

1. The GeneChip Wheat Genome Array.
2. RNeasy Micro Kit.
3. One-cycle cDNA Synthesis Kit.
4. GeneChip IVT Labeling Kit.
5. GeneChip Sample Cleanup Module.
6. Non-stringent wash buffer: 6× SSPE, 0.01% Tween-20.
7. Stringent wash buffer: 100 mM MES, 0.1 M [Na+], 0.01% Tween-20.
8. GeneChip hybridization oven 640.
9. GeneChip fluidics station 450.
10. GeneArrayTM scanner 3000.
11. GeneChip Operating Software.
12. Pipettes.
13. Centrifuge.

Table 1
Selected oligonucleotide primers used in qRT-PCR analysis

Gene	Forward primer (5′ → 3′)	Reverse primer(5′ → 3′)
CA731144	CCCGCGAGTGGTACTTCTTC	GCCTTCCAGTAGCCGTTGC
BQ838257	CGGGCAGAAGGAGATACAGAA	CCATGCACTGCTGGTCGTACT
TaSAG3	CGACATCCGACGATTCAAC	TCGCACCACCATCCATTC
TaSAG5	CGGCTGCGAAGTTGGTTAC	ACCTGCTCCTGAGATAATGGC
β-actin	TGCTATCCTTCGTTTGGACCTT	AGCGGTTGTTGTGAGGGAGT

2.5 Quantitative RT-PCR Analysis

1. Reverse transcriptase.
2. Oligo (dT) 18 primers (*see* **Note 4**).
3. Oligonucleotide primers of *SAGs* (Table 1).
4. SYBR Green I master mix for qPCR.
5. RNase-free DNase.
6. RNase-free water.
7. Real-Time PCR system ABI7500.
8. qPCR plates and optical foils.
9. PCR tubes.
10. Pipettes.
11. Centrifuge.

2.6 Construct Generation and Transformation

1. Binary vectors (e.g., pCAMBIA-2300, Cambia, Canberra, Australia).
2. Restriction endonucleases.
3. T4 DNA ligase.
4. Competent cells of *Escherichia coli*.
5. Plasmid miniprep kit.
6. DNA gel extraction kit.
7. Reagents for tissue culture: Murashige and Skoog salt, sucrose, mannitol, inositol, glutamine, proline, phytagel, 2,4-dichlorophenoxyacetic acid, 6-benzylaminopurine, antibiotics, etc.
8. Facilities for tissue culture: Petri dishes, flasks, scissors, forceps, tweezers, laminar flow hood, etc.
9. PDS1000/He particle bombardment system.
10. Environmentally controlled incubators/growth chambers.

2.7 Phenotype Analyses of Transgenic Plants

1. DNA extraction kit.
2. RNA extraction kit.

3. Reagents and facilities for chlorophyll determination (*see* Subheading 2.2).

4. Reagents and facilities for qRT-PCR (*see* Subheading 2.5).

5. Oligonucleotide primers of marker genes (Table 1).

6. Digital camera.

3 Methods

3.1 Plant Growth and Sample Harvest

1. Wheat seeds are sowed on soil in the experimental field in early October, in Shijiazhuang, Hebei, China (*see* **Note 5**). After germination, let the plants grow through the winter time under normal cultivation conditions. Starting from March of the next year, plants are watered and fertilized normally.

2. Flag leaves are excised at different growth stages through April to May (Fig. 1A). Harvested samples are put into liquid nitrogen immediately and stored at −80 °C for future use.

3.2 Measurement of Chlorophyll Content (See Note 6)

1. Excise flag leaves from different plants, weigh with a fine balance, and record the weight.

2. Put the leaves in to 10 mL glass tubes and add 5.0 mL of 80% acetone each.

3. Wrap the glass tubes with aluminum foil, and incubate the samples in dark at 22 °C for 16 h.

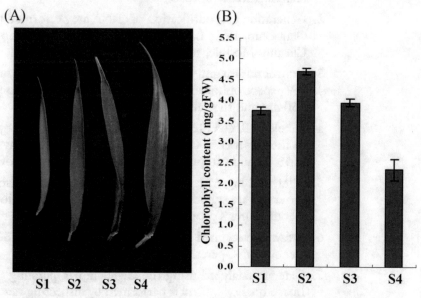

Fig. 1 Flag leaves used in microarray analysis. (**A**) Four developmental stages of wheat flag leaf. S1, young leaves; S2, mature leaves; S3, early senescent leaves; S4, late senescent leaves. These four samples were harvested on April 15 and May 1, 15, and 30 in 2011, respectively. (**B**) Chlorophyll contents of leaves at different developmental stages. The mean ± SE of four repeats are shown

4. Mix samples well, transfer the solution to 1.0 mL cuvettes, and determine the absorption using 80% acetone as control in a UV spectrophotometer at 645 and 663 nm, respectively.

5. Determine the concentration of chlorophyll using the following equation (Fig. 1B):

Chlorophyll a (mg/L) = $12.7 \times A_{663} - 2.59 \times A_{645}$

Chlorophyll b (mg/L) = $22.9 \times A_{645} - 4.67 \times A_{663}$

Total chlorophyll (mg/g) = (Ca + Cb) × 10/weight (g)

3.3 RNA Extraction

1. Total RNAs are isolated from flag leaves with a plant RNA extraction kit according to the manufacturer's instructions.

2. Concentration of RNA samples is determined by measuring the absorption at 260 nm in a UV spectrophotometer.

3. Use RNase-free DNase to remove contaminating genomic DNA in the RNA samples. Incubate the reactions at 37 °C for 1 h.

4. RNA samples are purified using the RNeasy Mini Kit (*see* **Note 7**).

5. Quality and integrity of total RNA samples are assessed by denaturing formaldehyde gel electrophoresis, where the presence of sharp 28S and 18S ribosomal RNA bands at an intensity ratio of ~2:1 indicates good integrity (*see* **Note 8**).

3.4 Microarray Analysis (See Note 9)

1. Five milligram of total RNA is used to generate labeled cRNA using the One-Cycle cDNA Synthesis Kit according to the manufacturer's protocol.

2. Generation and purification of cRNA are performed using the GeneChip IVT Labeling Kit and the GeneChip Sample Cleanup Module, respectively.

3. Concentration and qualification of cRNAs are determined in a UV spectrophotometer by measuring the absorption at 260 nm and the value of A_{260}/A_{280}.

4. Labeled cRNAs (15 μg) are then fragmented and hybridized to the Affymetrix GeneChip® Wheat Genome Array for 16 h at 45 °C and 60 rpm in a GeneChip hybridization oven.

5. After washing with non-stringent and stringent wash buffers, the arrays are stained in the Fluidics Station 450 following the manufacturer's protocol.

6. Arrays are scanned using the GeneChip® Scanner 3000 with GeneChip® Operating Software.

7. Data are analyzed using the GeneChip® Operating Software. Image quality control is performed by inspecting raw intensity (DAT) files for scratches/smears and uniform performance of the B2 oligo around the border of each image.

8. Log-transformed expression values are subjected to analysis of variance (ANOVA) where filters are set to $p < 0.01$ and differential expression (DE) either less than −3 or greater than +3.

3.5 Quantitative RT-PCR Analysis (See Note 10)

1. Using purified total RNAs from Subheading 3.3, cDNAs are generated by reverse transcription reaction following the manufacturer's protocols.

2. Prepare quantitative real-time PCR mixes. A typical reaction mix (20 µL): 10.0 µL 2× SYBR Premix Ex Taq, 2.0 µL of 0.5 mM primer pairs, 2.0 µL cDNA, 0.4 µL of ROX Reference Dye II.

3. Run quantitative real-time PCR reactions in a 96-well PCR plate with an Applied Biosystems 7500 Real-Time PCR system using a SYBR Green I master mix. A typical cycler condition: Initial denaturation for 5 min at 94 °C, followed by 40 cycles each comprising denaturation for 10 s at 94 °C, annealing for 10 s at 60 °C, and elongation for 20 s at 72 °C.

4. Calculate the fold changes in gene expression using the comparative threshold cycle (Ct) method. PCR experiments are repeated at least three times (Fig. 2).

3.6 Construct Generation and Transformation

1. For overexpression constructs, coding sequences of specific *SAG* genes are amplified by PCR from cDNA templates and inserted into a binary vector by conventional molecular cloning protocols. To make RNAi constructs, around 300 bp gene-specific fragments are amplified and then inserted into the binary vector in both forward and reverse orientations (*see* **Note 11**).

2. Embryogenic calli of wheat variety Kenong 199 are bombarded with the overexpression or RNAi constructs together with a UBI::BAR selectable marker plasmid at a 2:1 molar ratio. A PDS1000/He particle bombardment system is used with a target distance of 6.0 cm from the stopping plate at the helium pressure 1100 psi, as described previously [13].

3. Transformants are selected on MS medium containing appropriate antibiotics. After screening and examining by

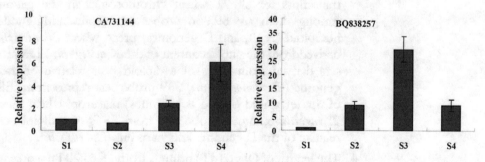

Fig. 2 Two examples showing the qRT-PCR results of candidate *SAGs*. The relative transcription levels of two candidate *SAGs*, *CA731144* and *BQ838257*, were determined through qRT-PCR using RNA from wheat flag leaves (the bars represent mean ± SE, and β-actin was used as an internal control)

conventional PCR and Southern blot assay, the positive transgenic plants are transplanted into soil. The plants are grown
under normal conditions until harvest.

**3.7 Phenotype
Analyses
of Transgenic Plants
(See Note 12)**

1. Total RNA of transgenic plants (both overexpression lines and
   RNAi lines) and wild-type Kenong 199 are isolated and
   reverse-transcribed. Then qRT-PCR is performed to examine
   the transcription levels of the target *SAGs* and senescence
   marker genes *TaSAG3* and *TaSAG5* (*see* **Note 13**).

2. Senescence-related phenotypes are recorded with a digital
   camera.

3. Depending on research purposes, senescence parameters
   including chlorophyll content are determined and compared
   between transgenic plants and wild type (*see* **Note 14**).

4 Notes

1. Common wheat "Shiluan 02-1" is one of the most cultivated
   varieties in Northern China, which shows high protein and
   wheat gluten contents. This variety, however, shows a little
   early senescence in flag leaves during the grain filling stage.
   "Kenong 199" is a winter wheat variety commercially released
   in 2006 and is one of the best recipients for transformations by
   bombardment methods.

2. RNA isolation could be done using an RNA extraction kit or
   other extraction reagents, such as TRIzol. We use a TRIzol
   Max™ RNA Isolation Kit (Invitrogen, Carlsbad, USA) in this
   experiment.

3. There are a couple of commercial microarray chips available.
   We use the Affymetrix (Santa Clara, California, USA) products because there is an Affymetrix GeneArray facility on cam-
   pus of Hebei Normal University. The GeneChip Wheat
   Genome Array includes 61,127 probe sets representing 55,052
   transcripts for all 21 wheat chromosomes in the genome.
   Among them are 59,356 probes sets representing modern
   hexaploid (A, B, and D genomes) bread wheat (*T. aestivum*)
   derived from the public content of the *T. aestivum*, 1215 probe
   sets derived from ESTs of a diploid near relative of the A
   genome (*T. monococcum*), 539 probe sets representing ESTs
   of the tetraploid (A and B genomes) macaroni wheat species
   T. turgidum, and 5 probe sets from ESTs of a diploid near
   relative of the D genome known as *Aegilops tauschii*.

4. The length of Oligo (dT) primers from 15 to 20 bp works well
   for reverse transcription reaction.

5. "Shiluan 02-1" is a winter wheat, so it should be sowed in
   autumn, around October 5–15 in Northern China. And it is

watered once before the snowing season, around November 10. In a regular growing season, wheat shoots in early April, flag leaves emerge in mid-April, grain filling starts from early May, and wheat is harvested around June 1.

6. Measurement of chlorophyll content could be performed by different methods. We use the classical acetone extraction protocol. The detailed procedures and steps have been described previously [15].

7. Up to 45 μg total can be used for the RNeasy Mini Kit. In the case of more than 45 μg, total RNA is used, and the RNeasy Micro Kit is recommended.

8. The quality and integrity of total RNA are crucial for microarray analysis, so extra caution is needed for the RNA purification procedure. We recommend using denaturing formaldehyde gel to check purified RNAs before performing labeling and hybridization.

9. Array analysis should be performed by well-trained technicians or specialized person under the manufacturer's instructions, and quality control should be monitored at all times during the processes of labeling, hybridization, washing, scanning, etc.

10. To validate the data of microarray analysis, quantitative RT-PCR analysis or RNA gel blot array is generally performed. Around 20 genes/ESTs are necessary for confirmation. For the experiment presented in Fig. 2, we used qRT-PCR analysis, and all the 22 genes examined showed the same expression pattern as that from the microarray analysis.

11. For overexpression analysis, to avoid potential complications in interpreting pleiotropic phenotypes caused by constitutive overexpression of SAGs, a plant native promoter-drove coding sequence construct or hormone-regulated transcriptional induction system can be used for functional analysis of SAGs [16]. For RNAi constructs, the specific sequences used in the vector should be carefully selected. The sequence should be unique, that means it will not target other sites except for the target SAG.

12. To confirm and validate the functions of SAGs with transgenic plants, genetic, physiological, biochemical, molecular, and cellular evidences are indispensable. To save space and avoid overlapping with the other chapters, only determination of gene expression and chlorophyll contents is discussed in this section.

13. It is expected to see both target SAGs and marker genes show high expression levels in overexpression lines, and low levels or no expression in RNAi lines, compared to wild-type plants. To avoid accidental transgenic events occurring during

transformation, at least three independent transgenic lines should be examined.

14. Besides chlorophyll data, Fv/Fm data, malondialdehyde content data, electrolyte leakage assay, trypan blue staining assay, etc., are often used for assessment of senescence status of the plants.

Acknowledgments

This work was supported by the grants from the National Science Foundation of China (Grant No. 31671694) and the Support Program for Hundreds of Outstanding Innovative Talents in Higher Education Institutions of Hebei Province (III) (Grant No. SLRC2017043) to CZ.

References

1. Gan S, Amasino RM (1997) Making sense of senescence (molecular genetic regulation and manipulation of leaf senescence). Plant Physiol 113:313–319

2. Lim PO, Kim HJ, Nam HG (2007) Leaf senescence. Annu Rev Plant Biol 58:115–136

3. Guo Y, Gan S (2005) Leaf senescence: signals, execution, and regulation. Curr Top Dev Biol 71:83–112

4. Zhang H, Zhou C (2013) Signal transduction in leaf senescence. Plant Mol Biol 82:539–545

5. Wu X, Kuai B, Jia J, Jing H (2012) Regulation of leaf senescence and crop genetic improvement. J Integr Plant Biol 54:936–352

6. Zhou C, Gan S (2009) Senescence. In: Pua EC, Davey MR (eds) Plant developmental biology: biotechnological perspectives. Springer, Berlin Heidelberg, pp 151–169

7. Guo Y, Gan S (2014) Translational researches on leaf senescence for enhancing plant productivity and quality. J Exp Bot 65:3901–3913

8. Guo Y, Cai Z, Gan S (2004) Transcriptome of Arabidopsis leaf senescence. Plant Cell Environ 27:521–549

9. Breeze E, Harrison E, McHattie S, Hughes L, Hickman R et al (2011) High-resolution temporal profiling of transcripts during Arabidopsis leaf senescence reveals a distinct chronology of processes and regulation. Plant Cell 23:873–894

10. Liu L, Zhou Y, Zhou G, Ye R, Zhao L et al (2008) Identification of early senescence-associated genes in rice flag leaves. Plant Mol Biol 67:37–55

11. Ay N, Clauss K, Barth O, Humbeck K (2008) Identification and characterization of novel senescence-associated genes from barley (*Hordeum vulgare*) primary leaves. Plant Biol (Stuttg) 10(Suppl 1):121–135

12. Gregersen PL, Holm PB (2007) Transcriptome analysis of senescence in the flag leaf of wheat (*Triticum aestivum* L.) Plant Biotechnol J 5:192–206

13. Uauy C, Distelfeld A, Fahima T, Blechl A, Dubcovsky J (2006) A NAC gene regulating senescence improves grain protein, zinc, and iron content in wheat. Science 314:1298–1301

14. Liang C, Wang Y, Zhu Y et al (2014) OsNAP connects abscisic acid and leaf senescence by fine-tuning abscisic acid biosynthesis and directly targeting senescence-associated genes in rice. Proc Natl Acad Sci U S A 111:10013–10018

15. Zhao L, Zhang H, Zhang B, Bai X, Zhou C (2012) Physiological and molecular changes of detached wheat leaves in responding to various treatments. J Integr Plant Biol 54:567–576

16. Zhou C, Cai Z, Guo Y, Gan S (2009) An arabidopsis mitogen-activated protein kinase cascade, MKK9-MPK6, plays a role in leaf senescence. Plant Physiol 150:167–177

Chapter 20

Isolation, Purification, and Detection of Micro RNAs in Plant Senescence

Yujun Ren and Ying Miao

Abstract

Micro RNAs (miRNAs) are small noncoding RNA molecules that function in transcriptional level to regulate gene expression both in plants and animals. Increasing researches have shown that miRNAs are key regulators in plant development and stress responses, and emerging evidence indicates the potential role of miRNAs on plant senescence. In this chapter we summarize the daily methods used for identification and study of miRNAs in plants, including the isolation of total RNA, the purification of miRNAs, and the methods used to detect miRNAs in plants. The committed steps or modifications of these methods used in plant senescence research are noted.

Key words Micro RNAs, Plant senescence, Total RNA isolation, Micro RNA purification, Micro RNA detection

1 Introduction

Plant senescence is a gene-controlled cell death procedure that can be affected or induced by internal and external environmental factors. Cross talks between the developmental and stress-related signal pathways in plants facilitate the optimization and fine-tuning of leaf senescence in response to the challenging environmental conditions, which have important impacts in agricultural industry [1, 2]. The regulatory mechanisms of plant senescence have been widely characterized in the past decades, and increasing researches have emphasized the importance of transcription factors in signal transduction networks related to the plant senescence process [1, 3]; among them, the WRKY and NAC transcription factor families play central roles in regulating leaf senescence [3]. Recently, emerging evidence declared the effects of small RNA in mediating gene expression related to leaf senescence [4], which presents a new window for researchers to decipher the secret of senescence in plants.

Micro RNAs (miRNAs) are a class of small endogenous noncoding RNAs that are predominantly 21–23 nucleotides (nt) in

Yongfeng Guo (ed.), *Plant Senescence: Methods and Protocols*, Methods in Molecular Biology, vol. 1744,
https://doi.org/10.1007/978-1-4939-7672-0_20, © Springer Science+Business Media, LLC 2018

length [5, 6]. Encoded by eukaryotic nuclear DNA in plants and animals, miRNAs function via base-pairing with complementary sequence within messenger RNA (mRNA) molecules [5–7]. As a result, these mRNA strands are silenced because they cannot be translated into proteins by ribosome [6–8]. Plant miRNAs usually have perfect pairing with their targets and induce gene repression through degradation of their target transcripts; the pairing regions to mRNA can be both coding regions and untranslated regions [6–8]. A given miRNA may have multiple mRNA targets, and a given target might similarly be targeted by multiple miRNAs, which shows a combinatorial regulation of miRNA to control gene expression [5–7]. Mature miRNAs are derived from single-stranded RNA precursors that possess an imperfect stem-loop secondary structure [5, 6]. These hairpins are processed by a Dicer-like enzyme and are loaded into the RNA-induced silencing complexes (RISC), where they direct cleavage of complementary mRNAs in the cytoplasm [6]. Plant miRNAs play important roles in multiple biological processes [8], such as leaf morphogenesis and polarity [9, 10], floral organ identity [11], and stress responses [12]. Recently, several studies showed that miRNAs also play roles in regulating leaf senescence, such as *miR164* [13], *miR319* [14], and *miR390* [15, 16]. And in our research, we found that micro RNA *miR840a* regulates leaf senescence by controlling the expression of several downstream senescence-related genes (Ren et al., unpublished data). These results indicate the possibility to use miRNA to regulate leaf senescence in the future for agricultural production improvements.

The appropriate manipulation of micro RNAs is important for their functional study in plants and animals. In this chapter, we outline several methods for isolation and detection of micro RNAs in plants, including the extraction of total RNA, the purification of micro RNAs, and the detection of specific micro RNA in total RNA or purified miRNAs pools. The committed experiment steps and modifications of these methods specifically used in micro RNA study in plant senescence are emphasized.

2   Materials

All the reagents and solutions used for RNA experiments should be RNase-free. Disposable plastic gloves should be worn and changed frequently to avoid RNase contamination. Handle organic compounds in a fume hood and avoid splashing them to skin or eyes.

2.1   Total RNA Isolation

2.1.1   General Materials

1. RNaseZap® RNase Decontamination Solution (Life Technologies).

2. RNase-free double-distilled water: Treated with 0.1% DEPC.

3. 75% ethanol: Diluted from absolute ethanol with RNase-free water.

4. Nuclease-free water.

5. 1× TAE buffer: 40 mM Tris, 20 mM acetic acid, 1 mM EDTA, pH 8.0, prepared fresh from 50× TAE stock solution.

6. 1.2% agarose gel: Buffered with 1× TAE buffer.

7. RNA loading buffer: 98% deionized formamide, 10 mM EDTA, 1 mg/mL xylene cyanol, 1 mg/mL bromophenol blue.

8. RNA staining solution.

9. Liquid nitrogen.

10. ND1000 spectrophotometer (NanoDrop Technologies).

11. 1.5 mL RNase-free polypropylene microcentrifuge tubes.

12. RNase-free pipette tips.

13. Benchtop centrifuges (at 4 °C and RT).

14. RNase-free mortar and pestle (Heated at 200 °C for at least 4 h).

15. Nucleic acid electrophoresis devices.

16. Water bath.

17. Vortex.

18. PCR machine.

2.1.2 Total RNA Isolation Using TRIzol Reagent

1. TRIzol reagent (Life Technologies).

2. Chloroform.

3. Isopropyl alcohol.

4. 100% deionized formamide.

2.1.3 Total RNA Isolation Using LiCl/SDS/ Phenol

1. LiCl extraction buffer: 100 mM LiCl, 1% SDS, 100 mM Tris–HCl, pH 9.0, 10 mM EDTA.

2. Chloroform.

3. Phenol, pH 8.0.

4. 8 M LiCl.

2.1.4 Total RNA Isolation Using CTAB/Phenol

1. CTAB extraction buffer: 2% CTAB, 25 mM EDTA, pH 8.0, 2 M NaCl, 4% PVP-40 (v/v), 100 mM Tris–HCl, pH 8.0, 0.05 g/L spermidine.

2. β-mercaptoethanol (β-ME).

3. Acid-saturated phenol.

4. Isopropyl alcohol.

5. Chloroform-isoamyl alcohol (24:1; v/v).

6. 15 mL RNase-free polypropylene centrifuge tubes.

2.2 Micro RNA Purification

2.2.1 PEG Method

1. 3 M NaCl/30% polyethylene glycol (PEG8000) solution.

2. Absolute ethanol.

3. Phenol-chloroform-isoamyl alcohol: 25:24:1 (v/v/v).

4. 3 M NaAOc, pH 5.2.

5. RNase-free DNase I with 10× reaction buffer (1 U/μL, ThermoFisher).

6. RiboLock RNase Inhibitor (40 U/μL, ThermoFisher).

7. Chloroform.

2.2.2 LiCl Method

1. Isopropyl alcohol.

2. 8 M LiCl.

3. RNase-free DNase I with 10× reaction buffer (1 U/μL, ThermoFisher).

4. RiboLock RNase Inhibitor (40 U/μL, ThermoFisher).

5. Chloroform.

6. 3 M NaAOc, pH 5.2.

2.3 Micro RNA Detection by Northern Blotting

2.3.1 General Materials

1. RNase-free double-distilled water (treated with 0.1% DEPC).

2. 0.5× TBE buffer: 44.5 mM Tris-borate, 44.5 mM boric acid, 10 mM EDTA, pH 8.0, prepare fresh from 10× TBE stock solution.

3. 1× TBE buffer: 89 mM Tris-borate, 89 mM boric acid, 20 mM EDTA, pH 8.0, prepare fresh from 10× TBE stock solution.

4. TEMED: Tetramethylenediamine.

5. 10% APS: Ammonium persulfate, fresh made each time when use.

6. RNA loading buffer: 98% deionized formamide, 10 mM EDTA, 1 mg/mL xylene cyanol, 1 mg/mL bromophenol blue.

7. 17% denatured polyacrylamide gel solution: prepare 25 mL by adding 10.63 mL of 40% polyacrylamide (acrylamide:bis, 19:1; w/v), 10.5 g of urea, 2.5 mL of 10× TBE buffer, and 4.43 mL RNase-free water.

8. Hybridization buffer: 50 mL of 100% deionized formamide, 25 mL of 1 M Na-PO_4 pH 7.2, 5 mL of 5 M NaCl, 200 mL of 0.5 M EDTA, 7 g SDS, 20 mL of RNase-free water.

9. 1% SDS solution.

10. Wash solution: 2× saline sodium citrate (SSC)/0.2% SDS.

11. Stringent wash solution: 0.5× SSC/0.1% SDS.

12. Nuclease-free water (1 L, Ambion).

13. Water bath (at 50 and 100 °C) and ice bath.

14. Hybond-N⁺ nylon membrane.

15. Bio-Rad Protean II vertical electrophoresis system.

16. UV cross-link machine.

17. Nucleic acid hybridization oven and tubes.

18. PhosphorImager and screens.

19. Geiger counter.

2.3.2 Micro RNA Detection Using α-³²P dCTP Labeling Probe

1. Terminal deoxynucleotidyl transferase and 5× reaction buffer (Promega or Ambion).

2. α-^{32}P *dCTP* (50 µCi).

3. G-25 MicroSpin columns (GE Healthcare Life Science).

2.3.3 Micro RNA Detection Using α-³²P UTP Labeling Probe

1. rATP/rGTP/rCTP (2.5 mM each) .

2. UTP (1 mM).

3. α-^{32}P *UTP* (50 µCi).

4. T7 or Sp6 RNA polymerase with 5× transcription buffer (Promega or Ambion).

5. 0.1 M DTT.

6. RNase inhibitor (Promega or Ambion).

7. RNase-free DNase (Promega or Ambion).

8. 3 M NaAOc (pH 5.2).

9. Absolute ethanol.

10. 200 mM carbonate buffer: 0.672 g $NaHCO_3$, 1.277 g Na_2CO_3, mix with Nuclease-free water to 100 mL.

2.4 Micro RNA Detection by Stem-Loop RT-PCR

1. Micro RNA specific RT-PCR primers.

2. Nuclease-free water.

3. Reverse transcriptase with 10× reaction buffer (Life Technologies).

4. 100 mM dNTPs: ATP/CTP/GTP/TTP, 25 mM each.

5. RNase Inhibitor (Life Technologies).

6. 20× TaqMan Micro RNA Assay (Life Technologies).

7. 2× TaqMan Universal PCR Master Mix (Life Technologies).

8. 0.2 mL RNase-free thin wall PCR tubes.

9. Ice bath.

10. PCR machine and real-time PCR system.

3   Methods

We outline three RNA isolation methods and two following micro RNA purification methods here. The TRIzol method is commonly used for total RNA and micro RNA isolation from most plant species [17]. For isolation of total RNA and micro RNAs from materials with high polysaccharide content, the LiCl/SDS/Phenol and CTAB/Phenol methods are strongly recommended. The LiCl/

SDS/Phenol method has been successfully used in isolation of total RNA and micro RNAs from *Arabidopsis* (*Arabidopsis thaliana*), rice (*Oryza sativa*), tobacco (*Nicotiana tabacum*), barley (*Hordeum vulgare*), wheat (*Triticum aestivum*), prickly pear (*Opuntia robusta*), banana (*Musa paradisiaca*), tomato (*Solanum lycopersicum*), and so on [13, 18, 19]. The CTAB/Phenol method has been used in total RNA and micro RNA isolation from *Arabidopsis* (*Arabidopsis thaliana*), narcissus species, rhododendron (*Rhododendron Linn.*), rubber tree (*Hevea brasiliensis*), cassava (*Manihot esculenta*), banana (*Musa paradisiaca*), pine tree (*Pinus densiflora*), and jasmine (*Jasminum sambac*) [20–22].

3.1 Total RNA Isolation

3.1.1 TRIzol Method

1. Spray 1:10 (v/v) diluted RNase*Zap*® solution (with RNase-free water) to the surface of the experiment table to destroy RNase contamination, and then clean with pre-wetted RNase-free paper towel.

2. Grind 50–100 mg plant tissue into fine powder in liquid nitrogen using an RNase-free mortar and pestle set. Transfer the tissue powder to a precooled 1.5 mL RNase-free microcentrifuge tube, and add 1.0 mL of TRIzol reagent. Vigorously shake the tube to completely homogenize the sample, and then put the tube into liquid nitrogen to wait for other samples. When all the samples are processed, take them out from liquid nitrogen, and lay them down on the table to let them thaw to room temperature (RT) and continuously stay for another 5 min (*see* **Notes 1** and **2**).

3. Centrifuge the samples at 12,000 × *g* for 10 min at 4 °C, and transfer the cleared supernatant to a new tube.

4. Add 200 μL of chloroform, and vigorously shake the tube by hand for 15–20 s, and then keep the samples at RT for 2–3 min.

5. Centrifuge the tubes at 12,000 × *g* for 15 min at 4 °C, and transfer 400 μL upper aqueous phase to a new tube (*see* **Note 3**).

6. Add 400 μL of isopropyl alcohol to each tube, and invert the tubes 20 times by hand. Incubate the samples at RT for 10 min or keep at −20 °C overnight.

7. Centrifuge at 16,000 × *g* for 15 min at 4 °C; remove the supernatant.

8. Add 1 mL of 75% ethanol, cap the tube and invert by hand for 10 s, and then stay at RT for 5 min (*see* **Note 4**).

9. Centrifuge at 9000 × *g* for 5 min at 4 °C. Discard the wash.

10. Pulse spin and pipette off the residual ethanol; air-dry the RNA pellet in the tubes for 5 min (*see* **Note 5**).

11. Resuspend the RNA pellet in 30 μL RNase-free water or 100% deionized formamide, and incubate in a water bath at 60 °C for 10 min (*see* **Note 6**).

12. Run 2.0 μL of each total RNA sample on a 1.2% agarose gel in 1× TAE buffer to check the quality of the isolated total RNA and determine the RNA concentration of each sample on a NanoDrop ND1000 spectrophotometer (*see* **Note 7**).

3.1.2 LiCl/SDS/ Phenol Method

1. Use 1:10 (v/v) diluted RNase*Zap*® solution to clean the table surface for RNA isolation, and rinse with pre-wetted RNase-free paper towel.

2. In an RNase-free mortar and pestle set, grind 0.1 g fresh or prefrozen tissue (stored at −80 °C) into fine powder in liquid nitrogen, transfer the powder to a precooled 1.5 mL RNase-free microcentrifuge tube, and add 500 μL of LiCl extraction buffer and 500 μL of phenol (pH 8.0) (*see* **Note 8**).

3. Vigorously shake the sample on a vortex for 30 s. Store each sample in liquid nitrogen until all samples are processed (*see* **Note 2**).

4. Let all samples return to room temperature (RT), and then incubate at 60 °C in a water bath for 5 min.

5. Centrifuge the samples at 12,000 × *g* for 10 min at 4 °C.

6. Carefully transfer the supernatant to a new 1.5 mL tube.

7. Add 500 μL of chloroform to each tube, and mix on a vortex for 1 min; then keep the samples at RT for 5 min.

8. Centrifuge at 12,000 × *g* for 15 min at 4 °C, and transfer the cleared upper phase to a new tube (*see* **Note 9**).

9. Add 1/3 volume of 8 M LiCl to the tube and mix by inverting samples 20 times. Incubate overnight (O/N) at 4 °C.

10. Centrifuge at 4000 × *g* for 10 min at RT, and transfer the supernatant containing micro RNAs (LMW RNAs) to a 1.5 mL tube, and store at 4 °C until use (for **step 1** in Subheading 3.2.2); the remaining pellet contains high molecular weight RNAs (HMW RNAs).

11. Add 1 mL of ice-cold 75% ethanol to the pellet, invert tubes 10–15 times, and then centrifuge at 9000 × *g* for 5 min at 4 °C (*see* **Note 4**).

12. Remove the supernatant, pulse spin, and pipette off the residual ethanol in the tube; air-dry the cleared pellet for 5 min (*see* **Note 5**).

13. Resuspend the RNA pellet in 30 μL RNA-free water, and incubate at 60 °C in a water bath for 10 min.

14. Run 2.0 μL of each HMW RNA sample on a 1.2% agarose gel in 1× TAE buffer to check the quality of the isolated HMW RNAs and determine the RNA concentration of each sample on a NanoDrop ND1000 spectrophotometer (*see* **Note 7**).

3.1.3 CTAB/
Phenol Method

1. Decontaminate the table surface for RNA extraction with 1:10 (v/v) diluted RNaseZap solution, and then clean with RNase-free pre-wetted paper towel.

2. Grind 0.5 g fresh or prefrozen tissue (stored at −80 °C) to fine powder with liquid nitrogen in a RNase-free mortar and pestle set. Transfer the tissue powder to an 15 mL RNase-free polypropylene centrifuge tube, and add 5 mL of extraction buffer pre-warmed to 65 °C and 200 μL of β-ME to the tube (*see* **Note 8**).

3. Mix vigorously on a vortex for 1 min, and then let it stay at room temperature (RT) for 5 min.

4. Add an equal volume of acid-saturated phenol and vortex vigorously for 1 min.

5. Centrifuge the sample at 12,000 × g for 15 min at 4 °C, and transfer the clean supernatant to a new 15 mL tube.

6. Add an equal volume of chloroform:isoamyl alcohol (24:1, v/v) to the tube and vortex vigorously for 1 min.

7. Centrifuge at 12,000 × g for 10 min at 4 °C, and transfer the supernatant to a new 15 mL tube (*see* **Note 9**).

8. Repeat **steps 6** and **7** once, respectively.

9. Add an equal volume of isopropyl alcohol to the tube, invert 20 times, and let it stay at RT for 10 min or −20 °C overnight (O/N).

10. Centrifuge at 16,000 × g for 10 min at 4 °C, remove the supernatant, and wash RNA pellet with 5 mL of ice-cold 75% ethanol (*see* **Note 4**).

11. Centrifuge at 9000 × g for 5 min at 4 °C. Discard the wash solution.

12. Pulse spin the tube, pipette off the residual ethanol, and air-dry the RNA pellet for 5–10 min (*see* **Note 5**).

13. Resuspend RNA pellet in 100 μL RNase-free water, and then incubate at 60 °C in a water bath for 15 min.

14. Run 2.0 μL of total RNA on a 1.2% agarose gel in 1× TAE buffer to check the quality of total RNA and quantify RNA concentration on a NanoDrop™ ND1000 spectrophotometer (*see* **Note 6**).

3.2 Micro RNA Purification

3.2.1 PEG Method (See Note 10)

1. Begin from the collected supernatant from **step 5** in Subheading 3.1.1 or from **step 8** in Subheading 3.1.2 or from **step 8** in Subheading 3.1.3.

2. Add 1/3 volume of 3 M NaCl/30% polyethylene glycol (PEG8000) solution to the tube, vortex for 30 s, and incubate on ice for 30 min.

3. Centrifuge at 13,000 × g for 10 min at 4 °C.

4. Transfer the supernatant containing micro RNAs (LMW RNAs) to a new tube. Store the pellet containing high molecular weight RNAs (HMW RNAs) in absolute ethanol until use (*see* **Note 11**).

5. Add an equal volume of phenol-chloroform-isoamyl alcohol (25:24:1; v/v/v) to the collected supernatant, vortex for 1 min, and centrifuge at max speed in a microfuge at 4 °C for 10 min.

6. Transfer supernatant to a new tube, and precipitate micro RNAs (LMW RNAs) by adding 1/10 volume (the final volume after adding ethanol) of 3 M NaAOc (pH 5.2) and 2.5 volume (the volume before adding ethanol) of absolute ethanol to the tube. Mix by inverting 10–20 times, and incubate at −20 °C for at least 30 min or overnight (O/N).

7. Centrifuge at max speed in a microfuge for 15 min at 4 °C.

8. Discard supernatant, wash the pellet with ice-cold 75% ethanol, and air-dry the pellet at RT for 2 min.

9. Add 20–100 μL nuclease-free water to dissolve the pellet, and incubate at 60 °C for 10 min.

10. Set up a DNA digestion reaction to remove DNA contamination to the isolated micro RNAs (below), mix by gently pipetting, and incubate at 37 °C for 40 min.

Reagent	Volume (μL)
Isolated micro RNAs	20
Nuclease-free water	31
DNase reaction buffer (10×)	6
RNase-free DNase I (1 U/μL)	2
RNase inhibitor (40 U/μL)	1

11. Add 90 μL of nuclease-free water to the reaction tube to adjust the total volume to 150 μL, and then add an equal volume of chloroform and mix well by vortexing the tube for 30 s. Incubate at RT for 3 min and centrifuge at 13,000 × g for 10 min at 4 °C.

12. Transfer the upper aqueous phase to a new 1.5 mL tube.

13. Add 1/10 volume (the final volume after adding ethanol) of 3 M NaAOc (pH 5.2) and 3 volume (the volume before adding ethanol) of absolute ethanol, mix by inverting the tube 20 times, and incubate at −20 °C for at least 30 min.

14. Centrifuge at 16,000 × g for 20 min at 4 °C; discard the supernatant.

15. Add 1 mL of ice-cold 75% ethanol to wash the pellet, and then centrifuge at $13,000 \times g$ for 5 min at 4 °C.

16. Discard the supernatant, pulse spin the tube, and pipette off the residual ethanol; air-dry the pellet for 2 min at RT.

17. Dissolve with 20 μL nuclease-free water, and store the purified micro RNAs at −80 °C or directly proceed to downstream application (*see* **Note 12**).

3.2.2 LiCl Method
(See Note 13)

1. Begin from the collected supernatant of **step 10** in Subheading 3.1.2, add an equal volume of isopropyl alcohol to the tube, invert for 10–20 times, and keep at −20 °C for 4 h or overnight (O/N).

2. Centrifuge at $14,000 \times g$ for 10 min at 4 °C and discard the supernatant.

3. Add 1 mL of ice-cold 75% ethanol, invert the tube 10 times, and then centrifuge at $13,000 \times g$ for 5 min at 4 °C.

4. Remove the supernatant, pulse spin, and pipette off the residue liquid.

5. Air-dry the pellet for 2 min at RT.

6. Add 500 μL of 8 M LiCl to the pellet, and shake overnight on a benchtop shaker at 4 °C.

7. Centrifuge at $4000 \times g$ for 30 min at 4 °C.

8. Transfer the supernatant to a new tube, mix with an equal volume of isopropyl alcohol, and keep at −20 °C for 4 h or O/N.

9. Centrifuge at $14,000 \times g$ for 30 min at 4 °C and discard the supernatant.

10. Add 1 mL of ice-cold 75% ethanol to the tube, shake by hand for 10 s, and then centrifuge at $14,000 \times g$ for 5 min at 4 °C.

11. Discard the supernatant, pulse spin, and pipette off the residual ethanol. Air-dry the pellet for 2 min at RT.

12. Resuspend the pellet in 20 μL nuclease-free water.

13. Set up a DNA digestion reaction to eliminate DNA contamination to the isolated micro RNAs.

Reagent	Volume (μL)
Isolated micro RNAs	20
Nuclease-free water	31
DNase reaction buffer (10×)	6
RNase-free DNase I (1 U/μL)	2
RNase inhibitor (40 U/μL)	1

14. Mix by pipetting and incubate at 37 °C for 40 min.

15. Add 90 μL of nuclease-free water to the reaction tube to adjust the total volume to 150 μL, and then add an equal volume of chloroform and mix well by vortexing the reaction tube for 30 s. Incubate at RT for 3 min, and centrifuge at $16,000 \times g$ for 10 min at 4 °C.

16. Transfer the upper aqueous phase to a new 1.5 mL tube.

17. Add 1/10 volume (the final volume after adding ethanol) of 3 M NaAOc (pH 5.2) and 3 volume (the volume before adding ethanol) of absolute ethanol, mix by inverting the tube 20 times, and incubate at −20 °C for at least 30 min.

18. Centrifuge at $16,000 \times g$ for 20 min at 4 °C; discard the supernatant.

19. Add 1 mL of ice-cold 75% ethanol to wash the pellet, and then centrifuge at $16,000 \times g$ for 5 min at 4 °C.

20. Discard the supernatant, pulse spin, and pipette off the residual ethanol; air-dry the pellet for 2 min at RT.

21. Dissolve the pellet with 20 μL nuclease-free water, and store the purified micro RNAs at −80 °C or directly proceed to downstream application (*see* **Note 12**).

3.3 Micro RNA Detection

3.3.1 Micro RNA Detection by α-³²P dCTP Labeled DNA Probe

1. Prepare 25 mL 17% denatured polyacrylamide gel solution. Mix thoroughly by gently swirling. Incubate at 40 °C for 20 min to help the urea dissolve completely.

2. Add 12.5 μL of TEMED and 150 μL of 10% APS to the above solution, and mix by gently swirling (*see* **Note 14**).

3. Quickly pour the gel solution into a preassembled RNase-free gel rig (Bio-Rad Protean II vertical electrophoresis system or similar), place the gel comb into the gel rig, and incubate at RT for 1.5 h to ensure the complete polymerization of the gel.

4. Transfer the gel rig to an electrophoresis tank containing 0.5× TBE buffer, remove the gel comb, and rinse the sample wells by using a syringe to flush out the undissolved urea from each well.

5. Pre-run the gel at 140 V for 1 h to warm the gel, and let the undissolved urea migrate to the top of the gel; use a syringe to flush out each well again.

6. Prepare RNA samples by mixing the purified micro RNAs (5–10 μg, about 10 μL) with an equal volume of RNA loading buffer, incubate at 80 °C for 5 min, and then immediately put on ice for 5 min.

7. Briefly centrifuge the tubes, directly load each sample to the pre-warmed gel, and run at 200 V at RT for 1 h and then 400 V until the dye has migrated to the bottom of the gel (*see* **Note 15**).

8. Prepare a Hybond-N⁺ nylon membrane which has the same size with the gel and pre-wetted with 1x TBE buffer.

9. Carefully disassemble the gel rig, and transfer the gel to an electroblotting apparatus containing 1× TBE buffer. Place the pre-wetted nylon membrane directly above the gel, and transfer micro RNAs from the gel to the membrane at 40 V for 2 h at RT.

10. Carefully disassemble the electroblotting apparatus, use a UV cross-link machine to cross-link micro RNAs to the membrane (the RNA side up toward the UV source) (*see* **Note 16**).

11. Prepare hybridization solution, set the nucleic acid hybridization oven to 42 °C, and pre-warm the hybridization solution to 42 °C.

12. Place the membrane RNA-side face up in the hybridization tube using tweezers, and add 25 mL of pre-warmed hybridization solution. Incubate by rotating the hybridization tube in the oven to pre-hybridize the RNA-blotted membrane at 42 °C for at least 4 h while preparing the DNA or RNA probe used for hybridization.

13. Prepare α-^{32}P dCTP-labeled DNA probe (reverse complement with specific micro RNA) in a 50 µL labeling reaction. The probe is 5′ end-labeled by α-^{32}P dCTP with terminal deoxynucleotide transferase.

Reagent	Volume (µL)
DNA oligonucleotide (10 µM)	2.0
Reaction buffer (5×)	10.0
α-^{32}P dCTP (50 µCi)	5.0
Terminal deoxynucleotide transferase	1.0
Nuclease-free water	32.0

14. Mix and incubate the reaction at 37 °C for 60 min, and purify the probe by using a G-25 MicroSpin column following the manufacturer's instruction.

15. Add the purified labeling probe to the pre-hybridized membrane, and hybridize for 16–20 h at 42 °C with constant rotation.

16. Discard hybridization solution, and wash the membrane with 30 mL of pre-warmed wash solution in the tube for 20 min at 55 °C with constant rotation.

17. Discard the wash solution; repeat **step 16** two times.

18. Discard the wash solution, add 50 mL of pre-warmed stringent wash solution in the tube, and incubate by rotating for 5 min.

19. Discard the stringent wash solution, place the membrane on a piece of Whatman filter paper, and measure the radioactivity; seal the membrane in a heat-sealed bag, and transfer to a PhosphorImager screen system to expose for overnight at RT (*see* **Note 17**).

20. Wash the membrane with 50 mL of 1% SDS for 30 min at 90 °C two times to strip the probe by rotation in the hybridization oven. Check for residual radioactive counts on the membrane; if necessary repeat this operation one more time.

3.3.2 Micro RNA Detection by α-³²P UTP Labeled RNA Probe

1. Follow the gel preparation and pre-hybridization **steps** from **1** to **12** in Subheading 3.3.1.

2. Prepare α-^{32}P UTP labeled RNA probe in a 20 μL labeling reaction. The radiolabeled RNA is transcribed from linearized plasmid DNA containing inserted sequence of specific target gene. The plasmid has both the T7 and SP6 transcription start sites at the two ends of the insert in opposite direction.

Reagent	Volume (μL)
Linearized plasmid DNA (~0.2 μg)	2.0
Transcription buffer (5×)	4.0
rATP/rCTP/rGTP (2.5 mM each)	4.0
UTP (1 mM)	0.2
α-^{32}P UTP (50 μCi)	3.8
DTT (0.1 M)	2.0
RNase inhibitor (40 U/μL)	1.0
RNA polymerase (T7 or SP6)	1.0
Nuclease-free water	2.0

3. Mix and incubate at 37 °C for 1 h.

4. Add 1.0 μL of RNase-free DNase to the reaction, mix, and incubate at 37 °C for 15 min.

5. Add 1/10 volume (the final volume after adding ethanol) of 3 M NaAOc and 3 volumes (the volume before adding ethanol) of absolute ethanol, gently mix, and incubate on ice for 30 min.

6. Centrifuge at max speed in a microfuge for 15 min at 4 °C; carefully remove the supernatant.

7. Add 1 mL of ice-cold 75% ethanol to wash the pellet, and then centrifuge at $16,000 \times g$ for 5 min at 4 °C.

8. Discard the supernatant, pulse spin, and pipette off the residual solution; air-dry the pellet for 1 min.

9. Resuspend the transcribed α-^{32}P UTP-labeled RNA probe in 20 μL of nuclease-free water; check the approximate radiolabel incorporation by measuring radioactive counts with a Geiger counter.

10. Add 300 μL of 200 mM carbonate buffer to 20 μL of radiolabeled probe. Incubate at 60 °C for as long as it takes to hydrolyze the probe to an average length of ~50 nt (*see* **Note 18**).

11. After the carbonate reaction is completed, add 20 μL of 3 M NaAOc to terminate the reaction.

12. Add the probes to the pre-hybridized RNA-blotted membrane (follow **step 12** in Subheading 3.3.1). Hybridize at 4 °C for 16–20 h with constant rotation in the hybridization oven.

13. Remove the hybridization buffer; wash the membrane in 25 mL of pre-warmed wash solution for 20 min at 55 °C with constant rotation.

14. Discard the wash solution; repeat **step 13** two times.

15. Discard the wash solution, add 50 mL of pre-warmed stringent wash solution in the tube, and incubate by rotating for 5 min.

16. Discard the stringent wash solution, place the membrane on a piece of Whatman filter paper, and measure the radioactivity. Seal the membrane in a heat-sealed bag, and transfer to a PhosphorImager screen system to expose for overnight at RT (*see* **Note 17**).

17. Wash the membrane with 100 mL of 1% SDS for 30 min at 90 °C to strip the probe by rotation in the hybridization oven. Check for residual radioactive counts on the membrane; if necessary repeat this operation one more time.

3.3.3 Micro RNA Detection by Stem-Loop RT-PCR (See Note 19)

1. Prepare a reverse transcription (RT) reaction by adding the following components in a RNase-free 0.2 mL thin-wall PCR tube on ice (*see* **Note 20**):

Reagent	Volume (μL)
Nuclease-free water	12.6
Reverse transcription buffer (10×)	2.0
dNTPs mix (ATP/CTP/GTP/TTP, 25 mM each)	0.2
RNase inhibitor (40 U/μL)	0.2
Reverse transcriptase (50 U/μL)	2.0
Total RNA (10 ng/μL)	2.0
Micro RNA specific primer (1 μM)	1.0

2. Mix by gently pipetting the solution to ensure reaction components are completely mixed, pulse spin the tube, and incubate on ice for 5 min.

3. Transfer the reaction tube to a PCR machine to perform the reverse transcription reaction using the following settings: 30 min at 16 °C, followed by pulsed RT of 60 cycles at 30 °C for 30 s, 42 °C for 30 s, and 50 °C for 1 s, and terminate the reaction by heating at 85 °C for 5 min. Keep the completed reaction on ice.

4. Dilute 1 volume of RT product with 9 volume of cold nuclease-free water; keep the dilution on ice.

5. Set up a real-time PCR reaction in an RNase-free PCR tube on ice (*see* **Note 21**).

Reagent	Volume (µL)
1:10 diluted RT product (cDNA template)	1.0
TaqMan micro RNA assay (20×)	1.0
TaqMan Universal PCR Master mix (2×)	10.0
Nuclease-free water	8.0

6. Mix by gently pipetting the solution to ensure reaction components are completely mixed; pulse spin to collect the solution to the bottom of the tube.

7. Transfer the reaction to a real-time PCR system, and amplify the specific micro RNA by using the following settings: 1 cycle of 95 °C for 10 min (to activate the enzyme and completely denature the template), 40 cycles of 95 °C for 15 s, and 60 °C for 1 min. The micro RNA expression level in plant can be calculated and compared with the endogenous micro RNA controls by using the software associated with the real-time PCR system. Also the real-time PCR products can be analyzed via electrophoresis to check the specificity of the PCR amplification.

4   Notes

1. Do not exceed 100 mg of fresh material per 1.0 mL of TRIzol. The best total RNA isolation efficacy is achieved when the starting material is 50–100 mg. In this range, both RNA quantity and integrity are assured.

2. When dealing with big number of samples (more than ten) at one time, we suggest to keep the homogenized samples in liquid nitrogen or on dry ice before all the samples are processed; this will decrease the browning of samples in TRIzol buffer when the samples are just kept on ice waiting for the other samples. This modification is strongly recommended when process senescing tissues in plants such as senescing leaves.

3. After centrifugation, the mixture separates into three phases. RNA remains exclusively in the upper aqueous phase. Carefully transfer the aqueous phase by angling the tube at 45°, and pipette the solution out. In order to avoid the contamination of DNA and proteins in the interphase, we usually pipette only 400–450 μL of the aqueous solution.

4. The RNA can be stored in 75% ethanol at least 1 year at −20 °C or at least 1 week at 4 °C. For longer storage, keep RNA in 75% ethanol at −80 °C.

5. Be careful not to dry the pellet for too long. If so, the pellet will lose solubility and cannot be completely dissolved in water.

6. The total RNA pellet can be dissolved in RNase-free water as well as 100% deionized formamide for different purposes. If the total RNA is to be used as templates for cDNA synthesis via RT-PCR, real-time PCR, or stem-loop RT-PCR, it should be dissolved in RNase-free water. If the total RNA is to be used directly for Northern blotting, it can be suspended in 100% deionized formamide.

7. The nucleic acid electrophoresis device used for RNA detection should be treated with 1:10 (v/v) diluted RNase_Zap®_ solution or rinsed in 0.5% SDS solution and washed with RNase-free water to remove RNase contamination.

8. The additional fine powder of plant material can be saved in an RNase-free tube at −80 °C for future use. Be careful not to let the material melt when transferring to the freezer.

9. Carefully transfer the upper aqueous phase by angling the tube at 45°, and pipette the solution out.

10. This is a modified one-step micro RNA isolation protocol that omits total RNA preparation based on the early protocols [17, 18, 20]. It can integrate the downstream micro RNAs isolation program directly from three major total RNA extraction methods (TRIzol, LiCl/SDS/Phenol, CTAB/Phenol; _see_ Subheading 3.1). After the release of total RNA and genome DNA from crushed tissues, the nucleic acid mixture in lysis buffer is extracted with chloroform or chloroform-containing organic solvent for several times, and then the cleared upper aqueous phase is directly proceeded for micro RNA purification with PEG (_see_ Subheading 3.2.1). Alternatively, if researchers want to purify micro RNAs from a previously isolated total RNA (in Subheadings 3.1.1 and 3.1.3), they can also refer to the two-step micro RNA isolation method using PEG as a separating agent [17, 22].

11. The HMW RNAs can be used for other experiments, such as reverse transcription using Oligo(dT)n primers or gene-specific primers or Northern blotting.

12. After dissolving micro RNAs in RNase-free water, be careful not to let the solution stay at RT. Always keep the tube on an ice bath, and pipette with RNase-free pipette tips. Divide the micro RNA solution into several parts in 200 μL RNase-free PCR tubes, and keep them in a −80 °C freezer; use only one or several tubes one time.

13. This is also a one-step micro RNA isolation protocol that uses 8 M LiCl as a separating agent. It is integrated with the LiCl/SDS/Phenol total RNA extraction method in application (in Subheading 3.1.2). However, if researchers want to purify micro RNAs from a previously isolated total RNA (in Subheadings 3.1.1 and 3.1.3), they can also refer to the two-step micro RNA isolation method using 8 M LiCl as a separating agent [19].

14. The 10% APS solution should be made fresh and kept at −20 °C. Don't use this solution prepared more than 3 months ago.

15. The settings of the voltage used for micro RNA electrophoresis in a denatured polyacrylamide gel are based on the operation instructions of the Bio-Rad Protean II vertical electrophoresis system; if the researchers use other systems, they need to refer to the supplier instructions for voltage settings.

16. UV cross-link is not compellingly required if the nylon membrane is positively charged (such as Hybond-N+ or XL), but for other type of uncharged membrane, remember to do the UV cross-linking.

17. RNA-blotted membranes are washed until the counts-per-minute (CPM) is below 5. This usually requires three 20-min long washes at 55 °C with wash solution and a short wash with stringent wash solution for 5 min. If the membrane counts are still high after these washes, more stringent washes should be applied until the CPM is below 5.

18. The time of carbonate treatment is calculated using the mathematic program: $T_{min} = (L_i - L_f)/(K \times L_i \times L_f)$, where T_{min} is the incubation time (min), L_i represents the initial length of probe in kilobases (kb), and L_f is the final length of probe in kilobases (kb). K is the rate constant of 0.11 kb/min.

19. This method is a supplement to the traditionally used Northern blotting methods for micro RNA detection. It can quantitatively assess the expression levels of known specific small RNAs via quantitative RT-PCR reaction from total isolated RNAs using a small RNA-specific stem-loop primer [23, 24]. This method has been successfully used for known micro RNA detection in different plant species such as *Arabidopsis* (*Arabidopsis thaliana*), pumpkin (*Cucurbita maxima*), cucumber (*Cucumis sativus*), and so on [24, 25].

20. For known specific micro RNA sequences, the micro RNA-specific forward primer and the TaqMan MGB probe can be ordered from Life Technologies.

21. For statistic calculation, at least three biological and three technical replications should be applied for each small RNA analysis in different genetic backgrounds or different time points. At the same time, endogenous small RNA control should be set up and used to normalize the expression levels of assayed micro RNAs.

References

1. Lim PO, Kim HJ, Nam HG (2007) Leaf senescence. Annu Rev Plant Biol 58:115–136

2. Gregersen PL, Culetic A, Boschian L et al (2013) Plant senescence and crop productivity. Plant Mol Biol 82:603–622

3. Balazadeh S, Riaño-Pachón DM, Mueller-Roeber B (2008) Transcription factors regulating leaf senescence in *Arabidopsis thaliana*. Plant Biol 10:63–75

4. Humbeck K (2013) Epigenetic and small RNA regulation of senescence. Plant Mol Biol 82:529–537

5. Carthew RW, Sontheimer EJ (2009) Origins and mechanisms of miRNAs and siRNAs. Cell 136:642–655

6. Vionnet O (2009) Origin, biogenesis, and activity of plant microRNAs. Cell 136: 669–687

7. Carrington JC, Ambros V (2003) Role of microRNAs in plant and animal development. Science 301:336–338

8. Jones-Rhoades MW, Bartel DP, Bartel B (2006) MicroRNAs and their regulatory roles in plants. Annu Rev Plant Biol 57:19–53

9. Gandikota M, Birkenbihl RP, Höhmann S et al (2007) The miRNA156/157 recognition element in the 3'UTR of the Arabidopsis SBP box gene SPL3 prevents early flowering by translational inhibition in seedlings. Plant J 49: 683–693

10. Palatnik JF, Allen E, Wu X et al (2003) Control of leaf morphogenesis by microRNAs. Nature 425:257–263

11. Aukerman MJ, Sakai H (2003) Regulation of flowering time and floral organ identity by a microRNA and its APETALA2-like target genes. Plant Cell 15:2730–2741

12. Sunkar R, Zhu JK (2004) Novel and stress-regulated microRNAs and other small RNAs from Arabidopsis. Plant Cell 16:2001–2019

13. Kim JH, Woo HR, Kim J et al (2009) Trifurcate feed-forward regulation of age-dependent cell death involving *miR164* in Arabidopsis. Science 323:1053–1057

14. Schommer C, Palatnik JF, Aggarwal P et al (2008) Control of jasmonate biosynthesis and senescence by miR319 targets. PLoS Biol 6:1991–2001

15. Adenot X, Elmayan T, Lauressergues D, Boutet S, Bouche N, Gasciolli V, Vaucheret H (2006) DRB4-dependent TAS3 trans-acting siRNAs control leaf morphology through AGO7. Curr Biol 16:927–932

16. Fahlgren N, Montgomery TA, Howell MD et al (2006) Regulation of AUXIN RESPONSE FACTOR3 by TAS3 ta-siRNA affects developmental timing and patterning in Arabidopsis. Curr Biol 16:939–944

17. Chomczynski P, Sacchi N (2006) The single-step method of RNA isolation by acid guanidinium thiocyanate-phenol chloroform extraction: twenty-something years on. Nat Protoc 1:581–585

18. Rosas-Cardenas FD, Duran-Figueroa N, Vielle-Calzada JP et al (2011) A simple and efficient method for isolating small RNAs from different plant species. Plant Methods 7:4

19. Verwoerd TC, Dekker BM, Hoekema A (1989) A small-scale procedure for the rapid isolation of plant RNAs. Nucleic Acids Res 17:2362

20. An Z, Li Y, Lili X et al (2013) A rapid and economical method for low molecular weight RNA isolation from a wide variety of plant species. Biosci Biotechnol Biochem 77: 1599–1601

21. Carra A, Gambino G, Schubert A (2007) A cetyltrimethylammonium bromide-based method to extract low-molecular-weight RNA from polysaccharide-rich plant tissues. Anal Biochem 60:318–320

22. Carra A, Mica E, Gambino G et al (2009) Cloning and characterization of small non-coding RNAs from grape. Plant J 59:750–763

23. Chen CF, Ridzon DA, Broomer AJ et al (2005) Real-time quantification of microR-NAs by stem-loop RT-PCR. Nucleic Acids Res 33:e179

24. Wu RM, Wood M, Thrush A et al (2007) Real-time PCR quantification of plant miRNAs using universal ProbeLibrary technology. Biochemica 2:12–15

25. Varkonyi-Gasic E, Wu RM, Wood M et al (2007) Protocols: a highly sensitive RT-PCR method for detection and quantification of microRNAs. Plant Methods 3:12

Chapter 21

In Situ Detection of Programmed Cell Death in Senescing *Nicotiana tabacum* Leaves Using TUNEL Assay

Branka Uzelac, Dušica Janošević, and Snežana Budimir

Abstract

Leaf senescence constitutes a highly regulated final phase of leaf development, leading to cell death that is recognized as a type of programmed cell death (PCD). Degradation of nuclear DNA into oligonucleosomal fragments (DNA ladder) and terminal deoxynucleotidyl transferase-mediated dUTP nick end labeling (TUNEL) assay are methods commonly used to detect PCD-specific DNA cleavage. TUNEL reaction in situ labels free 3′-OH DNA strand breaks (nicks), thus allowing histological localization of nuclear DNA degradation during PCD. Here we describe in situ labeling of PCD-specific nuclear DNA fragmentation on conventional histological sections of senescing tobacco leaves. Incorporation of fluorescein-labeled dUTPs is detected by fluorescence microscopy, which enables in situ visualization of PCD at the single-cell level in the leaf mesophyll tissues undergoing senescence.

Key words DNA fragmentation, Histology, Leaf senescence, *Nicotiana tabacum*, Programmed cell death, TUNEL assay

1 Introduction

Leaf senescence is the age-dependent deterioration process that constitutes the final phase of leaf development [1]. This highly regulated active process, during which leaf cells undergo orderly changes in their structure, metabolism, and gene expression, results in massive cell death, the type of which is referred to as programmed cell death (PCD). A very specific hallmark of PCD is the nonrandom, internucleosomal fragmentation of nuclear DNA (DNA laddering) that occurs as a result of specific endonuclease activation prior to condensation of nuclear chromatin [2, 3]. However, in plant PCD the occurrence of oligonucleosome-sized nuclear DNA fragments does not seem to be an obligatory stage in the processing of the initial large DNA fragments. That the absence of a DNA ladder does not necessarily rule out PCD was shown in the case of plant PCD in cultured cells containing only a small number of dead cells [4] or in ozonated tobacco cv Bel W3 leaves

Yongfeng Guo (ed.), *Plant Senescence: Methods and Protocols*, Methods in Molecular Biology, vol. 1744,
https://doi.org/10.1007/978-1-4939-7672-0_21, © Springer Science+Business Media, LLC 2018

where cell death occurred in a small group of cells buried in a bulk of surrounding healthy cells [5]. DNA laddering has rarely been reliably detected in senescing leaves, possibly due to low population of mesophyll cells undergoing internucleosomal DNA fragmentation [6].

DNA fragmentation corresponding to the DNA ladder can be detected using a method for the in situ visualization of PCD at the single-cell level, developed by Gavrieli et al. [7]. The terminal deoxynucleotidyl transferase (TdT)-mediated dUTP nick end labeling (TUNEL) assay is based on the specific binding of TdT to 3′-OH DNA strand breaks (nicks). TdT catalyzes polymerization of dUTPs conjugated to a detectable marker to free 3′-OH DNA ends in a template-independent manner, which results in a synthesis of labeled nucleotide polymers that are subsequently visualized depending on the introduced label [7, 8]. To avoid the loss of fragmented DNA and to allow enzyme and nucleotide entrance, cells need to be fixed and subsequently permeabilized prior to the labeling reaction. Although considered as a good specific criterion of PCD in animals, in some plant systems, TUNEL reaction has been reported to produce false-positive staining due to the sample preparation procedures that caused sufficient nicking of nuclear DNA [9]. However, adapting the TUNEL reaction protocols to plant cells eliminates artifacts and together with appropriate control treatments, allows only the detection of nuclei where DNA fragmentation occurred in vivo [10].

Here we describe in situ labeling of PCD-specific nuclear DNA fragments in the mesophyll cells of leaves of in vitro grown *Nicotiana tabacum* plants, the system in which DNA ladder could not be detected. Nuclear DNA on conventional histological sections was nick end labeled using TdT for the incorporation of fluorescein-dUTP at the sites of DNA breaks, enabling their detection by fluorescence microscopy (Fig. 1). Our TUNEL experiments demonstrated that nuclear DNA fragmentation characteristic for PCD occurred early during tobacco leaf development and in subsequent developmental stages but that the level of DNA degradation was much higher in the cells of senescent leaves [11].

2   Materials

Prepare all solutions with MilliQ water (the resistivity close to 18 MΩ cm) and analytical grade reagents. Use fresh solvents for the preparation of Paraplast-embedded tissue. Water, glassware, Eppendorf tubes, and pipette tips should be sterilized by autoclaving at 114 °C for 25 min. Frozen aliquots should be thawed immediately before usage and centrifuged briefly.

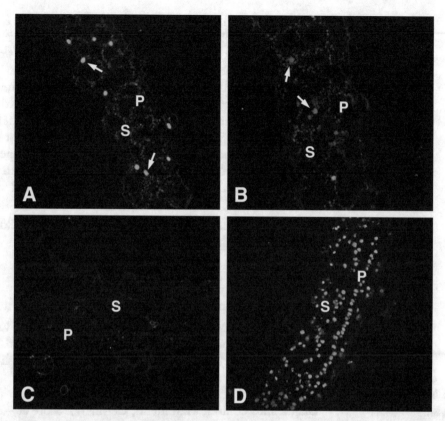

Fig. 1 In situ detection of nuclear DNA fragmentation in tobacco leaf tissues at different stages of senescence by fluorescence microscopy. (**a, b**) TUNEL stained leaf cross sections at the onset of senescence (**a**) and late senescence stage (**b**), show nuclear DNA cleavage as yellowish-green fluorescence in TUNEL positive nuclei (arrows) under blue excitation. Bright green fluorescent nuclei of the cells in palisade and spongy mesophyll are indicative of high level of DNA fragmentation. The weaker TUNEL staining signal in (**b**) suggests significant loss of DNA at the late stage of senescence. Note strong endogenous autofluorescence of plant tissues, particularly of chloroplasts and vascular bundles. (**c, d**) TUNEL staining of leaf cross sections used as labeling control. The negative labeling control (**c**), used to monitor background, was prepared without terminal deoxynucleotidyl transferase in the reaction mixture. The positive labeling control (**d**), used to monitor the signal strength, was pretreated with DNase I (1000 U mL^{-1}) to induce DNA strand breaks, prior to TUNEL reaction. Note intensive staining of all nuclei in the section. *P* palisade mesophyll tissue, *S* spongy mesophyll tissue

2.1 Plant Tissue Sectioning

1. Plant material: *Nicotiana tabacum* L. cv. Wisconsin 38 (*see* **Note 1**).

2. 1× Phosphate buffered saline (PBS): 0.14 M NaCl, 2.7 mM KCl, 8.0 mM Na$_2$HPO$_4$, and 1.5 mM KH$_2$PO$_4$, pH 7.4. Dissolve 8.18 g NaCl, 0.2 g KCl, 1.42 g Na$_2$HPO$_4$, and 0.2 g KH$_2$PO$_4$ (*see* **Note 2**) in 800 mL of water by stirring in a 1 L glass beaker. Adjust pH to 7.4 with 1 *N* HCl (*see* **Note 3**). Transfer the solution to a 1 L graduated cylinder and adjust volume to 1 L with additional water. Sterilize by autoclaving and store at room temperature.

3. Fixation solution: 4% paraformaldehyde in PBS, pH 7.4. Weigh 4 g paraformaldehyde and transfer to a glass beaker (*see* **Note 4**). Add 100 mL 1 × PBS, followed by 3–4 drops of 8 *N* (or 10 *N*) NaOH. Heat the mixture on a hot plate and stir slowly until paraformaldehyde is completely dissolved (when the liquid is clear and no powder precipitates on the bottom of the beaker). Monitor the temperature of the mixture using thermometer (optimal is 55–57 °C) and make sure it does not exceed 60 °C! Cool the solution down to the room temperature. Adjust pH to 7.4 with H_2SO_4. Pour into glass jar with screwcap, label properly (name, concentration, date), and store at 4 °C, for up to 1 month (*see* **Note 5**).

4. Xylene and ethanol: absolute, 96%, 85%, 70%, 50%, 30%, and 10%, diluted with purified water.

5. Eosin.

6. Paraplast (Tm = 56 °C).

7. Fume hood, thermostat, microtome, and flotation bath.

8. Glass vials, paper, pen, duct tape, forceps, scalpel, vacuum, Parafilm, ceramic evaporating dishes, paper (cardboard) embedding molds, and SuperFrost®Plus microscopic slides.

2.2 TUNEL Assay Components

1. In Situ Cell Death Detection Kit: Fluorescein (Roche Diagnostics GmbH, Mannheim, Germany). Store at −20 °C.

2. Washing buffer: 1 × PBS: Prepare a 1 L solution, as described in Subheading 2.1.

3. Xylene and ethanol: absolute, 96%, 85%, 70%, 50%, 30%, and 10%, diluted with purified water.

4. 150 mM Tris–HCl (pH 7.4): Weigh 9.085 g Tris base and transfer to a glass beaker. Add water to a volume of 400 mL and mix well (*see* **Note 6**). Gradually add concentrated HCl, drop by drop, using a Pasteur pipette. Stir thoroughly and check the pH (*see* **Note 3**). When desired pH is reached (stable at pH 7.4), use a 500 mL graduated cylinder to bring the volume up to 500 mL with water. Pour into labeled Duran bottle and sterilize by autoclaving. Store at room temperature.

5. 100 mM Tris–HCl (pH 7.4): Weigh 1.211 g Tris base and prepare 100 mL solution as described above. Autoclave and store at room temperature.

6. 50 mM Tris–HCl (pH 7.4): Mix 5 mL 100 mM Tris–HCl with 5 mL water, autoclave and store at room temperature.

7. 20 mM Tris–HCl (pH 7.4): Pour 133.3 mL 150 mM Tris–HCl (pH 7.4) into 1 L graduated cylinder and bring the volume up to 1 L with water. Prepare 100 or 200 mL aliquots, autoclave, and store at room temperature.

8. Tris–$CaCl_2$ buffer: 50 mM Tris–HCl, 5 mM $CaCl_2$, pH 7.4. Weigh 0.61 g Tris base and dissolve in 80 mL water in a glass

beaker. Add 0.055 g anhydrous $CaCl_2$ (*see* **Note 7**) to the solution and mix well. Adjust pH with concentrated HCl. Adjust volume to 100 mL with water. Autoclave and store at room temperature.

9. Proteinase K: nuclease-free, recombinant, PCR grade. Store at 4 °C.

10. Proteinase K stock solution (20×): 400 µg mL^{-1} Proteinase K, 25 mM Tris–HCl, 2.5 mM $CaCl_2$, pH 7.4. Weigh 0.010 g Proteinase K, lyophilized powder, aseptically (*see* **Note 8**) and add to 12.5 mL Tris–$CaCl_2$ buffer (pH 7.4) in a plastic test tube. Mix gently (by inverting the tube), without vortexing, until dissolved. Then add 12.5 mL sterile glycerol and mix well. Spin down for a second. Aliquot and store at −20 °C.

11. Proteinase K working solution (1×): 20 µg mL^{-1} Proteinase K in 20 mM Tris–HCl (pH 7.4) containing 2 mM $CaCl_2$. Add 5 mL Proteinase K stock solution to 37.5 mL Tris–$CaCl_2$, in a 100 mL graduated cylinder. Make up to 100 mL with water. Mix well. Prewarm to 37 °C before use (*see* **Note 9**).

12. DNase I: Grade I. Store lyophilized powder at 4 °C.

13. 25 mM $MgCl_2$ solution: Store at −20 °C.

14. Albumin bovine serum (BSA): Store at 4 °C.

15. BSA stock solution (10/1): 10 mg mL^{-1} BSA in 50 mM Tris–HCl, pH 7.4. Weigh 10 mg BSA in sterile Eppendorf tube. Add 1 mL sterile 50 mM Tris–HCl aseptically and mix well. Prepare 100 µL aliquots and store at 4 °C.

16. DNase I stock solution: 10,000 U mL^{-1} DNase I in 20 mM Tris–HCl, 1 mM $MgCl_2$, 50% (v/v) glycerol, pH 7.4. Mix 400 µL 50 mM Tris–HCl (pH 7.4), 40 µL 25 mM $MgCl_2$, 60 µL water, and 500 µL sterilized anhydrous glycerol in a sterile 1.5-mL Eppendorf tube (*see* **Note 10**). Add 1000 µL of this mixture to 10,000 U DNase I (grade I), that is, the lyophilizate contained in the original vial, and mix gently by pipetting. Centrifuge briefly. Prepare 100-µL aliquots and store at −20 °C (*see* **Note 11**).

17. DNase I reaction buffer: 1000 U mL^{-1} DNase I in 50 mM Tris–HCl, 10 mM $MgCl_2$, 1 mg mL^{-1} BSA, pH 7.4. Mix 100 µL DNase I stock solution, 287 µL 150 mM Tris–HCl (pH 7.4), 396 µL 25 mM $MgCl_2$, 100 µL BSA stock solution, and 117 µL water in a sterile 1.5 mL Eppendorf tube. Mix gently by pipetting and spin down shortly.

18. Mounting medium: PBS/glycerol mixture (1:1 PBS/glycerol). Pour 3–5 mL PBS in a beaker, add equal volume of glycerol, and mix well.

19. Humidified chamber or plastic cytology slide folder, staining jars or slide staining set, liquid-repellent slide marker pen (optional), forceps, cloth towel, Parafilm or duct tape, coverslips, and coverslip sealant.

3 Methods

3.1 Preparation of Tissue Sections

1. Label the vials and fill them with at least 5 mL chilled fixation solution (*see* **Note 4**).

2. Cut the leaves on ice with a sharp blade on a glass plate into small (~5 mm × 5 mm) pieces (*see* **Note 12**).

3. Transfer the specimens into labeled vials containing chilled fixation solution using forceps. Keep the specimens in fixation solution at 4 °C overnight (*see* **Note 13**).

4. Using Pasteur pipette, replace the fixation solution with cooled PBS. Keep specimens in PBS for at least 1 h. Perform minimum three washes with PBS at room temperature (*see* **Note 14**).

5. Replace PBS stepwise with 10, 30, 50, and 70% ethanol. Each passage should last for 1 h, with two washes with alcohol at room temperature (*see* **Note 15**).

6. Continue dehydration by passing specimens through 85% ethanol (two washes), at room temperature for 2 h. Replace 85% ethanol with 96% ethanol containing 0.1% (w/v) eosin and store at 4 °C overnight (*see* **Note 16**).

7. Exchange alcohol in the vials with absolute ethanol containing 0.1% (w/v) eosin. Perform two washes (30 min each) at 4 °C and then one last wash at room temperature, for another 30–60 min (*see* **Note 17**).

8. Replace absolute ethanol with graded series of absolute ethanol/xylene mixture, with increasing concentrations of xylene: 2:1, 1:1, 1:2 (v/v). Passage through absolute ethanol/xylene mixture should be carried out in a fume hood, at room temperature, and should last 1 h for each new xylene concentration (*see* **Note 18**).

9. Replace the absolute ethanol/xylene mixture with pure xylene. Perform three washes with xylene for 1 h.

10. Pour out specimens with the last xylene wash, into labeled ceramic evaporating dishes (*see* **Note 19**). Quickly add about the same amount of melted Paraplast (to obtain 1:1 xylene/Paraplast mixture) and incubate in thermostat for 1 h.

11. Replace the xylene/Paraplast mixture with pure Paraplast and incubate in thermostat overnight.

12. Replace with fresh Paraplast and incubate in thermostat overnight. After this step, Paraplast should be exchanged with fresh one and kept in thermostat overnight.

13. Pour the specimens (in Paraplast) into embedding mold (*see* **Note 20**).

14. Using flamed needles, orient the specimens in the block in a desired direction (*see* **Note 21**).

15. Paraplast will solidify within few hours or overnight. Solidified blocks should be kept in a refrigerator until use.

16. Before sectioning, remove Paraplast block out of the mold.

17. Trim the block with flamed scalpel to obtain smaller rectangular blocks, each containing one specimen in the center (*see* **Note 22**).

18. Put small amount of Paraplast on metal spatula and heat it above alcohol lamp until Paraplast completely melts. Pour the melted Paraplast over the specimen holder of the microtome. Quickly press the cut off Paraplast block on the melted layer, in a desired orientation, to fuse it with the specimen holder.

19. With heated spatula, shape the block so that its upper surface, instead of a rectangle, appears as an isosceles trapezoid.

20. Cool warmed Paraplast blocks attached to the specimen holders down to the room temperature. Keep them in refrigerator until use (*see* **Note 23**).

21. Affix the specimen holder with a chilled Paraplast block to the microtome object clamp and orient it in a desired position toward the knife (*see* **Note 24**).

22. Insert a disposable microtome blade in a knife holder and place it to a knife holder base. Adjust the knife clearance angle (usually between 1° and 5°).

23. Section the specimen to obtain a ribbon of 5-μm thick sections (*see* **Note 25**).

24. Using forceps or brushes, transfer the ribbon to flotation bath and place it with the smooth side down onto the water surface. Allow the sections to stretch gradually (they must not melt), but leave them just long enough to flatten (*see* **Note 26**).

25. Mount sections onto SuperFrost®Plus glass slides by immersing the slide carefully into water under the ribbon. Once the ribbon is floated on the slide, remove the slide vertically from the water.

26. Drain the slides and leave them to air dry overnight. Make sure each slide is labeled accurately. Store in a cytology slide folder until use, protected from dust.

3.2 TUNEL Assay

1. Prepare a humidified chamber (*see* **Note 27**) in which slides will be incubated during the TUNEL assay. Each experimental set up must include one positive and two negative controls (*see* **Note 28**). For DNase I treatment of positive control slides, fill an appropriate jar with ~100 mL 20 mM Tris–HCl (pH 7.4) and put the lid on (*see* **Note 29**). Use forceps to handle the slides, holding them by the label end.

2. Prepare fresh Proteinase K working solution and warm it up in thermostat (*see* **Note 9**).

3. Insert chosen slides into baskets of the slide staining set. Fill the slide containers with the appropriate amount (100–150 mL) of xylene; absolute ethanol; 96%, 85%, 70%, 50%, and 30% ethanol; and 20 mM Tris–HCl (pH 7.4). Use fresh solvents for dewaxing, rehydration, and rinsing!

4. Dewax tissue sections by placing them in xylene, for 10 min. Repeat this step with fresh xylene.

5. Rehydrate tissue sections by passing them through a series of decreasing concentrations of alcohol, starting with absolute ethanol, through 96%, 85%, 70%, 50%, and 30% (10 min each, *see* **Note 30**).

6. Rinse the slides twice with 20 mM Tris–HCl (pH 7.4).

7. Pour 100 mL Proteinase K working solution into the staining jar. Remove slides, one by one, from 20 mM Tris–HCl with forceps and drain excess fluid. Immerse the slides into Proteinase K working solution. Add lid, seal with duct tape or Parafilm, and incubate in thermostat at 37 °C for 30 min (*see* **Note 31**).

8. Rinse the slides twice with PBS. First wash (PBS I) should last for about 5 min. Second wash (PBS II) should last another 5 min for the positive control slide, and longer (until DNase I treatment of the positive control slide is finished) for all other slides.

9. After the second wash, remove only the positive control slide from PBS (all other slides remain in PBS II). Place it on a cloth towel (loaded side up) to remove excess buffer. Dry the area around the sections. Pipette 50–100 μL of DNase I working solution onto the tissue sections (*see* **Note 32**). Transfer the slide carefully into the staining jar (or slide folder) containing 20 mM Tris–HCl (pH 7.4) and place it horizontally *above* the buffer—do not immerse into the buffer! Add lid and incubate at room temperature for 15 min.

10. During DNase I treatment, prepare TUNEL reaction mixture *on ice* (*see* **Note 33**). From the In Situ Cell Death Detection Kit, transfer 100 μL of the label solution (for two negative controls) into a sterile Eppendorf tube and keep on ice until use. To obtain 500 μL TUNEL reaction mixture (*see* **Note 34**), add total volume (50 μL) of the enzyme solution to the remaining 450 μL label solution and mix well (by inverting the vial). Centrifuge briefly at 6–7000 rpm (~$3,000 \times g$) at 4 °C with a microfuge to collect the reaction mixture at the bottom of the tube and keep on ice until use.

11. After the DNase I treatment is finished, remove the DNase I working solution from the positive control slide by pouring it

off onto the cloth towel. Gently rinse the slide with a few mL of Tris–HCl (over the treated tissue sections). Immerse the slide into the buffer contained in the jar where it has been incubated, to thoroughly wash all the remaining DNase I working solution. Then rinse the positive control slide with fresh PBS.

12. For all slides: remove slides from PBS II using forceps. Let excess PBS drain off. Dry area around chosen tissue sections, but make sure that sections do not dry (*see* **Note 35**). If available, use liquid-repellent slide marker pen to draw a circle around the sections (*see* **Note 36**). Place a slide horizontally (in a humidified chamber or an open plastic cytology slide folder with moist filter paper on its bottom).

13. For samples and positive control: with a sterile pipette tip, apply 50 μL of TUNEL reaction mixture onto 3–4 adjacent (encircled) tissue sections. For negative control: apply 50 μL of label solution (removed from the original vial and stored in a sterile Eppendorf tube) on each of the two negative control slides.

14. Seal the boxes with duct tape or Parafilm. Make sure to do it carefully to prevent any spillage of TUNEL reaction mixture or label solution from the slides.

15. Place the boxes in thermostat and incubate at 37 °C for 60 min.

16. Collect the supernatants from labeling reactions in a tightly closed bottle (or non-breakable container) and discard according to the local regulations for toxic waste disposal!

17. Rinse slides in fresh PBS (three washes, 5 min each). Slides should remain in the last PBS until coverslipping is performed (*see* **Note 37**).

18. Remove slide from PBS, one at a time, and let excess PBS drain off. Set the slide on a cloth towel. Using glass stirring rod, apply the mounting medium to the slide. Place the coverslip onto the slide as quickly as possible to avoid air-drying (*see* **Note 38**). Wipe the edges and the bottom surface of the slide with a cloth towel to remove excess mounting medium. Seal the coverslip with coverslip sealant. Place the slide on the slide tray. Repeat this step for all the slides.

19. Examine the slides under a fluorescence microscope, in excitation range of 450–500 nm and detection range of 515–565 nm (*see* **Note 39**). Nuclear DNA cleavage can be detected as yellowish-green fluorescence in the nuclei of the cells, as shown in Fig. 1 for tobacco leaf mesophyll. If not examined immediately after mounting, slides can be stored in a refrigerator in the dark for a few days until viewing under microscope.

4 Notes

1. *Nicotiana tabacum* L. cv. Wisconsin 38 plants are germinated and grown in vitro, on half-strength Murashige and Skoog [12] culture medium at 25 ± 2 °C (16-h photoperiod), for 8–9 weeks. Leaf samples are obtained from the seedlings with 9–10 leaves. The middle part of the leaf near midvein is used for assay.

2. Make sure that Na_2HPO_4 is anhydrous, as the formula weights will be different. Otherwise, the weight of the available hydrate needs to be calculated to ensure the desired molarity.

3. Adjust the final pH using a sensitive pH meter. Do not stir the solution while taking the reading. Unless specified otherwise, use 1 N HCl rather than concentrated HCl when adjusting the pH, to prevent its sudden drop below the desired value. When preparing Tris buffers, it is recommended to start with concentrated HCl and then switch to 1 N HCl as you get close to the desired pH.

4. Perform all further actions in the fume hood. If weighing cannot be done in a fume hood, care should be taken to minimize the exposure to paraformaldehyde powder. Wear protective gloves (nitrile, rather than latex), safety glasses, or face shields, and avoid breathing dust and fumes. Fixation solution must be prepared and all the tasks that include handling it must be carried out in the fume hood, to limit the exposure to toxic and otherwise harmful vapors of paraformaldehyde.

5. Paraformaldehyde is a white powder, containing water insoluble higher polymers of formaldehyde. To be useful as a fixative, a solution must contain monomeric formaldehyde as its major solute. This is achieved by dilution with a buffer solution at physiological pH. Hydrolysis of the polymers is catalyzed by the hydroxide ions present in the slightly alkaline solution. Heating is also necessary, but temperature must be kept below 60 °C (temperatures above 60 °C will make it useless as a fixative due to denaturation). Paraformaldehyde dissolves slowly (over period of 30–60 min) and the solution appears cloudy. Keep the solution covered whenever possible! Label the fixation solution in a manner to include the word "formaldehyde" and its concentration. We find it best to prepare the fresh fixation solution each time. Alternatively, it can be aliquoted in 50-mL falcon tubes, frozen and stored at −20 °C. Chill before use!

6. Water can be warmed up (~37 °C) for Tris to dissolve faster. Stir on magnetic stirrer with a magnetic stirring bar. Allow the solution to cool down to room temperature before adjusting pH with concentrated HCl.

7. If $CaCl_2$ is available as a hydrate, its weight needs to be calculated to ensure the desired molarity. When using dihydrate for preparing the buffer, 0.074 g $CaCl_2 \times 2H_2O$ should be used.

8. Weigh Proteinase K lyophilized powder with sterilized spatula, into a sterile Eppendorf tube. Use previously sterilized buffer and glycerol. Perform actions aseptically, in a laminar flow hood.

9. Proteinase K working solution is stable for a month at 4 °C. However, we prefer preparing a fresh working solution each time before starting a pretreatment of Paraplast-embedded tissue, then warming it up to 37 °C in thermostat until use.

10. All components of the DNase dissolving buffer should be handled aseptically, using sterile solutions and pipette tips. Sterilize 50 mM Tris–HCl and glycerol (anhydrous) by autoclaving. We have used sterile 25 mM $MgCl_2$ contained in the AmpliTaq Gold® PCR Kit (Applied Biosystems, Roche, Branshburg, New Jersey, USA).

11. Do not vortex the enzyme while dissolving. Appropriate aliquots can be stored at −20 °C for up to 18 months.

12. To ensure that fixation solution penetrates deep into the tissue, the piece of tissue should not be too large (~ 5 mm³). Immediately before taking the tissue samples, vials should be labeled appropriately (date, type of fixation solution used, plant species) and filled with chilled fixation solution.

13. The tissue is properly fixed when the specimens settle to the bottom of the fixative containing vial. If the specimens float on the surface of the fixative, as is the case with most plant tissues that have cuticle, they should be vacuum infiltrated to remove the trapped air and to allow penetration of the fixative. Care should be taken to ensure that the fixation solution does not boil during vacuum infiltration and that the vacuum is released slowly to avoid tissue disruption. Since formaldehyde vapor is volatile, the fixative in the vial should be replaced with fresh solution after the vacuum treatment.

14. Fixation solution waste must be placed in tightly sealed and labeled bottles. Used fixative along with chemical waste collected throughout the dehydration, clearing, embedding, rehydration, and TUNEL staining processes must be disposed of according to local regulations.

15. It is important that dehydration process is performed gradually. Stepwise dehydration minimizes overall tissue shrinkage, and prevents difficulties during sectioning. Dehydration steps should take at least 60 min each. For better tissue preservation, lower starting ethanol concentrations (e.g., 10%) and

smaller increments in alcohol series are recommended. Material can be stored in 70% ethanol at 4 °C overnight (up to several weeks).

16. At this point of the procedure, it is recommendable to place Paraplast granules in a glass beaker to melt in thermostat at 58–60 °C overnight.

17. Dehydration in absolute ethanol should be repeated at least three times in order to efficiently remove water molecules from the tissue.

18. Xylene clearing must be performed in a fume hood, to limit exposure to toxic xylene fumes! To minimize the exposure to xylene, a graded series of mixtures can be replaced by only one passage through ethanol/xylene (1:1). Note that improper clearing (incomplete removal of alcohol by clearing fluid) of the specimens can cause difficulties during sectioning, making sections crumble, and specimen tear out.

19. Before pouring out the liquid with specimens into ceramic evaporating dish, make sure that appropriate label, corresponding to that on the vial, is placed in each ceramic dish.

20. Make sure to do it quickly, so that specimens do not remain trapped in solidifying Paraplast and glued to the base or walls of the ceramic dishes. Alternatively, this can be done by pouring fresh Paraplast into embedding mold and then transferring the specimens into liquid Paraplast contained in the mold, using flamed forceps. We avoid embedding this way, because of the rapid Paraplast solidification and, more importantly, because of the possible damage to the leaf tissue when handled with flamed forceps.

21. Embedding should be performed quickly (before Paraplast starts solidifying) and carefully (to avoid formation of air bubbles in the solidifying Paraplast, especially around the specimen). Should the air bubbles appear around the specimen, remove them with the hot needle. Consider the orientation of the specimens already at this stage, as this will later influence the orientation of the specimen to the blade, which will directly influence section quality.

22. When trimming the block with flamed scalpel, care should be taken that the cutting line is positioned minimum 3–5 mm away from the specimens. Trimming should be performed carefully, to avoid frictions in the Paraplast block. Usually one cut with the flamed scalpel is not sufficient. Rather, not too deep cut, along the same line, should be repeated few times, and scalpel should be heated before each cut is made.

23. Ensure blocks are cold before sectioning. Avoid placing them in a freezer, as this can cause surface cracking. The best way to

chill the blocks is to place them on the surface of melting ice, for a few minutes. Re-chilling of the block may be required if the block face becomes warm during sectioning, especially if the room is too warm.

24. For security reasons, always affix the specimen holder first, before affixing the knife holder with the blade.

25. Ensure that the blade is held firmly in place, as a loose blade can destroy a block. Use the blade systematically, working from one end to the other, thus maximizing the blade life. Use one part of the blade for trimming and another (new) part for final sectioning, or use separate blades for these two procedures. Make sure that the block edge is parallel to the blade edge and that upper and lower edges of the block are parallel, to prevent obtaining bent ribbons. For best results, a slow uniform cutting stroke is recommended. When the tissue is properly fixed, dehydrated, cleared, and infiltrated with Paraplast, the section quality is determined more by the condition of the blade and blade clearance angle, than by any other factors. Debris adhering to upper and lower edges of the block, or to the back of the blade, makes it difficult to obtain cohesive ribbons and causes the ribbon to lift off the blade on the upstroke. If debris is present, clear it away with a piece of xylene-soaked cloth and re-chill the block before starting again. Frequently, lifting sections off the blade on the upstroke is caused by the angle of the blade, which should be readjusted, or by a dull blade, which should be replaced with new one. If sections tear, remove the dirt from the blade with a piece of xylene-soaked cloth or try using a different part of the blade. If sections fail to form a ribbon, allow the block to warm up to the room temperature (if it has been too thoroughly chilled before sectioning). Make sure that block edge is parallel to the blade edge. We also suggest to try decreasing the clearance angle, clean the blade with a piece of xylene-soaked cloth, or replace it with a sharp one. Split ribbon or lengthwise scratches in the ribbon are indicative of the nicks in the blade or dirty blade edges. If cleaning both sides of the blade or even replacement with the sharp one do not help, it is advised to decrease the knife tilt. If the sections are excessively compressed, replace the blade with the sharp one, or cool the trimmed block and the blade by placing ice chunks on each just before sectioning. Another reason for this could be a loose knife (knife tilt should be increased) or too rapid sectioning (cutting should be slowed, to approximately one revolution of the microtome drive wheel per second). Commonly faced difficulty during sectioning is static electricity, which can be reduced by increasing the humidity in the room (e.g., by boiling water in a beaker placed near the microtome).

26. Flotation ensures that sections are completely flat and expanded to original dimensions. The temperature of the water in flotation bath needs to be ~10 °C below the melting point of the Paraplast. Care should also be taken that the flotation bath is not too cold or contaminated and that there are no bubbles adhering to the base and sides of the bath. Before removing slides from the water, make sure no air bubbles are trapped under the section.

27. For this purpose, plastic cytology slide folders can be used. Place filter paper on the bottom of the slide folder. Pour about 100 mL sterilized water over the paper, add lid, and keep the box closed during the pretreatment of Paraplast-embedded tissue, so that the atmosphere gets saturated with water vapor.

28. The amount of label and enzyme solutions in one pair of vials is sufficient for processing nine samples, one positive and two negative controls.

29. Staining jar (or slide folder) in which the positive control slide is treated with DNase should not be used for TUNEL reaction. Instead, Tris–HCl solution contained in this jar should be used later to rinse thoroughly with DNase working solution from the positive control slide (to avoid accidental contamination of the samples and negative control with DNase working solution).

30. Prepare 90% and lower percentage alcohols with sterilized water. Rehydration time in 50% and 30% ethanol can be shortened to 5 min each.

31. Upon start of the proteolytic treatment, vials containing label and enzyme solutions (from the In Situ Cell Death Detection Kit, stored at −20 °C) should be taken out of the freezer and placed on ice. While slides are incubated with Proteinase K, prepare DNase I working solution.

32. Choose several (e.g., four) adjacent sections depending on their number and arrangement on the slide. We find it better not to coverslip the treated sections, as these might suffer damage when removing the coverslip after the incubation is finished. The humidity in the incubation jar and proper administration of the working solution will ensure that sections do not run dry during DNase I treatment. To prevent accidental spillage of the drop of the solution, sections can be encircled with liquid-repellent slide marker pen, prior to applying DNase I working solution (*see* **Note 36**).

33. The TUNEL reaction mixture should be prepared immediately before use. Minimize warming up of the solutions and exposure to the light. Keep TUNEL reaction mixture on ice

until use. Wear safety gloves and avoid exposure to toxic and carcinogenic substances contained in the label solution. Dispose of pipette tips immediately after use, as regulated for toxic waste.

34. We have prepared the TUNEL reaction mixture according to manufacturer's instructions. However, high TdT and nucleotide concentrations used for staining animal cells can lead to high background and non-specific staining in the plant cells. In such cases, a substantial reduction of enzyme and/or nucleotide concentrations is required to avoid background interference [10].

35. Care should be taken that at no point sections are let to dry completely!

36. To obtain more uniform TUNEL staining, liquid-repellent slide marker pen can be used to draw a water-repelling circle (~1 cm diameter) around desired sections. Pen can also be used when treating the positive control with DNase I. Make sure that the part of the slide where circle will be drawn is completely dry. Let the circle dry for approximately 1 min, before applying solutions to encircled sections.

37. Collect used PBS containing traces of label solution in a tightly closed bottle and discard according to the local regulations for toxic waste disposal.

38. Slides should be removed from PBS one at a time, to avoid drying of the specimen surface. Minimize specimen exposure to the light. Be careful to avoid formation of large air bubbles when placing down the coverslip. Clear nail polish can also be used to seal the coverslip.

39. We do not recommend direct observation of specimens that are mounted only in a few drops of PBS, as PBS quickly evaporates due to the heat produced by the lamp. Consequently, the image appears rather grainy due to the lowered signal level. Instead, we advise the use of PBS/glycerol mixture, which provides longer lasting, high quality image. Or, if available, a mounting medium that is optimized for fluorescence microscopy (nondrying medium with fluorochrome-preserving agents) can be used.

Acknowledgments

This work was supported by the Serbian Ministry of Education, Science and Technological Development, Grant No 173015.

References

1. Lim PO, Kim HJ, Nam HG (2007) Leaf senescence. Annu Rev Plant Biol 58:115–136
2. Yen CH, Yang CH (1998) Evidence for programmed cell death during leaf senescence in plants. Plant Cell Physiol 39:922–927
3. Simeonova E, Sikora A, Charzyńska M et al (2000) Aspects of programmed cell death during leaf senescence of mono- and dicotyledonous plants. Protoplasma 214:93–101
4. Mc Cabe PF, Levine A, Meijer PJ et al (1997) A programmed cell death pathway activated in carrot cells cultured at low cell density. Plant J 12:267–280
5. Pasqualini S, Piccioni C, Reale L et al (2003) Ozone-induced cell death in tobacco cultivar Bel W3 plants. The role of programmed cell death in lesion formation. Plant Physiol 133:1122–1134
6. Lee R-H, Chen S-CG (2002) Programmed cell death during rice leaf senescence is nonapoptotic. New Phytol 155:25–32
7. Gavrieli Y, Sherman Y, Ben-Sasson SA (1992) Identification of programmed cell death in situ via specific labeling of nuclear DNA fragmentation. J Cell Biol 119:493–501
8. Gorczyca W, Gong J, Darzynkiewicz Z (1993) Detection of DNA strand breaks in individual apoptotic cells by the in situ terminal deoxynucleotidyl transferase end nick translation assays. Cancer Res 53:1945–1951
9. Danon A, Delorme V, Mailhac N et al (2000) Plant programmed cell death: a common way to die. Plant Physiol Biochem 38:647–655
10. Wang H, Li J, Bostock RM, Gilchrist DG (1996) Apoptosis: a functional paradigm for programmed cell death induced by a host-selective phytotoxin and invoked during development. Plant Cell 8:375–391
11. Uzelac B, Janošević D, Budimir S (2008) *In situ* detection of programmed cell death in *Nicotiana tabacum* leaves during senescence. J Microsc-Oxford 230:1–3
12. Murashige T, Skoog F (1962) A revised medium for rapid growth and bioassays with tobacco tissue cultures. Physiol Plant 15:473–497

Chapter 22

Activities of Vacuolar Cysteine Proteases in Plant Senescence

Dana E. Martínez, Lorenza Costa, and Juan José Guiamét

Abstract

Plant senescence is accompanied by a marked increase in proteolytic activities, and cysteine proteases (Cys-protease) represent the prevailing class among the responsible proteases. Cys-proteases predominantly locate to lytic compartments, i.e., to the central vacuole (CV) and to senescence-associated vacuoles (SAVs), the latter being specific to the photosynthetic cells of senescing leaves. Cellular fractionation of vacuolar compartments may facilitate Cys-proteases purification and their concentration for further analysis. Active Cys-proteases may be analyzed by different, albeit complementary approaches: (1) in vivo examination of proteolytic activity by fluorescence microscopy using specific substrates which become fluorescent upon cleavage by Cys-proteases, (2) protease labeling with specific probes that react irreversibly with the active enzymes, and (3) zymography, whereby protease activities are detected in polyacrylamide gels copolymerized with a substrate for proteases. Here we describe the three methods mentioned above for detection of active Cys-proteases and a cellular fractionation technique to isolate SAVs.

Key words Cysteine protease, Senescence, Protease activity, Protease labeling, Zymogram, Vacuole, Senescence-associated vacuoles (SAVs), Cellular fractionation

1 Introduction

Cys-proteases have emerged among the main classes of proteases (serine, cysteine, metallo-, and aspartic proteases) as the most frequently associated to senescence, according to gene expression and enzyme activity studies [1–3]. Most Cys-proteases are predicted to locate to the central vacuole (CV) [3, 4]. On the other hand, the senescence-specific Cys-protease SAG12 locates to senescence-associated vacuoles (SAVs), which were shown to harbor the most intense Cys-protease activities within the cell of senescing leaves [5, 6]. Cys-protease activation may be regulated by different posttranslational mechanisms, including cleavage of N-terminal auto-inhibitory domains, inhibitors, and pH [7–9]. Indeed, these mechanisms might also be regulatory steps for activation of constitutively expressed Cys-proteases involved in senescence. Hence, activity-based protease detection (ABPD) represents the most

Yongfeng Guo (ed.), *Plant Senescence: Methods and Protocols*, Methods in Molecular Biology, vol. 1744,
https://doi.org/10.1007/978-1-4939-7672-0_22, © Springer Science+Business Media, LLC 2018

straightforward approach for the study of Cys-proteases involved in senescence, allowing also for their purification [10, 11]. Active Cys-proteases may be indirectly detected in vivo through the use of particular substrates that become detectable upon specific cleavage. The rhodamine 110-based substrate R-6502® becomes fluorescent after Cys-protease catalyzed cleavage and its signal can be monitored by fluorescence microscopy, thereby evidencing specific cellular compartment/s containing active Cys-proteases [5]. In addition, R-6502® fluorescence may be quantified by spectrofluorometry. One drawback of using R-6502® is that it has to be incubated with isolated cells or protoplasts instead of leaf tissue as R-6502® is probably unstable in the apoplast environment.

Active Cys-proteases can be labeled with specific tags often derived from protease-specific inhibitors that react irreversibly with the active-site thiol of these enzymes [10]. These tags are linked to reporter molecules allowing for the detection and purification of the tagged protease. The most used tag in plants has been DCG-04, a biotinylated derivative of the Cys-protease inhibitor E-64. Based on biotin-streptavidin binding systems and using conjugated streptavidin (i.e., streptavidin-peroxidase, streptavidin-coated beads), DCG-04 has been applied for labeling and identification of senescence-associated Cys-proteases by mass spectrometry in *Arabidopsis* [11], for analyzing the senescence-associated profile of vacuolar Cys-proteases in wheat [3], and for labeling Cys-proteases located to senescence-associated vacuoles in tobacco [6].

Zymography is a versatile ABPD technique that allows for the detection of proteolytic activities based on the protease molecular weight and/or isoelectric point with nanoscale sensitivity [12]. The technique uses polyacrylamide gels, commonly SDS-PAGE, copolymerized with a protease substrate such as casein or gelatin. Zymography can also be combined with mass spectrometry techniques to identify the detected protease/s [13]. *In-gel* activity assays also allow for studying the effect of particular variables (i.e., the effect of cofactors, inhibitors, pH), on protease activities. One drawback of zymography is that protease precursors or inactive proteases might become activated during zymogram development [12].

Here we describe protocols for detection of active Cys-proteases by using artificial substrates and by tagging them. The use of the commercial probe R-6502® is presented as a model procedure for in vivo proteolysis detection with a rhodamine 110-based Cys-protease substrate. The use of DCG-04 exemplifies a procedure for active Cys-protease tagging. In addition, as senescence-associated Cys-protease activities seem to concentrate to SAVs, a cellular fractionation protocol to isolate SAVs is described.

2 Materials

Due to the variety of specific chemicals employed in the techniques described here, those reagents used for standard recipes such as for SDS-PAGE (i.e., how to prepare acrylamide/bis-acrylamide stock solution) are omitted. All reagents should be prepared with MiliQ or similar grade water.

2.1 General Reagents and Equipments

1. Laemmli sample buffer (LSB): To make 100 mL 4× LSB, add 15 mL glycerol, 15 mL 0.5 M Tris–HCl, pH 6.8, 10 mL 20% (w/v) sodium dodecyl sulfate (SDS), 15 mg bromophenol blue, and 60 mL MQ water. Before use, mix 1 vol of 4× LSB with 3 vol of sample, and add 100 μL of β-mercaptoethanol in 900 μL of this mixture. To prepare 1× LSB, mix 1 vol 4× LSB with 2.6 vol water and 0.4 vol β-mercaptoethanol.

2. SDS-PAGE resolving buffer mix: To make 100 mL of 12% acrylamide buffer, add 23 mL 1.5 M Tris–HCl, pH 8.8, 40 mL acrylamide/bis-acrylamide 30:1, 2 mL 20% (w/v) SDS, 34.4 mL MQ water, 450 μL 10% (w/v) ammonium persulfate (APS), and 100 μL N,N,N′,N′ tetramethylethylenediamine (Temed) (*see* **Note 1**).

3. SDS-PAGE stacking buffer mix: To make 100 mL, add 25 mL 0.5 M Tris–HCl, pH 6.8, 20 mL acrylamide/bis-acrylamide 30:1, 2 mL 20% (w/v) SDS, 52 mL MQ water, 100 μL 10% (w/v) APS, 100 μL Temed (*see* **Note 1**).

4. Running buffer: 0.2 M Tris, 1.5 M glycine, 0.4% (w/v) SDS.

5. Coomassie Brilliant Blue: 0.1% (w/v) Coomassie Brilliant Blue, 50% (v/v) methanol, 10% (v/v) glacial acetic acid, and 40% (v/v) MQ water. Dissolve the CBB in methanol; add glacial acetic acid and then MQ water. Stir the solution for 3–4 h and then filter through filter paper. Store at room temperature.

6. Destaining solution: 20% (v/v) methanol, 7% (v/v) glacial acetic acid, and 73% (v/v) MQ water.

7. Electrophoresis equipment.

8. Electrotransfer equipment.

9. Refrigerated centrifuge.

2.2 In Vivo Detection of Cys-Protease Activities Using Fresh Cells or Protoplasts

1. Suspension buffer: 0.6 M mannitol, 0.1% (w/v) bovine serum albumin (BSA), 20 mM MES, and pH 6.0 (*see* **Note 2**).

2. Leaf tissue digestion buffer (*see* **Note 3**): 0.5 M mannitol, 0.1% (w/v) BSA, 20 mM MES, and pH 5.8. Add the following digestion enzymes to final concentrations (w/v) of 1.0% cellulase, 0.3% driselase. For protoplasts preparation also add 1.0% pectinase (final concentration).

3. R-6502®: Rhodamine 110, bis-(CBZ-L-phenylalanyl-L-arginine amide), dihydrochloride (Molecular Probes, Eugene, OR, USA). Prepare a stock solution of 5 mM R-6502 in dimethylformamide (DMF) and keep at −20 °C. This stock solution is stable for several months.

4. Fresh isolated cells or protoplasts.

5. Fluorescence microscope: Epifluorescence or laser scanning confocal microscope (LSCM).

2.3 Cysteine Protease Labeling

1. DCG-04: Prepare a stock solution of 5 mM DCG-04 in dimethyl sulfoxide (DMSO) [10]. DCG-04 has not been licensed for commercial use. Professor Matthew Bogyo, Department of Pathology, School of Medicine, Stanford University, USA (http://bogyolab.stanford.edu) has been courteously providing DCG-04 to several labs around the world.

2. Cys-protease irreversible inhibitor E-64: N-(trans-Epoxysuccinyl)-L-leucine 4-guanidinobutylamide. Prepare a stock solution of 5 mM E-64 in DMSO (*see* **Note 4**).

3. Incubation buffer: 25 mM Na-acetate, 10 mM cysteine, pH 5.0, 0.05% Triton ×100 (*see* **Note 5**).

4. Ultrasensitive streptavidin-horseradish peroxidase conjugate: For DCG-04 tagged Cys-protease detection in Western blots by chemiluminescence.

5. Streptavidin-coated beads: For DCG-04 tagged Cys-proteases purification.

6. Acetone.

7. Chemiluminescence detection reagents.

8. PBS-T buffer: 0.8% (w/v) NaCl, 0.02% (w/v) KCl, 0.144% (w/v) Na_2HPO_4, 0.024% (w/v) KH_2PO_4, and 0.2% (v/v) Tween 20. Dilute in MQ water and adjust to pH 7.2.

2.4 In-gel Protease Activity Detection (Zymography)

1. 5.0% gelatin: To make 100 mL of 5% (w/v) gelatin stock solution, prepare a beaker with 100 mL of pre-warmed MQ water at 40 °C with a stirrer bar, and add 5 g of gelatin slowly while stirring (*see* **Note 6**).

2. Leupeptin (N-acetyl-L-leucyl-L-leucyl-L-arginal) protease inhibitor: Prepare a stock solution of 10 mM leupeptin in MQ water (*see* **Note 7**).

3. Extraction buffer: 25 mM Tris, pH 7.5, 2 mM cysteine, 1.0% insoluble polyvinylpyrrolidone.

4. Incubation buffer: 85 mM Na-acetate, pH 4.0, 2 mM cysteine.

5. Wash buffer: 85 mM Na-acetate, pH 4.0, 2 mM cysteine, 2.0% (v/v) Tritón ×100 (*see* **Note 8**).

2.5 Senescence-Associated Vacuoles (SAVs) Isolation

1. Homogenization buffer: 25 mM HEPES, pH 7.0, 0.6 M mannitol, 5 mM cysteine, 2 mM ethylenediaminetetraacetic acid (EDTA), 1 mM PMSF, 2 µg mL^{-1} leupeptin, 1.0% (w/v) polyvinylpolypyrrolidone (*see* **Note 9**).

2. Neutral Red (NR): Make a stock solution of 0.01% (w/v) NR in HEPES buffer (*see* **Note 10**).

3. Sucrose discontinuous gradient solutions: 5, 25, 35, 45, and 60% (w/v) sucrose prepared in 25 mM HEPES, pH 7, 0.6 M mannitol (*see* **Note 11**).

4. Thermolysin stock solution: 3 mg mL^{-1} thermolysin in 25 mM HEPES, pH 7.0.

5. 25 mM CaCl$_2$: Make a stock solution of 25 mM CaCl$_2$ in 25 mM HEPES, pH 7.0.

6. Stop buffer: 7.5 mg mL^{-1} ethylene glycol-bis(2-aminoethylether)-N,N,N′,N′-tetraacetic acid (EGTA) in 25 mM HEPES, pH 7.0.

7. Ultracentrifuge: With a swinging-bucket rotor capable of generating 100,000 × g.

3 Methods

Due to the variety of specific procedures described in this section, standard protocols such as SDS-PAGE and Western blot analysis are omitted but can be found elsewhere [14, 15].

3.1 Preparation of Fresh Cells or Protoplasts from Leaf Tissue

1. Place the plant material in a Petri dish, and cut it to small pieces (around 2–3 mm) with a razor blade under suspension buffer.

2. Expose the Petri dish to a vacuum for cycles of 5 min each. Repeat the vacuum cycle until bubbling appears on the surface of the leaf pieces. Usually 2–4 cycles are sufficient.

3. Replace the suspension buffer with digestion buffer that contains the corresponding digestion enzymes (choose the corresponding enzymes for protoplast or cell isolation). The suspension buffer can be removed by using a Pasteur pipette. A ratio of around 200 mg of fresh weight leaves *per* mL of digestion buffer is suitable for *Arabidopsis* and wheat leaves.

4. Incubate the leaf pieces in digestion buffer at room temperature in the dark with constant gentle shaking for about 1.5 h. Optimal incubation time may vary with plant materials and species.

5. Separate released cells or protoplasts from undigested tissue fragments by gently aspirating the digestion buffer (containing the cells or protoplasts at this step) with a Pasteur pipette.

6. Pellet down isolated cells and protoplasts by centrifuging at 50 × g for 15 min at room temperature, and resuspend isolated cells and protoplasts in suspension buffer.

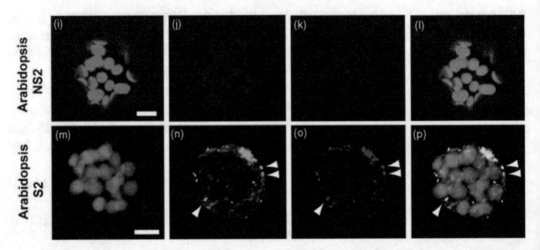

Fig. 1 In vivo detection of Cys-protease activities in isolated protoplasts from *Arabidopsis*. Protoplasts from senescing (S2) and non-senescing (NS2) leaves were labeled with R-6502 and with the acidotropic dye Neutral Red (NR) (k and o), which concentrates in senescence-associated vacuoles (SAVs). Note that R-6502 fluorescence signal is intense in S2 (n), whereas it is almost undetectable in NS2 (j). R-6502 fluorescence localizes predominantly to SAVs (o). Taken from Otegui et al. [5]

3.2 In Vivo Detection of Cys-Protease Activities

1. Dilute R-6502® from stock solution (5 mM R-6502 in DMF) into suspension buffer to a final concentration of 50 μM R-6502®. R-6502® is light sensitive; therefore, keep R-6502®-containing buffers in the dark.

2. Incubate freshly isolated cells or protoplasts in suspension buffer containing 50 μM R-6502® for 20–40 min in the dark (*see* **Note 12**). Small-scale experiments can be carried out in 1.5 mL polypropylene tubes.

3. Wash the cells or protoplasts by spinning down at 50 g for 10 min at room temperature, and resuspend in suspension buffer without R-6502®. Repeat this step at least twice to remove the excess of probe not taken up by cells (*see* **Note 12**).

4. Examine R-6502® fluorescence in cells or protoplasts by confocal microscopy (excitation/emission settings are 488/505–550 nm, although, depending on signal intensity, the emission window can be narrowed down to around 520–530 nm with a spectral detector, such as those of the LEICA SP5 LSCM). Alternative, R-6502® fluorescence might be examined in an epifluorescence microscope equipped with an FITC filter. R-6502® fluorescence can be quantified with a spectrofluorometer using the same ex/em settings as for confocal microscopy. Figure 1 shows a typical image of a senescing leaf protoplast from *Arabidopsis* displaying a punctuate pattern of R-6502 fluorescence in senescence-associated vacuoles.

3.3 Activity-Based Cys-Protease Labeling

1. Homogenize the sample (e.g., leaves, an organelle fraction) in chilled MQ water, and centrifuge at $13,000 \times g$ for 15 min at 4 °C. If leaf tissue is the starting material, grind the tissue directly in chilled MQ water (*see* **Note 13**). Put aside a small aliquot of the homogenate to run an E64 control.

2. For the E-64 control, preincubate one aliquot of the sample with E-64 by adding 5 mM E-64 stock solution to the sample to make a final concentration of 100 μM E-64. After at least 1 h of pre-incubation with E-64, proceed to **step 3** (*see* **Note 14**).

3. Mix 1 vol of sample with 3 vol of incubation buffer containing 5 μM DCG-04, and incubate at room temperature with gentle shaking for 3–4 h.

4. After incubation with DCG-04, samples can be processed for further DCG-04 tagged Cys-protease detection (**steps 5–7** below). Alternatively, DCG-04 bound proteases can be purified from the sample by using streptavidin-coated beads (**steps 8–10** below).

5. For detection in Western blots, precipitate DCG-04-labeled samples (**steps 2** and **3** above) with chilled (−20 °C) acetone (*see* **Note 15**).

6. Centrifuge at $13,000 \times g$ for 15 min at 4 °C, discard the acetone, and let the pellet dry out at room temperature. Resuspend the pellet in 1× LSB and boil for 1 min (*see* **Note 16**).

7. Analyze the DCG-04 bound proteases contained in the sample by Western blot using streptavidin-horseradish peroxidase conjugate (go to **step 11**).

8. For Cys-proteases purification, incubate the DCG-04-labeled sample with 100–200 μL of streptavidin-coated beads at room temperature with shaking for about 2–4 h. Small-scale experiments can be performed in 1.5 mL polypropylene tubes.

9. Wash the beads by spinning down in a centrifuge, and resuspend in incubation buffer. Repeat at least five times to wash out DCG-04 unbound proteins.

10. Elute DCG-04 bound proteins by boiling the streptavidin-coated beads in 1× LSB for 1 min (*see* **Note 16**). Spin down the streptavidin-coated beads and recover the supernatant. This contains purified Cys-proteases.

11. Load the protein sample resuspended in LSB (from **steps 6** or **10**) directly on a 12% acrylamide SDS-PAGE gel, run the gel, and electro-transfer to a nitrocellulose membrane (*see* **Note 17**).

12. Block the transferred nitrocellulose membrane with 10% (w/v) nonfat milk dissolved in PBS-T buffer.

13. Incubate the nitrocellulose membrane with (1:2000 v/v) ultrasensitive streptavidin-peroxidase conjugate for 1 h.

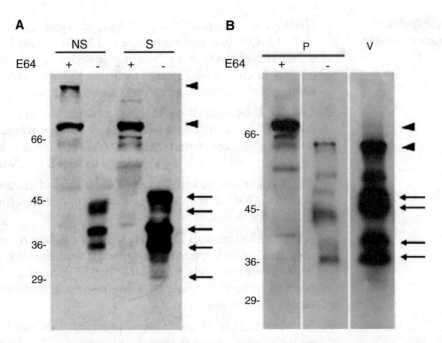

Fig. 2 DCG-04 tagged Cys-protease profiling in senescing (S) and non-senescing (NS) wheat leaves. E-64 pre-incubated samples were loaded on the left of each sample (E64 +). (**a**) DCG-04 tagged proteases (marked with arrows) increased in S respect to NS samples. Note that DCG-04 tagged proteases are not detected when the sample is preincubated with E-64, whereas unspecific DCG-04 tagged proteins are detected also when preincubated with E-64 (arrowheads). (**b**) DCG-04-labeled Cys-proteases locate to the central vacuole (V isolated vacuoles, P protoplast). Taken from Martinez et al. [3]

14. Wash the membrane with PBS-T buffer for 5–6 times.

15. Incubate the membrane with the chemiluminescence reagents according to the detection system and develop.

16. Figure 2 shows a Western blot of DCG-04 tagged Cys-proteases from senescing leaf extracts of wheat. These proteases are detected in isolated central vacuoles.

3.4 In-Gel Protease Activity Detection (Zymography)

1. Cast activity gels. Use conventional SDS-PAGE resolving and stacking buffer mix recipes with the addition of 5% (w/v) gelatin to a final concentration of 0.04% (*see* **Note 18**).

2. Homogenize the sample in extraction buffer, and centrifuge at $13,000 \times g$ for 15 min at 4 °C. If leaf tissue is the starting material, grind the tissue directly in extraction buffer (*see* **Note 19**).

3. Place the polymerized activity gel in the electrophoresis chamber, and fill it with chilled running buffer. Set up the electrophoresis system (electrophoresis chamber plus power supply unit) at 4 °C with enough time in advance to allow for temperature equilibration.

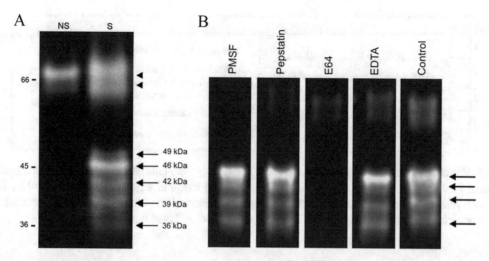

Fig. 3 Senescence-associated Cys-protease activities detected by zymography. (**a**) Protease activities from senescing (S) and non-senescing (NS) leaf extracts of wheat were examined in activity gels. A group of active bands between 36 and 50 kDa is specific to senescing leaves. (**b**) The mechanistic type of proteases responsible for the senescence-related proteolytic activities was determined by using a battery of specific protease inhibitors. The addition of the Cys-protease inhibitor E-64, but not other inhibitors, prevented Cys-protease activity detection. Taken from Martinez et al. [3]

4. Right before running the gel, mix 1 vol of sample with 0.25 vol of 4× LSB, and load immediately onto the gel (*see* **Note 20**).

5. Once the electrophoresis is completed, place the gel in wash buffer with gentle shaking at 4 °C for at least 1 h. Gel incubation in wash buffer for longer than 3 h is not recommended (*see* **Note 21**).

6. Replace the wash buffer with incubation buffer, and place the gel at 37 °C for 14–16 h (*see* **Notes 21** and **22**).

7. Incubate the gel in Coomassie Brilliant Blue with gentle shaking. Gel staining may take from a few hours. Replace Coomassie Brilliant Blue with distaining solution. Active proteases are detected as clear bands on a blue background. Figure 3 shows senescence-associated Cys-protease activities detected by zymography. Protease activities from senescing (S) and non-senescing (NS) leaf extracts of wheat were examined in activity gels.

3.5 Senescence-Associated Vacuoles (SAVs) Isolation (see Note 23)

Carry out all procedures at 4 °C unless otherwise specified.

1. Prepare sucrose fractions (25, 35, 45 and 60% w/v) according to Table 1 (*see* **Note 24**).

2. Prepare a discontinuous sucrose concentration gradient in centrifuge tubes appropriate for a swinging-bucket rotor. For a 5 mL tube, pipette the volumes of each sucrose solution shown in Fig. 4. Start from the heaviest (60%) to the lightest (5%) solution.

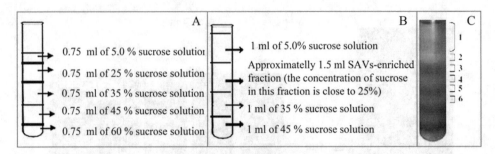

Fig. 4 SAVs isolation with discontinuous sucrose gradients. (**a, b**) Diagrams depicting the volume and position of each fraction in the first and second sucrose gradients. (**c**) Image of a typical sucrose gradient for SAVs isolation. After centrifugation, a reddish band corresponding to Neutral Red-labeled SAVs can be observed. The SAV-containing fraction locates to around 25% sucrose (fraction 3). Taken from Martinez et al. [16]

Table 1
First discontinuous sucrose concentration gradient fraction preparation
(*see* Note 23)

Sucrose fraction (%)	25 mM HEPES pH 7.0 (mL)	60% sucrose stock (mL)	Mannitol (g)
5	9.2	0.8	1.09
25	5.8	4.2	1.09
35	4.1	5.9	1.09
45	2.5	7.5	1.09
60	–	10	1.09

3. Cut approximately 25 g of tobacco leaves into small pieces and transfer immediately to a glass beaker with 100 mL of homogenization buffer; disrupt the leaf tissue using a homogenizer. Five up-and-down strokes, 1 s each, are sufficient for good leaf tissue homogenization.

4. Add 1 mL of Neutral Red (NR) stock solution to the homogenate in order to get a final concentration of 0.01% (w/v) NR. Stir gently for 15 min at room temperature.

5. Filter through a nylon mesh, and centrifuge at 2500 × *g*, 4 °C for 5 min to pellet chloroplasts and chloroplast membranes. Discard the pellet and recover the supernatant (leaf extract).

6. Layer approximately 1.7 mL of the leaf extract on top of the gradient. Centrifuge at 100,000 × *g* for 1 h at 4 °C in swinging-bucket rotor. After centrifugation, several bands will be noticeable (Fig. 4). A red band located at approximately 25% sucrose corresponds to the SAV-enriched fraction. This band is red because SAVs concentrate the acidotropic dye NR.

Table 2
Second discontinuous sucrose concentration gradient fraction preparation (*see* Note 23)

Sucrose fraction (%)	25 mM HEPES pH 7.0 (mL)	60% sucrose stock solution (mL)	Mannitol (g)
5	9.2	0.8	1.09
35	4.1	5.9	1.09
45	2.5	7.5	1.09

7. Remove the fractions carefully with a pipette, and place the SAV-enriched red fraction in a separate tube.

8. The SAV-enriched fraction may also contain contaminating proteins. To remove contaminating soluble proteins present outside SAVs, incubate the SAV-enriched fraction with thermolysin. To this end add 2 μL thermolysin stock solution and 2 μL 25 mM $CaCl_2$ per 100 μL of SAV-enriched fraction. Final concentrations in the reaction mixture should be 60 μg mL^{-1} thermolysin and 0.5 mM $CaCl_2$. Mix the contents by gently inverting the tube. Incubate for 30 min at 4 °C, and stop the thermolysin treatment by adding 10 μL EGTA stop buffer stock in 100 μL reaction (*see* **Note 25**).

9. Re-isolate SAVs in a density gradient as performed before, layering SAVs on top of a second sucrose gradient (Table 2). Centrifuge at $100,000 \times g$ for 1 h at 4 °C in swinging-bucket rotor. The SAVs fraction is the red fraction floating on 35% sucrose. Recover the SAVs fraction and store at −20 or −80 °C.

4   Notes

1. APS catalyzes acrylamide polymerization; therefore, it should be added to the SDS-PAGE resolving or stacking mix right before pouring the mix into the gel mold.

2. Mannitol might show some auto fluorescence under R-6502® ex/em settings (488/505–530).

3. Driselase is a mixture of polysaccharide hydrolases, i.e., cellulase, pectinase, beta-xylanase, and beta-mannanase. Its use improves the yield of isolated cells and protoplasts, although, we found that it is not essential for the procedure described here. In our experience either 0.3% (w/v) pectinase or 0.3% (w/v) pectolyase is effective for *Arabidopsis* and wheat leaf protoplasts isolation. However, it has been shown that the efficiency of the digestion enzymes on cell wall degradation might vary across species. Macerozyme represents an alternative to pectinase.

4. Either E-64c or E-64d, an ethyl ester of E-64c, may be used.

5. Na-acetate buffer pH 5.0 is suggested. However other buffers around pH 5.0 or at other pH values might be suitable or even optimal depending on the plant species and types of sample.

6. It is important to keep the water warm during gelatin dissolution. Keep the beaker in a hot water bath while stirring. Gelatin does not dissolve properly in cold water, whereas it becomes sticky when it is poured rapidly into hot water. Gelatin stock solution can be stored at −20 °C. We have tested gelatins provided by different suppliers, and no substantial differences between them were observed when used as zymogram substrate.

7. Leupeptin is a cysteine, serine, and threonine reversible protease inhibitor. The addition of low concentrations (10 μM) of leupeptin to the extraction buffer minimizes protease activities prior and during the gel run and is then washed out during the gel washing step. The use of leupeptin is recommended; however it is not essential for zymography.

8. Dithiothreitol (DTT) at the same concentration can be used instead of cysteine. DTT or cysteine should be added to the buffers right before use. In the protocol described here, wash buffer and incubation buffer are prepared at pH 4.0, based on the optimal Cys-protease *in-gel* activity determined for senescing leaves of wheat. However lower or higher pH values might be optimal for Cys-proteases from other plant species.

9. HEPES buffer can be stored at 4 °C for approximately a month, but pH should be checked before use. Homogenization buffer cannot be stored and must be prepared immediately before use.

10. Neutral Red solution is unstable. Make fresh immediately before use and keep away from light.

11. Dissolve sucrose in a considerably smaller volume than the final volume needed because sucrose solutions increase in volume. Adjust the volume after complete dissolution of sucrose.

12. The optimal incubation time of 20–40 min was determined for *Arabidopsis* protoplasts and soybean cells; plant materials from other species might require other incubation times that should be determined empirically. Prolonged incubation of the sample with R-6502® might lead to over staining.

13. If leaf tissue is the starting material, use chilled mortar and pestle, and add quartz sand to chilled MQ water to facilitate tissue grinding (use 1–2 mg tissue/mL MQ water). If a cellular fraction (i.e., CV, SAVs, other) is used as starting material, homogenize with chilled MQ water (use 1–2 mL cellular fraction/mL MQ water).

14. Some proteins might bind to the biotin moiety of DCG-04 (which is referred to as unspecific binding). Therefore E-64 competition experiments are essential to identify DCG-04 specific binding to Cys-proteases. Those bands that are observed (either in a Western blot or in a Coomassie Blue-stained SDS-PAGE gel of purified Cys-proteases) in DCG-04 tagged samples preincubated with E-64 are due to unspecific binding. Those bands that are not detected when the sample is E-64 preincubated (i.e., where the active site of the protease is irreversibly occupied by E-64 prior to incubation with DCG-04) correspond to Cys-proteases (Fig. 2).

15. Mix 1 vol of sample with 5 vol of chilled (−20 °C) acetone. Incubate at −20 °C for at least 6 h or overnight.

16. The use of small volumes of LSB is highly recommended in order to avoid excessive dilution of proteins in the sample. The addition of protease inhibitors to LSB is recommended, ideally an inhibitor "cocktail" or a mix of serine- and metallo-protease inhibitors (i.e., phenylmethanesulfonyl fluoride (PMSF) and 10-phenanthroline or ethylenediaminetetraacetic acid (EDTA) for serine and metallo-proteases, respectively).

17. It is convenient to load the sample in the SDS-PAGE alongside with the sample previously incubated with E-64 in order to facilitate the comparison of profiles and identification of unspecific bands (Fig. 2).

18. Prepare the resolving buffer mix in a beaker by adding the MQ water first, then add the gelatin stock (warm or at room temperature), and then the rest of the reagents. As a determined volume of 5.0% (w/v) gelatin stock will be added, the volume of water should be recalculated to keep the final volume needed for casting gels. For 100 mL of gelatin-containing resolving buffer, prepare a beaker with a stirrer in motion, and add 24.4 mL MQ water, 10 mL of 5.0% (w/v) gelatin, 40 mL of acrylamide/bis-acrylamide 30:1, 23 mL 1.5 M Tris–HCl, pH 8.8, and 2 mL 20% (w/v) SDS. Immediately before loading the mix, add 100 μL Temed and 450 μL 10% (w/v) APS (*see* **Note 1**). After the resolving gel polymerizes, continue by pouring the stacking gel mix according to the steps for conventional SDS-PAGE gels. Once the gel is polymerized, keep it at 4 °C. Activity gels should be used fresh or within no more than 2 days when stored at 4 °C.

19. If leaf tissue is the starting material, use chilled mortar and pestle, and add quartz sand to chilled extraction buffer to facilitate tissue grinding (use 1–2 mg tissue/mL buffer). If a cellular fraction (i.e., CV, SAVs, other) is used as starting material, homogenize with chilled extraction buffer (use 1–2 mL cellular fraction/mL buffer).

20. If the effect on protease activities of a given reagent or protease irreversible inhibitor is to be tested, incubate the sample for at least 1 h at 4 °C with the reagent or inhibitor before adding 4× LSB. Prior tests of the effects of β-mercaptoethanol in sample buffer should be done as resolution can be improved by omitting β-mercaptoethanol.

21. For small gels (8.6 × 6.7 cm, 1 mm width), use 50 mL of wash buffer and 50 mL of incubation buffer.

22. The incubation time should be determined empirically for each sample. Shorter time than optimal may not be enough to detect activities, whereas longer times may lead to excess of activities, reducing the sensitivity of the method (i.e., one thick activity band is detected instead of two activity bands of similar size). Ensure that the gel does not dehydrate during the prolonged exposure to 37 °C. If a plastic container is used for incubation, it can be wrapped in plastic bags with wet filter paper inside.

23. This procedure has been optimized for tobacco leaves and might need to be modified for other plant species.

24. Prepare the sucrose concentration gradient tubes on ice or in a cold room until you are ready to use them. The sucrose gradients should be used within 2 h after preparation.

25. EGTA chelates the Ca^{+2} that thermolysin requires for activity. Check that thermolysin works. Take a small volume of the SAV-enriched fraction, add BSA as a control, and perform the thermolysin treatment. Run this reaction along with an appropriate negative control (i.e., SAVs plus BSA before thermolysin addition). Run a standard SDS-PAGE gel, and examine changes in BSA band concentration (BSA is easily observed around 66 kDa).

References

1. Guo Y, Cai Z, Gan S (2004) Transcriptome of Arabidopsis leaf senescence. Plant Cell Environ 27:521–549

2. Gepstein S, Sabehi G, Carp MJ et al (2003) Large-scale identification of leaf senescence-associated genes. Plant J 36:629–642

3. Martínez DE, Bartoli CG, Grbic V, Guiamet JJ (2007) Vacuolar cysteine proteases of wheat (Triticum aestivum L.) are common to leaf senescence induced by different factors. J Exp Bot 58:1099–1107

4. Kinoshita T, Yamada K, Hiraiwa N et al (1999) Vacuolar processing enzyme is up-regulated in the lytic vacuoles of vegetative tissues during senescence and under various stressed conditions. Plant J 19:43–53

5. Otegui MS, Noh YS, Martínez DE et al (2005) Senescence-associated vacuoles with intense proteolytic activity develop in leaves of Arabidopsis and soybean. Plant J 41:831–844

6. Carrión CA, Costa ML, Martínez DE et al (2013) In vivo inhibition of cysteine proteases provides evidence for the involvement of 'senescence-associated vacuoles' in chloroplast protein degradation during dark-induced senescence of tobacco leaves. J Exp Bot 64:4967–4980

7. Yamada K, Matsushima R, Nishimura M et al (2001) A slow maturation of a cysteine protease with a granulin domain in the vacuoles of senescing Arabidopsis leaves. Plant Physiol 127:1626–1634

8. Rojo E, Zouhar J, Carter C et al (2003) A unique mechanism for protein processing and degradation in Arabidopsis thaliana. Proc Natl Acad Sci 100:7389–7394

9. Battelli R, Lombardi L, Picciarelli P et al (2014) Expression and localisation of a senescence-associated KDEL-cysteine protease from Lilium longiflorum tepals. Plant Sci 214:38–46

10. Greenbaum D, Medzihradszky KF, Burlingame A et al (2000) Epoxide electrophiles as activity-dependent cysteine protease profiling and discovery tools. Chem Biol 7:569–581

11. Van der Hoorn RA, Leeuwenburgh MA, Bogyo M et al (2004) Activity profiling of papain-like cysteine proteases in plants. Plant Physiol 135:1170–1178

12. Vandooren J, Geurts N, Martens E, Van den Steen PE et al (2013) Zymography methods for visualizing hydrolytic enzymes. Nat Methods 10:211–220

13. Martinez DE, Borniego ML, Battchikova N, Aro EM, Tyystjärvi E, Guiamét JJ (2015) SASP, a senescence-associated subtilisin protease, is involved in reproductive development and determination of silique number in Arabidopsis. J Exp Bot 66:161–174

14. Bollag DM, Edelstein S, Rozicky M (1996) Proteins methods. Jhon Wiley and Sons Publications, New York

15. Hames BD (1998) Gel electrophoresis of proteins: a practical approach, vol 197. Oxford University Press, New York

16. Martínez DE, Costa ML, Gomez FM et al (2008) 'Senescence-associated vacuoles' are involved in the degradation of chloroplast proteins in tobacco leaves. Plant J 56:196–206

Chapter 23

Study of Autophagy in Plant Senescence

Xuefei Cui, Jing Zheng, Jinxin Zheng, and Qingqiu Gong

Abstract

As a major intracellular degradation pathway, autophagy contributes to nutrient recycling and is indispensable during plant senescence. Here we describe methods used for investigating the autophagic process during leaf senescence. These include transcript analysis of core machinery autophagy genes, immunoblotting of ATG8, and microscopic observation of autophagosome formation.

Key words Autophagy, Leaf senescence, ATG8, Transcript analysis, Immunoblots, Confocal laser scanning microscopy

1 Introduction

Macroautophagy (autophagy) is an intracellular trafficking pathway that involves the formation of a double-membrane vesicle termed autophagosome and its subsequent fusion with the vacuole/lysosome [1]. Then the so-called autophagic body, which includes the inner membrane and the contents, is degraded by the vacuolar proteases to yield free amino acids. In plants, autophagy has critical roles in nutrient recycling, which is an important process during plant senescence [2, 3]. Nearly all null mutants of Arabidopsis *AuTophaGy (ATG)* genes undergo early senescence and produce less seeds under optimum growth conditions and are sensitive to nutrient starvation [4–7]. During leaf and petal senescence, expression of essential *ATG* genes is induced, accompanying elevated autophagic activities [8, 9]. Autophagy is also partly responsible for chloroplast protein and even chloroplast turnover during dark-induced leaf senescence and natural leaf senescence [10–12].

Many assays have been developed to monitor autophagy in plants. Some measure the numbers and/or volumes of autophagic elements at a given time point, such as doing quantitative, real-time PCR (Q-RT-PCR) of essential *ATG* genes [8, 9] and staining of acidic intracellular compartments with fluorescent dyes [13]. Others investigate time-dependent accumulation of a cargo marker or an

Yongfeng Guo (ed.), *Plant Senescence: Methods and Protocols*, Methods in Molecular Biology, vol. 1744,
https://doi.org/10.1007/978-1-4939-7672-0_23, © Springer Science+Business Media, LLC 2018

autophagosome/autophagic body marker [5, 14, 15], so that the relative rate of autophagy can be estimated and compared.

Here we introduce three commonly used methods in the study of autophagy in plant senescence, using dark-induced leaf senescence as an example. The first one is to document the mRNA level of *ATG* genes that are essential for autophagosome formation over the course of leaf senescence. The second is to quantify ATG8 protein levels. The third is to observe autophagosome formation with fluorescent microcopy by examining the presence and changing distribution of GFP-ATG8 inside the cell. For results, *see* Fig. 1.

2   Materials

2.1   Plant Growth

1. Seeds: For *Arabidopsis thaliana* (ecotype Columbia-0) and *Nicotiana benthamiana*.
2. Soil.

2.2   Quantitative Real-Time PCR

1. First-Strand cDNA Synthesis Kit.
2. Gene-specific primers (Table 1) (*see* **Note 1**).
3. SYBR Premix Ex Taq II (TaKaRa).

2.3   Immunoblots

1. Total protein extracted from senescing *Arabidopsis* leaves, 10 to 15 μg each.
2. Antibodies: anti-ATG8 antibody [16], anti-actin antibody (AbMart, Shanghai, China), peroxidase-labeled secondary antibodies.

2.4   Transient Transformation of N. benthamiana Leaves and Confocal Laser Scanning

1. Microscopy *Agrobacterium tumefaciens* strain GV3101.
2. Binary vector containing *p19* of tomato bushy stunt virus (TBSV) [17].
3. Binary vector containing *pUBQ10:GFP-ATG8a* or *pATG8a: GFP-ATG8a* (*see* **Notes 2** and **3**).
4. Rifampin stock: 30 mg/mL.
5. Kanamycin stock: 30 mg/mL.
6. Gentamicin stock: 80 mg/mL.
7. Agrobacterium selection medium: 10 g yeast extract, 10 g tryptone, 5 g NaCl in 1 L distilled water with appropriate antibiotics.
8. Infiltration medium: 10 mM $MgCl_2$, 10 mM MES, 200 μM acetosyringone.
9. 100 μM E-64d or 0.5 μM concanamycin A (final concentrations), stock solution prepared with DMSO.
10. 5% glycerol.
11. Confocal laser scanning microscope.

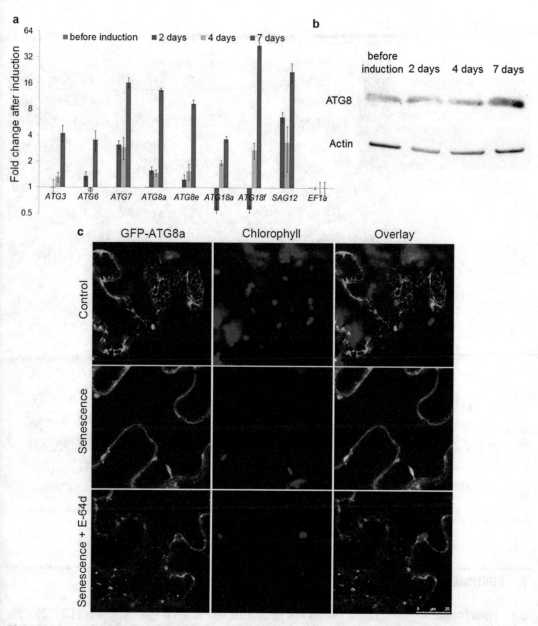

Fig. 1 Representative results. (**a**) Quantitative real-time PCR. Four-week-old soil-grown *Arabidopsis* plants are kept in the dark for 2, 4, and 7 days to induce senescence. Rosette leaves Nos. 3 to 6 are collected for Q-RT-PCR analysis. Control leaves are collected at day 0. Fold changes relative to control (set as 1) are plotted in log scale. Bars = SD. (**b**) ATG8 protein accumulation during dark-induced senescence. Plant materials used are the same as in (**a**). An antibody raised against GmATG8a is used. Anti-actin is used as a loading control. (**c**) ATG8-labeled structures in control leaves (top panel), leaves stayed in the dark for 24 h (middle panel), and leaves stayed in the dark for 23.5 h plus 0.5 h of E-64d treatment in the dark (bottom panel)

Table 1
Q-RT-PCR primers for Arabidopsis *ATG* genes

Names	Sequences
ATG8aAt4g21980F	GGGTTTGTTTCTCCCCCGAT
ATG8aAt4g21980R	TCCTTGCCTCGAGAGGGTTA
ATG8eAt2g45170F	AGCTGGAAGGATCAGGGAGA
ATG8eAt2g45170R	GGTCTGATGGCACAAGGTACT
ATG3At5g61500F	GGAGGAGTTCGATGAGGCTG
ATG3At5g61500R	CGTCCGAGTTCGGAGAATGT
ATG6At3g61710F	CTGGAAAGGCGATGGAGGAA
ATG6At3g61710R	TGGCCCACTTCAAGAGACAG
ATG7AT5G45900F	GTAATCCTGGCTGGCCACTT
ATG7AT5G45900R	CAGCGAAACCACGACTCTCT
ATG18aAt3g62770F	TGCAGTCGTGGAGATGCTTT
ATG18aAt3g62770R	GCCCTGGTGATCATCCCAAA
ATG18fAt5g54730F	TAACCCCGGAAGTGAAACCG
ATG18fAt5g54730R	CGAGAACAGCACCTAACCGA
EF1aAt5g60390F	AAGAGCGTGGACAAGAAGGA
EF1aAt5g60390R	AGTCTCATCATTTGGCACCC
SAG12At5g45890F	TGCTTTGCCGGTTTCTGTTG
SAG12At5g45890R	TGCTCCTTCAATAGCCGCAA

EF1a is used as internal control. *SAG12* is used as a positive control for leaf senescence

3   Methods

3.1   Plant Growth and Senescence Induction

1. All plants are grown in soil at 16 h (22 °C)/8 h (18 °C) with a photosynthetic photon flux density at 90 µE m^{-2} s^{-1}. For dark-induced senescence, plants were kept in the dark at the same temperature settings.

2. For dark-induced senescence of *Arabidopsis*, 4-week-old *Arabidopsis* plants are kept in the dark for 2, 4, and 7 days. Four-week-old plants grown at the same conditions are used for control. The 3–6 leaves are collected from more than ten plants each time for RNA and protein extractions (*see* **Note 4**).

3. For dark-induced senescence of *N. benthamiana*, Agrobacterium-transformed plants are kept in the dark for 1–2 days before microscopy.

3.2 Quantitative Real-Time PCR

1. Primers for the amplification are designed so that the PCR products are of similar lengths (70–140 bp) and of high specificity (*see* **Note 5**).

2. Follow the manufacturer's instructions for Q-PCR. Thaw all reagents at room temperature, mix gently, and keep on ice. Five technical repeats are done for the internal control gene, and three technical repeats are done for genes of interest (*see* **Note 6**).

   For each 25 μL reaction, mix on ice (*see* **Note 7**):

SYBR Premix Ex Taq II (2×)	12.5 μL
Forward primer (10 μM)	1 μL
Reverse primer (10 μM)	1 μL
cDNA (1:20 dilution)	2 μL
Distilled water	8.5 μL

   Cycler conditions (two-step PCR): 30 s at 95 °C, 40 cycles of 5 s at 95 °C, and 30 s at 60 °C. A melt curve is generated immediately after the PCR procedure. Reactions that only generate a single PCR product are used for further analysis (*see* **Notes 8** and **9**).

3. Gene expression is normalized to that of the internal control gene (*EF1a* in this case) by subtracting the Cycle$_{threshold}$ (Ct) value of *EF1a* from the Ct value of the gene of interest. Expression ratio is then generated from the calculation $2^{\Delta\Delta Ct}$, where $\Delta\Delta Ct$ represents ΔCt control minus ΔCt treatment.

3.3 Immunoblots

1. Protein samples are subjected to SDS-PAGE and then electrophoretically transferred onto polyvinylidene fluoride membranes for immunoblotting.

2. Perform standard SDS-PAGE immunoblotting. An antibody against ATG8 is used with 1:1000 dilution. An anti-actin antibody (AbMart, Shanghai, China) is used at 1:5000. Peroxidase-labeled goat anti-rabbit or goat anti-mouse secondary antibodies are used at 1:10,000 (*see* **Note 10**).

3.4 Transient Transformation of N. benthamiana Leaves and Confocal Laser Scanning Microscopy

1. Inoculate GV3101 and p19 containing Agrobacterium strain GV3101 in appropriate selection medium. For *pUBQ10:GFP-ATG8a* and *pATG8a:GFP-ATG8a* expression, medium containing 30 mg/L rifampin, 30 mg/L kanamycin, and 80 mg/L gentamicin is used for selection. Grow at 28 °C at 220 rpm overnight or until OD$_{600}$ = 1.0–1.2.

2. Spin down Agrobacterium cultures at 2000 g for 30 min. Discard supernatant.

3. Resuspend the GFP-ATG8a-expressing cultures with freshly prepared infiltration medium to OD$_{600}$ = 0.4–0.6. Resuspend the p19 culture to OD$_{600}$ = 0.8–1.0.

4. Incubate the cultures at room temperature for 3 h to overnight.

5. Mix GFP-ATG8a and p19 cultures in a 1:1 ratio. Prepare 20 mL mixture for each construct.

6. Use 1 mL syringes without needle to inject Agrobacterium mixture onto *N. benthamiana* leaves until the mixture is used up. Label the leaves (*see* **Note 11**).

7. Keep the plants under normal growth conditions for 2–3 days. Check the GFP expression daily with a UV light.

8. For dark-induced senescence, keep the plants in the dark for 1–2 days before microscopy (*see* **Note 12**).

9. For inhibitor treatments, collect leaves, and add E-64d or concanamycin A at appropriate time points (1–12 h) prior to microscopy.

10. Prepare leaf discs and incubate in 5% glycerol and scan with a confocal laser scanning microscope.

4   Notes

1. Not all plant *ATG* genes are transcriptionally regulated during leaf senescence. The genes used here in the example are selected based on publicly available, large-scale analyses and our own experience.

2. To monitor ATG8s with fluorescent proteins or tags, ATG8 should always be N-terminally tagged. Upon activation of autophagy, the C-terminus of ATG8 is processed by ATG4 to expose the glycine residue to facilitate ATG8 conjugation. Thus only N-terminally tagged ATG8s can serve as markers for autophagic flux.

3. We prefer UBQ10 promoter or native ATG8 promoter over 35S promoter to drive the expression of ATG8 because overexpression of ATG8 can induce autophagy and change plant development [16]. The intracellular distribution of GFP-ATG8 also appears different when driven by 35S promoter.

4. Expression of many *ATG* genes is under the influence of circadian rhythm. Therefore, always collect samples at the same time of the day.

5. Specificity is crucial when designing Q-RT-PCR primers since SYBR Green binds all double-stranded DNA to give fluorescent signals. Special attention must be paid to ATG8s and ATG18s since they are protein families. We recommend using NCBI Primer-BLAST (http://www.ncbi.nlm.nih.gov/tools/primer-blast/) for primer design.

6. When mixing reagents for Q-RT-PCR, the mixture must be kept on ice to prevent unspecific amplification. To reduce bleaching of SYBR Green, this step can be carried out under dim light. Caution should be taken to avoid formation of air bubbles, which hampers correct reading.

7. Primer concentration should range from 0.2 to 1.0 μM in the final mixture. 0.4 μM works best for us. cDNA dilution should be tested, too. 1:10 to 1:20 works in our hands.

8. The PCR program can be adjusted to improve amplification efficiency. Note that prolonged denaturing steps may be harmful to enzyme activities.

9. For specificity, it is necessary to generate melting curves for all PCR products upon completion of PCR reactions.

10. ATG8s have molecular weights of approximately 13 kD; hence appropriate concentration of SDS-PAGE gel should be selected. 6 M urea can be added to sharpen the bands.

11. For transient expression of GFP-ATG8, the age of plants is critical. Three to four-week-old, healthy plants work the best. If the plants are older, then transformation efficiency drops drastically, and the expression of GFP-ATG8 can also be significantly reduced.

12. With co-expression of the p19 protein, expression of GFP-ATG8 can be prolonged to up to 5 days. This is the time limit for experiment plans. In our hands, 24 to 48 h of dark treatment works best for the observation of autophagosome formation.

Acknowledgments

This work was supported by the National Natural Science Foundation of China (31401179) and the Fundamental Research Funds for the Central Universities to QG.

References

1. Mizushima N, Yoshimori T, Ohsumi Y (2011) The role of Atg proteins in autophagosome formation. Annu Rev Cell Dev Biol 27: 107–132

2. Diaz C, Lemaitre T, Christ A et al (2008) Nitrogen recycling and remobilization are differentially controlled by leaf senescence and development stage in Arabidopsis under low nitrogen nutrition. Plant Physiol 147: 1437–1449

3. Guiboileau A, Yoshimoto K, Soulay F et al (2012) Autophagy machinery controls nitrogen remobilization at the whole-plant level under both limiting and ample nitrate conditions in Arabidopsis. New Phytol 194: 732–740

4. Hanaoka H, Noda T, Shirano Y et al (2002) Leaf senescence and starvation-induced chlorosis are accelerated by the disruption of an Arabidopsis autophagy gene. Plant Physiol 129:1181–1193

5. Thompson AR, Doelling JH, Suttangkakul A et al (2005) Autophagic nutrient recycling in Arabidopsis directed by the ATG8 and ATG12

conjugation pathways. Plant Physiol 138:
2097–2110

6. Xiong Y, Contento AL, Bassham DC (2005)
AtATG18a is required for the formation of
autophagosomes during nutrient stress and
senescence in Arabidopsis thaliana. Plant
J 42:535–546

7. Phillips AR, Suttangkakul A, Vierstra RD (2008)
The ATG12-conjugating enzyme ATG10 is
essential for autophagic vesicle formation in
Arabidopsis thaliana. Genetics 178:1339–1353

8. Shibuya K, Niki T, Ichimura K (2013)
Pollination induces autophagy in petunia petals
via ethylene. J Exp Bot 64:1111–1120

9. Van Der Graaff E, Schwacke R, Schneider A
et al (2006) Transcription analysis of arabidopsis membrane transporters and hormone pathways during developmental and induced leaf
senescence. Plant Physiol 141:776–792

10. Wada S, Ishida H, Izumi M et al (2009)
Autophagy plays a role in chloroplast degradation during senescence in individually darkened
leaves. Plant Physiol 149:885–893

11. Ishida H, Yoshimoto K, Izumi M et al (2008)
Mobilization of rubisco and stroma-localized
fluorescent proteins of chloroplasts to the
vacuole by an ATG gene-dependent autophagic
process. Plant Physiol 148:142–155

12. Ono Y, Wada S, Izumi M et al (2013)
Evidence for contribution of autophagy to
rubisco degradation during leaf senescence in
Arabidopsis thaliana. Plant Cell Environ
36:1147–1159

13. Contento AL, Xiong Y, Bassham DC (2005)
Visualization of autophagy in Arabidopsis
using the fluorescent dye monodansylcadaver-
ine and a GFP-AtATG8e fusion protein. Plant
J 42:598–608

14. Yoshimoto K, Hanaoka H, Sato S et al (2004)
Processing of ATG8s, ubiquitin-like proteins,
and their deconjugation by ATG4s are essential for plant autophagy. Plant Cell
16:2967–2983

15. Xiong Y, Contento AL, Nguyen PQ et al
(2007) Degradation of oxidized proteins by
autophagy during oxidative stress in
Arabidopsis. Plant Physiol 143:291–299

16. Xia T, Xiao D, Liu D et al (2012) Heterologous
expression of ATG8c from soybean confers tolerance to nitrogen deficiency and increases
yield in Arabidopsis. PLoS One 7:e37217

17. Voinnet O, Rivas S, Mestre P et al (2003) An
enhanced transient expression system in plants
based on suppression of gene silencing by the
p19 protein of tomato bushy stunt virus. Plant
J 33:949–956

Chapter 24

Isolation of Chloroplasts for *In Organelle* Protein Degradation Assay

Scott W. Vande Wetering and Judy A. Brusslan

Abstract

A method for isolating intact chloroplasts from mature and senescent *Arabidopsis thaliana* leaves is described that utilizes two subsequent Percoll gradients. The isolated chloroplasts can be incubated in the dark to track *in organelle* protein degradation. To remove chloroplasts that lysed during the incubation period, a second Percoll gradient is performed prior to SDS-PAGE and immunoblot analysis.

Key words Intact chloroplasts, *Arabidopsis*, Percoll, Chlorophyll

1 Introduction

Percoll density gradients have been used extensively to separate cellular components. Percoll consists of 17-nm-diameter particles of colloidal silica that are coated with polyvinylpyrrolidone. Percoll has low viscosity and low osmolarity and is not toxic, thus permitting the separation of organelles with little shearing and minimal effect on function [1]. Percoll gradients were first used for chloroplast isolation from pea leaves [2]. The following chloroplast isolation protocol was adapted from techniques used in *Arabidopsis thaliana* [3], which was modified from a protocol developed in *Hordeum vulgare* [4]. Additional modifications were adopted from a protocol developed in pea [5, 6]. This protocol has been used to demonstrate degradation of glutamine synthetase 2 and both rubisco activase isoforms in isolated, intact chloroplasts [7].

2 Materials

For all solutions, use ultrapure water (18 MΩ) and analytical grade reagents.

Yongfeng Guo (ed.), *Plant Senescence: Methods and Protocols*, Methods in Molecular Biology, vol. 1744,
https://doi.org/10.1007/978-1-4939-7672-0_24, © Springer Science+Business Media, LLC 2018

2.1 Tissue Homogenization

1. 1 M HEPES buffer: 23.8 g of HEPES per 100 mL final volume. Adjust pH to 7.5 with KOH.

2. 0.5 M EDTA: 14.6 g of EDTA per 100 mL final volume, pH 8.0 (*see* **Note 1**).

3. 1 M MnCl$_2$: 12.6 g tetrahydrate per 100 mL.

4. 1 M MgCl$_2$: 9.5 g hexahydrate per 100 mL.

5. Grinding buffer: 50 mM HEPES-KOH, 2 mM EDTA, pH 8.0, 1 mM MnCl$_2$, 1 mM MgCl$_2$, 165 mM sorbitol, 5.7 mM ascorbic acid, 0.25% BSA (w/v), final pH 7.5 (*see* **Note 2**).

6. Omni TH tissue homogenizer: Omni, Marietta, GA.

2.2 Percoll Gradient

1. 40% Percoll solution: 40.0% Percoll (GE Healthcare Bio Sciences), 330 mM sorbitol, 2.1 mM MgCl$_2$, 1.6 mM MnCl$_2$, 50 mM HEPES-KOH pH 7.5, 2 mM EDTA-NaOH pH 8.0, 0.1% (w/v) BSA (*see* **Note 3**).

2. 85% Percoll solution: 85.0% Percoll, 330 mM sorbitol, 50 mM HEPES-KOH pH 7.5 (*see* **Notes 3** and **4**).

3. Step gradient: Prepare the gradient in a standard disposable polypropylene 15 mL centrifuge tube. Four milliliter of 85% Percoll solution is pipetted to the bottom of the tube. Three milliliter of 40% Percoll solution is then gently pipetted on top of the 85% Percoll solution. Be careful to minimize mixing of the two layers (*see* **Note 5**).

2.3 Incubation and Second Percoll Gradient

1. Incubation buffer: 50 mM HEPES-KOH, pH 7.5, 1 mM MnCl$_2$, 1 mM MgCl$_2$, 165 mM sorbitol, 5.7 mM ascorbic acid, 0.25% BSA(w/v), final pH 7.5 (*see* **Note 6**).

2. Dimethyl formamide.

3 Methods

3.1 Leaf Homogenization

1. 2.5–5.0 g of mature or senescent leaf tissue is minced with a scissors prior to homogenization with an Omni TH tissue homogenizer in increments of 1.0–2.0 g in 30.0 mL grinding buffer at 4 °C. To keep the sample cold, incubate the sample on ice after 30 s of homogenization (*see* **Note 7**).

2. Filter the homogenate through one layer of Miracloth in increments of 5.0 mL, clearing debris from the Miracloth with a spatula in between addition of more homogenate.

3. Centrifuge the filtered homogenate at $1000 \times g$ for 8 min at 4 °C in a fixed angle rotor.

3.2 Percoll Gradient

1. Resuspend the resulting pellet in 4.0 mL of grinding buffer.

2. Gently load the resuspension onto a freshly made 40–85% Percoll step gradient to minimize mixing of layers, and then centrifuge at $6000 \times g$ for 15 min in a swinging bucket rotor.

3. Gently remove the Percoll gradient from the rotor and the chloroplast layer can be seen at the intersection of the 40% and 85% step in Percoll concentration. Larger debris will be at the bottom of the tube.

4. Gently remove the green layer at the 40–85% Percoll interface with a 1 mL pipette, and add to 30 mL of cold incubation buffer in a centrifuge tube to wash the chloroplasts.

5. After gentle mixing, the chloroplast solution is centrifuged for 6 min at $1000 \times g$ in a fixed angle rotor. Resuspend the chloroplast pellet at the bottom of the tube in 1 mL of incubation buffer.

3.3 Quantification and Adjustment of Chlorophyll Concentration

1. The chlorophyll concentration is determined by pipetting 6 μL of resuspended chloroplasts in 794 μL of dimethyl formamide and reading absorbance at 647 and 664 nM. The following equation is used to determine the concentration of chlorophyll in μg/mL [8]:

    Total chlorophyll (μg/mL) = $(17.67 \times OD_{647}) + 7.17 \times OD_{664}) \times 800/6 \times 0.8$

2. Adjust chlorophyll concentration of all samples to 200 μg/mL by addition of incubation buffer (*see* **Note 8**).

3.4 Incubation and Second Percoll Gradient

1. Once diluted to 200 μg/mL chlorophyll, the chloroplasts are incubated in a foil-wrapped Oakridge tube to prevent light exposure and stored in a closed drawer at room temperature [9].

2. Collect samples of equal volume (usually 0.5–1 mL) at various time points (1–24 h) from the total volume.

3. These time point samples are immediately loaded onto a 40–85% Percoll gradient and centrifuged at $6000 \times g$ for 15 min. Intact chloroplasts are collected from the 40–85% solution interface, washed in 30 mL of incubation buffer as described above, and centrifuged at $1000 \times g$ for 6 min.

4. Resuspend the resulting pellet in incubation buffer to the initial time point sample volume (usually 0.5–1 mL), and store at −80 °C.

5. Remove equal volume of (10–20 μL) samples from each time point for standard SDS-PAGE and subsequent immunoblot analysis (*see* **Note 9**).

4 Notes

1. EDTA will not go into solution until pH is close to 8.0, so it is necessary to add NaOH while stirring and using a pH meter, to get the EDTA into solution.

2. For 100 mL: 5 mL 1 M HEPES, pH 7.5, 0.4 mL 0.5 M EDTA, pH 8.0, 0.1 mL of 1 M $MnCl_2$, 0.1 mL of 1 M $MgCl_2$, 3 g sorbitol, 100 mg ascorbic acid, 0.25 g BSA, qs to 100 mL with ultrapure water. Prepare and autoclave 1 M HEPES buffer, 0.5 M EDTA, 1 M $MnCl_2$, and 1 M $MgCl_2$ as stock solutions. For each sample to be ground, freshly prepare 30 mL of grinding buffer and keep at 4 °C.

3. For 50 mL: 20 mL Percoll, 3 g sorbitol, 105 μL 1 M $MgCl_2$, 80 μL 1 M $MnCl_2$, 2.5 mL 1 M HEPES, pH 7.5, 0.2 mL 0.5 M EDTA, pH 8.0, 50 mg BSA. Add reagents to about 20 mL of ultrapure water and qs to 50 mL with ultrapure water when sorbitol is in solution. Percoll solution should be made no earlier than the day before the chloroplast isolation and must be stored at 4 °C.

4. For 50 mL: 42.5 mL of Percoll, 3 g sorbitol, 2.5 mL HEPES, pH 7.5. Add reagents, allow sorbitol to dissolve, and qs to 50 mL with ultrapure water.

5. The step gradient should be made immediately prior to use.

6. For 100 mL: 5 mL 1 M HEPES, pH 7.5, 0.1 mL of 1 M $MnCl_2$, 0.1 mL of 1 M $MgCl_2$, 3 g sorbitol, 100 mg ascorbic acid, 0.25 g BSA, qs to 100 mL with ultrapure water.

7. Non-homogenized tissue is allowed to float to the top while the sample stays on ice, and then only the top 10.0–15.0 mL are re-homogenized to avoid disturbing existing contents.

8. If the initial concentration is 1850 μg/mL, then 8.25 mL of incubation buffer will be added to the 1 mL chloroplast suspension to give a final volume of 9.25 mL of 200 μg/mL chlorophyll.

9. Since the chlorophyll concentration is equivalent in all samples, and only a low level of chloroplast lysis can be observed, equal volumes of each sample have equivalent levels of chlorophyll.

References

1. Pertoft H, Laurent TC, Laas T et al (1978) Density gradients prepared from colloidal silica particles coated by polyvinylpyrrolidone (Percoll). Anal Biochem 88:271–282

2. Mills WR, Joy KW (1980) A rapid method for isolation of purified, physiologically active chlo-roplasts, used to study the intracellular distribution of amino acids in pea leaves. Planta 148:75–83

3. Schulz A, Knoetzel J, Scheller H et al (2004) Uptake of fluorescent dye as a swift and simple indicator of organelle intactness: import-

competent chloroplasts from soil-grown Arabidopsis. J Histochem Cytochem 52: 701–770

4. Brock I, Hazell L, Michl D et al (1994) Optimization of an in vitro assay for the import of proteins by isolated pea and barley thylakoids. Plant Mol Biol 23:717–725

5. Mitsuhashi W, Crafts-Brandner S, Feller U (1992) Ribulose-1,5-bis-phosphate carboxylase/oxygenase degradation by lysates of mechanically isolated chloroplasts from wheat leaves. Plant Physiol 40:504–514

6. Roulin S, Feller U (1998) Light-independent degradation of stromal proteins in intact chloroplasts isolated from *Pisum sativum* L. leaves: requirement for divalent cations. Planta 205: 297–304

7. Lee T, Vande Wetering S, Brusslan J (2013) Stromal protein degradation is incomplete in Arabidopsis thaliana autophagy mutants undergoing natural senescence. BMC Res Notes 6:17

8. Porra R, Thompson W, Kriedemannm P (1989) Determination of accurate extinction coefficients and simultaneous equations for assaying chlorophylls *a* and *b* extracted with four different solvents: verification of the concentration of chlorophyll standards by atomic absorption spectroscopy. Biochim Biophys Acta 975: 384–394

9. Roulin S, Feller U (1997) Light induced proteolysis of stromal proteins in pea (Pisum sativum L.) chloroplasts: requirement for intact organelles. Plant Sci 128:31–41

Chapter 25

Plant Cell Walls: Isolation and Monosaccharide Composition Analysis

Yingzhen Kong, Malcolm O'Neill, and Gongke Zhou

Abstract

Plant cell walls have important roles during all phases of plant growth and development. Polysaccharides are the major components of the primary walls surrounding growing plant cells, together with small amounts of protein and minerals. Secondary walls that are deposited when a cell has ceased to grow are also composed predominantly of polysaccharides, although lignin may account for up to 20% w/w of these walls. The types of polysaccharides and their structure and abundance often vary greatly in the cell walls of different plant species, different cell types, and different developmental stages. Significant changes in structure and composition of cell wall have been described in various types of plant senescence. Here we describe a general method for the isolation of cell wall polysaccharides as their alcohol-insoluble residues (AIR) and procedures for the determination of the neutral and acidic monosaccharides present in the wall.

Key words Cell wall, Polysaccharides, Extraction, Monosaccharides

1 Introduction

Plant cell walls are mainly composed of cellulose, hemicelluloses, and pectins. Primary cell wall composition in dicots and non-graminaceous monocots is approximately 15–40% cellulose, 20–30% xyloglucans, and 30–50% of pectin, with lesser amounts of heteroxylans and structural proteins [1]. The primary cell walls of graminaceous monocots contain much less pectin (~5%) and xyloglucan (5%) but much more heteroxylan (15–25%). These walls may also contain mixed-linkage glucans [2]. Most of the polysaccharides present in plant cell walls have been well characterized structurally. Cellulose consists of 1,4-linked β-D-glucan chains. Xyloglucan also has a 1,4-linked β-D-glucan backbone, but this backbone is partially substituted at O-6 with xylose and other neutral sugars [3]. Pectins are a family of polysaccharides including homogalacturonan and rhamnogalacturonans I and II [4] that all contain 1,4-linked-D-galacturonic acid (GalA) residues. It has been estimated that ~90% of the GalA in the wall derive from pectin [5].

Yongfeng Guo (ed.), *Plant Senescence: Methods and Protocols*, Methods in Molecular Biology, vol. 1744,
https://doi.org/10.1007/978-1-4939-7672-0_25, © Springer Science+Business Media, LLC 2018

The hemicellulosic and pectic polysaccharides are believed to be cross-linked together both ionically and covalently to form a wall matrix that help keep the shape of the cell and play important roles during plant growth and development.

In plants, senescence is typically accompanied with substantial changes in the composition and properties of the cell wall. Such changes have been studied almost exclusively in fruits, including plum, peach, and tomato [6, 7]. Numerous cell wall changes occur during plant organ senescence, such as deglycosylation, depolymerization, and solubilization of xyloglucan, pectin, and other matrix glycans. The loss of arabinose (Ara) and galactose (Gal), two major non-cellulosic sugars, is a common trait during most fruit senescence. However, Gal loss was not observed in plum or cucumber, and the decrease in Ara is minor or absent in apple, plum, and apricot [8]. Few studies have been performed on cell wall metabolism during senescence of other plant organs. Vetten and Huber [9] reported cell wall changes during the senescence of carnation petals, where the decrease in the amounts of cell wall is due largely to the loss of neutral sugars, primarily Gal and Ara. Miyamoto et al. [10] have examined changes in metabolism of cell wall polysaccharides in oat leaves during senescence, and they reported that the amounts of pectin polysaccharides are almost constant during oat leaf senescence, while the hemicellulosic and cellulosic polysaccharides decreased substantially.

In this chapter, we describe the isolation of cell walls using aqueous alcohol, methanol/chloroform, and acetone, which is straightforward and widely used to obtain walls suitable for glycosyl residue composition analyses and for structural analyses of the constituent polysaccharides. The isolated cell wall is referred to as the alcohol-Insoluble residue (AIR). This method is also quite efficient for deactivating cell wall enzymes and thus minimizes the fragmentation of cell wall polysaccharides [11].

2   Materials

2.1   Cell Wall Extraction

1. Alcohol.
2. Methanol.
3. Chloroform.
4. Acetone.

2.2   Cell Wall Fractionation

1. 50 mM sodium acetate (NaOAc) with 0.02% (w/v) thimerosal.
2. 100 mM sodium acetate (NaOAc).
3. 1 unit/mL amylase: thermostable.
4. 1 unit/mL pullulanase.
5. Acetone.

6. 50 mM cyclohexane-trans-1,2-diamine-NNN′N′-tetra acetate (CDTA).

7. 50 mM sodium carbonate (Na_2CO_3).

8. 1 N and 4 N potassium hydroxide (KOH).

9. Glacial acetic acid.

10. 20 mM sodium borohydride ($NaBH_4$).

11. Octanol.

2.3 Monosaccharide Analysis

1. Standards: include Ara, rhamnose (Rha), fucose (Fuc), xylose (Xyl), GalA, Gal, mannose (Man), glucose (Glc), and inositol.

2. 1 M methanolic 3 N HCl (Supelco).

3. Anhydrous methanol.

4. Tri-Sil (Pierce).

5. Hexane.

6. Heating block.

7. Drying apparatus.

8. Lyophilizer.

9. Gas chromatograph (Hewlett Packard 5890): equipped with capillary columns and with a flame ionization detector (FID) or mass selective detector (MSD).

3 Methods

3.1 Extraction of Alcohol-Insoluble Residue (AIR)

1. Plant material is ground using a mortar and pestle to a fine powder in liquid nitrogen (for vegetative tissues) or ground in aq. 80% ethanol directly (for fruits). The ratio of tissue to organic solvent is ~1:25. The suspension is then stirred for 2 h. The insoluble material (cell wall) is collected by centrifugation for 10 min at $6000 \times g$.

2. The insoluble material is then suspended in methanol/chloroform (1:1) and stirred for 2 h in a fume hood. Centrifuge at $6000 \times g$ for 10 min, decant the supernatant, and collect the insoluble material.

3. Resuspend the insoluble material in acetone and stir for 2 h in the hood, centrifuge at $6000 \times g$ for 10 min, decant the supernatant, and evaporate the solvent in the hood until dry. The dry cell wall material is called AIR (*see* **Note 1**).

3.2 Destarching of AIR

1. Resuspend about 20–30 mg AIR in 1.5 mL of 100 mM NaOAc solution, and then 1 unit/mL thermostable amylase is added, and the suspension is kept for 15 min at 85 °C.

The sample is then cooled to 25 °C, 1 unit/mL pullulanase is added, and the suspension is kept overnight at 37 °C in a shaker (*see* **Note 2**).

2. Centrifuge the suspension at 6000 × *g* for 10 min and collect the pellet (destarched AIR) and also the supernatant (*see* **Note 3**). Wash the destarched AIR three times with deionized water, with vortexing, centrifuging, and decanting of the washing water.

3. Resuspend the pellet in acetone, vortex, and centrifuge down and evaporate the solvent in the hood.

4. If needed the dried destarched AIR samples can be kept at room temperature until needed.

3.3 Fractionation of Cell Wall Polysaccharides

The sequential extraction procedure is modified from Selvendran and O'Neill et al. [5]. The initial treatments involve the use of aqueous calcium chelators to solubilize material enriched in pectic polysaccharides. Subsequent extractions use more harsh treatments with increasing concentrations of alkali to solubilize hemicelluloses and additional pectins.

1. Suspend 1 g destarched AIR in 100 mL 50 mM CDTA in 50 mM NaOAc buffer (pH 6) containing 0.02% (w/v) thimerosal. Stir the suspension for 24 h at room temperature. The supernatant is collected by centrifugation for 10 min at 6000 × *g* and labeled as CDTA-soluble fraction.

2. Resuspend the residual pellet in 50 mM Na_2CO_3 containing 20 mM $NaBH_4$, and stir for 24 h at room temperature. The supernatants is collected by centrifugation for 10 min at 6000 × *g* and labeled as Na_2CO_3 fraction.

3. Resuspend the Na_2CO_3-insoluble pellet in 1 N KOH containing 20 mM $NaBH_4$, and stir for 24 h at room temperature; the supernatant is collected by centrifuging at 6000 × *g* for 10 min. Add a few drops of octanol to prevent excessive foaming, and then adjust to pH 5.0 by dropwise addition of glacial acetic acid. The neutralized supernatant is labeled as 1 N KOH fraction.

4. Resuspend the 1 N KOH-insoluble pellet in 4 N KOH, and stir for 24 h; the supernatants is collected by centrifuging at 6000 × *g* for 10 min, and the pH is adjusted to 5.0 with acetic acid. The neutralized supernatant is labeled as 4 N KOH fraction. The major component of the residue is cellulose, which is known to be insoluble in the extractants used.

5. Dialyze the CDTA fraction, Na_2CO_3 fraction, 1 N KOH fraction, and 4 N KOH fraction (*see* **Note 4**) by using 3.5 kDa cutoff dialysis tubing against deionized water for

48 h, and change water every 12 h. Dialyzed fractions are lyophilized and can be used for the following monosaccharide analysis.

3.4 Monosaccharide Analysis of Cell Wall Fractions

Treating polysaccharides with hot anhydrous methanol containing HCl releases neutral and acidic monosaccharides as their methyl glycosides. These glycosides are then converted to volatile derivatives by silylation of their hydroxy groups. Each sugar typically gives several derivatives including the anomers and the pyranose and furanose ring forms. TMS methyl glycosides are readily separated by GC using capillary columns, although only sugars for which standards are available can be identified based on retention time alone. GC-EI-MS may provide data to identify a derivative as the EI mass spectra of hexoses, pentoses, 6-deoxyhexose, uronic acids, and keto sugars are sufficiently different [12].

The following TMS monosaccharide analysis method is modified from York et al. [13]. Perform all derivatization procedures in a hood.

1. Prepare 1 M methanolic HCl by placing 6 mL anhydrous methanol into a screw cap tube, add 3 mL of 3 N methanolic HCl, and vortex.

2. Make a standard mixture that contains 50 μg of Ara, Rha, Fuc, Xyl, GalA, Gal, Man, and Glc and 20 μg inositol in a glass screwtop tube with Teflon-lined cap.

3. Weight out 100–500 μg sample into a screw cap tube with Teflon-lined caps, and add 20 μg inositol as internal standard.

4. Place sample and standards on dry ice until frozen. Loosen cap and place on a lyophilizer (*see* **Note 5**).

5. Remove when dry and then add 30–50 drops of methanolic HCl and heat overnight (16–18 h) at 80 °C (*see* **Note 6**).

6. Concentrate methanolysis products to dryness under a flow of N_2 gas. Add about 10–20 drops anhydrous methanol and dry again. Repeat twice.

7. Add ten drops of Tri-Sil and heat for 20 min at 80 °C.

8. Concentrate reagents until just dry, add 2 mL hexane, and mix well (*see* **Note 7**).

9. Filter the sample through glass wool packed into a Pasteur pipette and concentrate to ~100 μL. Inject 1 μL onto the gas chromatograph interfaced to a 5970 MSD using a Supelco DB1-fused silica capillary column. The oven temperature program is kept at 80 °C for 2 min and ramped at 20 °C/min to 140 °C, then at 2 °C/min to 200 °C, and finally 30 °C/min to 250 °C.

10. To determine the amounts of each monosaccharide, known amounts of each standard sugar must be converted to it's TMS methyl glycoside to identify the elution times of their peaks. The combined areas of the peaks are then used to determine the molar response factor for each monosaccharide relative to inositol. The monosaccharide compositions are then expressed as mol% of total sugar. The amounts of each glycose in the cell wall or extract (the amounts of sample hydrolyzed must be known) are estimated using the sugar peak areas and response factors.

4   Notes

1. The AIR prepared using this method may be slightly green in color if leafy materials are used, but this will not affect monosaccharide analysis.

2. This step is to remove the starch from the tissue, especially when walls are prepared from leafy tissue. If starch is not removed, the large amounts of Glc generated may affect the quantification of Glc from cell wall polysaccharides.

3. The supernatant may contain water-soluble arabinogalactan. The supernatant can be dialyzed to remove Glc generated by amylase and then freeze-dried for the following monosaccharide composition analysis.

4. The CDTA fraction and the Na_2CO_3 fraction comprise primarily of pectins, the 1 N KOH fraction are comprised of mainly hemicelluloses and pectin polysaccharides, and the 4 N KOH fraction comprises primarily of hemicelluloses, such as glucomannans and xyloglucan. The major component of the residue after 4 N KOH extraction is cellulose.

5. Moisture may affect the accuracy of the monosaccharide quantification, so make sure that the standards and the sample are as dry as possible.

6. Close the cap tightly to avoid evaporation. Sonicate the tube if the cell wall form clumps after adding methanolic HCl.

7. TMS derivatives are volatile, and take precaution not to over-dry; otherwise you will lose some of your derivatives.

Acknowledgments

This work is supported by the National Natural Science Foundation of China (31470291, 31670302), the National Science and Technology Support Program (2015BAD15B03), and the Taishan Scholar Program of Shandong (to GZ).

References

1. Cosrove DJ, Jarvis MC (2012) Comparative structure and biomechanics of plant primary and secondary cell walls. Front Plant Sci **3**:204

2. Burke D, Kaufman P, McNeil M et al (1974) The structure of plant cell walls. VI. The survey of the walls of suspension-cultured monocots. *Plant Physiol* **54**:109–115

3. Pauly M, Gille S, Liu L et al (2013) Hemicellulose biosynthesis. *Planta* **238**:627–642

4. Harholt J, Suttangkakul A, Scheller HV (2010) Biosynthesis of pectins. *Plant Physiol* **153**: 384–395

5. Selvendran RR, O'Neill MA (1987) Isolation and analysis of cell walls from plant material. *Methods Biochem Anal* **32**:25–153

6. Ponce NMA, Ziegler VH, Stortz CA et al (2010) Compositional changes in cell wall polysaccharides from Japanese plum (Prunus salicina Lindl.) during growth and on-tre ripening. *J Agric Food Chem* **58**:2562–2570

7. Brummell DA, Dal Cin V, Crisosto CH et al (2004) Cell wall metabolism during maturation, ripening and senescence of peach fruit. *J Exp Bot* **55**:2029–2039

8. Gross KC, Sams CE (1984) Changes in cell wall neutral sugar composition during fruit ripening: a species survey. *Phytochemistry* **23**: 2457–2461

9. De Vetten NC, Huber DJ (1990) Cell wall changes during the expansion and senescence of carnation (*Dianthus caryophyllus*) petals. *Physiol Plant* **78**:447–454

10. Miyamoto K, Oka M, Uheda E et al (2013) Changes in metabolism of cell wall polysaccharides in oat leaves during senescence: relevance to the senescence-promoting effect of methyl jasmonate. *Acta Physiol plant* **35**:2675–2683

11. Coimbra MA, Delgadillo I, Waldron KW et al (1996) Isolation and analysis of cell wall polymers from olive pulp. In: Linskens HF (ed) *Plant cell wall analysis*. Springer Berlin Heidelberg, New York, pp 19–44

12. Doco T, O'Neill MA, Pellerin P (2001) Determi-nation of the neutral and acidic glycosyl-residue compo-sitions of plant polysaccharides by GC-EI-MS analysis of the trimethylsilyl methyl glycoside derivatives. *Carbohydrate Polym* **46**:249–259

13. York WS, Darvill AG, McNeil M et al (1986) Isolation and characterization of plant cell walls and cell wall components. *Methods Enzymol* **118**:3–40

Chapter 26

Visualizing Morphological Changes of Abscission Zone Cells in *Arabidopsis* by Scanning Electron Microscope

Chun-Lin Shi and Melinka A. Butenko

Abstract

Scanning electron microscope (SEM) is a type of electron microscope which produces detailed images of surface structures. It has been widely used in plants and animals to study cellular structures. Here, we describe a detailed protocol to prepare samples of floral abscission zones (AZs) for SEM, as well as further image analysis. We show that it is a powerful tool to detect morphologic changes at the cellular level during the course of abscission in wild-type plants and to establish the details of phenotypic alteration in abscission mutants.

Key words Abscission, Floral organ abscission, Cell separation, Scanning electron microscope, Abscission zone, Morphology, Critical point drying

1 Introduction

Abscission is a regulated process leading to the loss of plant organs, such as leaves, fruits, and flower petals. At the cellular level, the presence of an abscission zone (AZ) is a prerequisite for abscission to take place [1]. Organ detachment requires a cell separation process where cell wall-modifying and cell wall-hydrolyzing enzymes are involved in degrading the middle lamella between two adjacent cell files (reviewed by [2–8]). It has been discovered that the expression of genes encoding these enzymes is regulated by signaling pathways and plant hormones, such as the peptide ligand IDA [9], the receptor-like kinases HAE and HSL2 [10, 11], KNOX HD proteins [12], ethylene [13, 14], and auxin [15].

Scanning electron microscope (SEM) provides a method to visualize the cellular morphology of cells that make up an AZ. In *Arabidopsis* flowers, the AZ region is composed of morphological distinct AZ cells localized at the base of the sepals, petals, and stamens. The use of SEM to study abscission was first described in Sexton 1976 [16], describing the development of leaf AZ in *Impatiens sultani*. The cellular changes observed by SEM during

Yongfeng Guo (ed.), *Plant Senescence: Methods and Protocols*, Methods in Molecular Biology, vol. 1744,
https://doi.org/10.1007/978-1-4939-7672-0_26, © Springer Science+Business Media, LLC 2018

the abscission process corroborate well with the changes in petal breakstrength (pBS) (*see* Chapter 6) which measures the force required to remove a petal from the receptacle (in gram equivalents). Thus, these two methods together provide a way to both quantify and visualize the progression of cell wall degradation and remodeling during abscission. We describe here a detailed protocol for SEM to study floral abscission in *Arabidopsis*, as well as image analysis to quantify alterations in AZ cell size and number in abscission mutants. We also show some examples of SEM images from mutant and transgenic plants with altered cellular morphology compared to WT. The protocol described here is modified for floral AZs but can be adapted to study other regions containing AZ cells.

2   Materials

Prepare all solutions using distilled water and analytical grade reagents. Prepare and store all reagents at room temperature (unless indicated otherwise). Diligently follow all waste disposal regulations when disposing waste materials.

1. 1 M K_2HPO_4 stock buffer: weigh 174.18 g K_2HPO_4, add 900 mL water, and mix and make up to 1 L with water.

2. 1 M KH_2PO_4 stock buffer: weigh 136.09 g KH_2PO_4, add 900 mL water, and mix and make up to 1 L with water.

3. 0.5 M potassium phosphate (KPO_4) stock buffer, pH 7.2: combine 71.7 mL 1 M K_2HPO_4 stock buffer and 28.3 mL 1 M KH_2PO_4 stock buffer, and mix and make up to 200 mL with water. Dilute with water to 50 mM KPO_4 buffer before use.

4. Fixative: 4% glutaraldehyde (w/v) and 2% paraformaldehyde (w/v) in 50 mM KPO_4 buffer, pH 7.2 (*see* **Note 1**).

5. Graded ethanol series: 5%, 10%, 15%, 20%, 30%, 40%, 50%, 60%, 70%, 80%, 90%, 95%, and 100% ethanol (*see* **Note 2**).

6. Large petri dish: for tissue preparation.

7. Filter paper.

8. 10 mL glass vial: for fixation and storing.

9. Vacuum.

10. Critical point drying apparatus: Bal-Tec CPD030 critical point dryer.

11. Sputter coater.

12. Cressington Coating System: 308R.

13. 4800 field emission scanning electron microscope (Hitachi).

14. NIH Image J software.

3 Methods

Arabidopsis thaliana plants are grown in soil in growth chambers at 22 °C under long days (16 h day/8 h dark) at a light intensity of 100 mE·m22·s21.

3.1 Tissue Preparation and Fixation

The protocol described here is modified for floral organ abscission tissue [9, 10, 12v, 17]. Collect tissue for SEM from adult plants with 18–20 flower positions (*see* **Note 3**).

1. Collect individual flowers at positions 1, 2, 3, 4, 5, 6, 8, 10, and 12, respectively (*see* **Note 4**).

2. Pull off floral organs carefully, with forceps if necessary (*see* **Note 5**).

3. Transfer and cut AZ on the lid of a large petri dish covered with filter paper which is soaked with 50 mM KPO$_4$ buffer and place on ice (Fig. 1a).

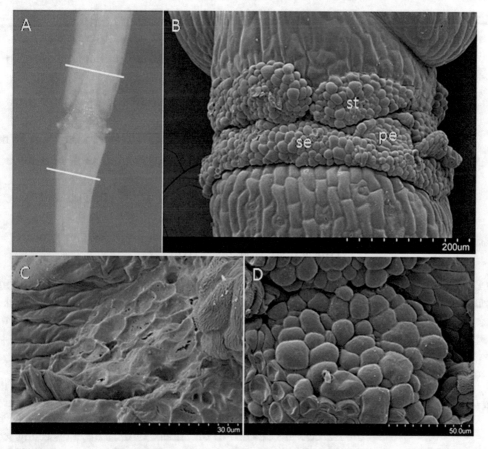

Fig. 1 Abscission zones (AZ) of different floral organs. (**a**) A mature silique. White lines indicate cutting positions. (**b**) SEM of floral AZs (*se* sepal AZ, *pe* petal AZ, *st* stamen AZ). (**c**) The petal AZ from pre-abscission stage. (**d**) The petal AZ from post-abscission stage

4. Transfer the AZ tissues into a glass vial (10 mL) containing 5 mL of fixative placed on ice. Up to 15 AZs can be placed into each vial (*see* **Note 6**).

5. Transfer the glass vials with fixative into the vacuum and vacuum at 20 Hg for 15 min.

6. Keep at 4 °C overnight.

3.2 Dehydration

1. Wash the AZ tissue in 50 mM KPO_4 buffer for 5 min by first removing the fixative with a glass pipette and subsequently adding buffer. Leave the buffer for 10 min, and repeat the washing with fresh buffer for three more times.

2. Remove the buffer and add 5% ethanol, and leave for 20 min.

3. Remove the 5% ethanol, add 10% ethanol, and leave for 20 min.

4. Remove the 10% ethanol, add 15% ethanol, and leave for 20 min.

5. Remove the 15% ethanol, add 20% ethanol, and leave for 20 min.

6. Remove the 20% ethanol, add 30% ethanol, and leave for 20 min.

7. Remove the 30% ethanol, add 40% ethanol, and leave for 20 min.

8. Remove the 40% ethanol, add 50% ethanol, and leave for 20 min.

9. Remove the 50% ethanol, add 60% ethanol, and leave for 20 min.

10. Remove the 60% ethanol, add 70% ethanol, and leave for 20 min.

11. Store at 4 °C overnight (*see* **Note 7**).

12. Remove the 70% ethanol, add 80% ethanol, and leave for 30 min (*see* **Note 8**).

13. Remove the 80% ethanol, add 90% ethanol, and leave for 30 min.

14. Remove the 90% ethanol, add 95% ethanol, and leave for 30 min.

15. Remove the 95% ethanol, add 100% ethanol, and leave for 30 min.

16. Store the tissue in 100% ethanol until ready to proceed with critical point drying (*see* **Note 9**).

3.3 Critical Point Drying

Critical point drying (CPD) is performed using a Bal-Tec CPD030 critical point dryer. Transfer the tissue into small plastic baskets, always having tissue covered in 100% ethanol. The baskets are then transferred to the CPD chamber filled with 100% ethanol. CPD is performed using CO_2.

3.4 Mounting and Coating

After CPD, the specimens should be stored dried in glass vials until mounting. Mount the specimens on aluminum stubs with double-sided carbon tape. Care should be taken to mount the specimens such that the AZ cells are exposed to the surface of the stub; a dissecting scope should be used for mounting. Mounted specimens are sputter coated with gold palladium using a Cressington Coating System 308R with a 3 nm thickness at 10^{-5} mbar employing nitrogen and argon gas.

3.5 Observation and Micrographs

View the samples at 5 K accelerating voltage on a 4800 field emission scanning electron microscope (Hitachi) (Fig. 1b–d).

3.6 AZ Measurements

AZ measurements are performed using the NIH Image J software. Cell elongation is verified by cell density, which is obtained by calculating cell no. within a certain area (*see* **Note 10**).

1. Select images with nice frontal view of petal AZ (Fig. 2) (*see* **Note 11**).

2. Define a 1600 µm² area in a petal AZ (*see* **Note 12**); count cells that fall completely or partially in the area (*see* **Note 13**).

3. Repeat the measurement with at least three additional images from the same position (*see* **Note 14**).

4. Import raw data into Microsoft Excel, average cell no., and calculate the standard deviation (SD) at interested positions. Do Student's t-test between mutant or transgenic plants and WT plants (Table 1).

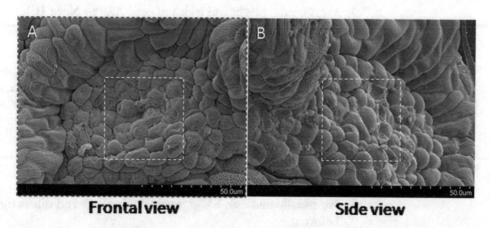

Fig. 2 Frontal view (**a**) and side view (**b**) of petal AZ. White square indicates a 1600 µM² area

Table 1
Petal AZ cell density (data from Shi et al. [12])

Genotype	p4	p7	p8	p12
WT	NA	40 ± 1.83	30 ± 1.26	19 ± 0.89
bp3	NA	30 ± 1.06*	26 ± 1.11*	19 ± 1.64
35S–IDA	29 ± 2.22	10 ± 1.21*	9 ± 0.63*	6 ± 1.29*

Petal AZ cell numbers in an area of 1600 µm²
NA Not applied
Asterisks indicate significant differences ($P < 0.002$) based on Student's *t*-test; *n* (flowers) ≥ 4

Table 2
Petal AZ cell number (data from Shi et al. [12])

Genotype	Height	Width
WT p8[a]	54.6 ± 4.55	87.8 ± 4.86
35S–IDA p4[a]	60.7 ± 4.01*	99.3 ± 9.49*
WT p12	59.9 ± 3.43	98.9 ± 8.42
bp3 p12	78.2 ± 4.04*	110.4 ± 6.59*

Asterisks indicate significant differences ($P < 0.001$) based on Student's t-test; n (flowers) ≥ 4
Positions where the WT and 35S–IDA AZs have the same cell density

3.7 Measurements of Cell Division

Cell division is verified by measurements of petal AZ height and width.

1. Select positions with similar cell density of petal AZ (*see* **Note 15**).

2. Measure the height and width of petal AZ (*see* **Note 16**).

3. Repeat the measurement with at least three additional images from the same position (*see* **Note 14**).

4. Import raw data into Microsoft Excel, average cell no., and calculate the standard deviation (SD) at interested positions. Do Student's t-test between mutant or transgenic plants and WT plants (Table 2).

4   Notes

1. We usually make an 8% glutaraldehyde stock and dilute right before use.

2. Make 70% and higher ethanol solutions with distilled water, since potassium phosphate buffer will precipitate in high-percentage ethanol solution.

3. Do not collect for more than 1 h.

4. WT plants normally shed their floral organs at position 6–8, the positions we choose here can be divided as three groups: pre-abscission stage (position 1–5), abscission stage (position 6 and 8), and post-abscission stages (position 10 and 12). For mutant or transgenic plants with premature abscission, abscission can occur as early as position 3 or 5 [12].

5. Pull off sepal and petals from young flowers to expose the petal AZ surface when performing scanning micrographs.

6. All work with fixative must be performed in a fume hood.

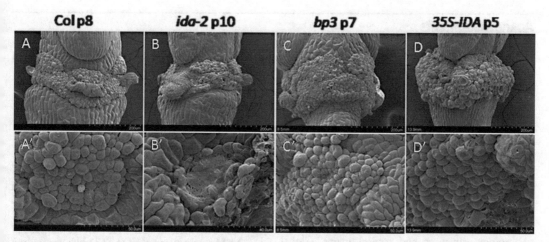

Col p8   **ida-2 p10**   **bp3 p7**   **35S-IDA p5**

Fig. 3 SEM of floral AZs (**a–d**) and petal AZs (**a'–d'**) from Col WTt, *ida-2* and *bp3* mutant, and *35S–IDA* transgenic plants. *p* Positions

7. Tissues can be stored up to 3 weeks in 70% ethanol at 4 °C. Seal the whole rack with parafilm to avoid evaporation.

8. Make 80% and higher ethanol solutions fresh.

9. Keep at least overnight before critical point drying.

10. Due to cell damage at pre-abscission stages and post-abscission stages from abscission-deficient mutants, images of post-abscission stages with good AZ cell patterns are used for petal AZ cell calculation (Fig. 3b, b').

11. The same area in images with side view of AZ will give relative more cells compared with that of front view of AZ, which could mimic a false positive of less cell elongation (Table 1).

12. For some plants, it is not possible to distinguish petal AZ due to an extreme enlargement of AZ, such as *35S:IDA* plants; the same defined area within the whole AZ region is used (Fig. 3d, d').

13. Move this defined region from side to side within petal AZ, repeat the cell counting twice, and obtain the average cell number from three independent samples.

14. This is to include variation from individual plants.

15. We have shown that cell density can be very different at certain positions between WT and mutant/transgenic plants (Table 1, position 7 and 8). Therefore, we choose positions with similar cell density between WT and mutant/transgenic plants for petal AZ height and width measurements. We believe that this will be more correct compared to simple measurement of the petal AZ height at certain position described in Leslie et al. [18].

16. Make a line vertical to AZ for height and a line horizontal to AZ for width.

Acknowledgments

This work was supported by Grants 13785/F20 to MAB and 348256/F20 to C-L.S from the Research Council of Norway.

References

1. Mckim SM, Stenvik GE, Butenko MA et al (2008) The BLADE-ON-PETIOLE genes are essential for abscission zone formation in Arabidopsis. Development 135:1537–1546

2. Aalen RB, Butenko MA, Stenvik G-E et al (2006) Genetic control of floral abscission. In: Floriculture, ornamnetal and plant biotechnology: advances and topical issues. Global Science books, London

3. Bleecker A, Patterson SE (1997) Last exit: senescence, abscission, and meristem arrest in *Arabidopsis*. Plant Cell 9:1169–1179

4. Roberts JA, Elliot KA, Gonzalez-Carranza ZH (2002) Abscission, dehiscence and other cell separation processes. Annu Rev Plant Biol 53:131–158

5. Patterson SE (2001) Cutting loose. Abscission and dehiscence in Arabidopsis. Plant Physiol 126:494–500

6. Lewis MW, Leslie ME, Liljegren SJ (2006) Plant separation: 50 ways to leave your mother. Curr Opin Plant Biol 9(1):59–65

7. Liljegren A (2013) Strategic collaboration councils in the mental health services: what are they working with? Int J Integr Care 13:e009

8. Estornell LH, Agusti J, Merelo P et al (2013) Elucidating mechanisms underlying organ abscission. Plant Sci 199–200:48–60

9. Butenko MA, Patterson SE, Grini PE et al (2003) INFLORESCENCE DEFICIENT IN ABSCISSION controls floral organ abscission in Arabidopsis and identifies a novel family of putative ligands in plants. Plant Cell 15:2296–2307

10. Stenvik GE, Tandstad NM, Guo Y et al (2008) The EPIP peptide of INFLORESCENCE DEFICIENT IN ABSCISSION is sufficient to induce abscission in Arabidopsis through the receptor-like kinases HAESA and HAESA-LIKE2. Plant Cell 20:1805–1817

11. Cho SK, Larue CT, Chevalier D et al (2008) Regulation of floral organ abscission in Arabidopsis thaliana. Proc Natl Acad Sci U S A 105:15629–15634

12. Shi CL, Stenvik GE, Vie AK et al (2011) Arabidopsis class I KNOTTED-like homeobox proteins act downstream in the IDA-HAE/HSL2 floral abscission signaling pathway. Plant Cell 23:2553–2567

13. Patterson SE, Bleecker AB (2004) Ethylene-dependent and -independent processes associated with floral organ abscission in Arabidopsis. Plant Physiol 134:194–203

14. Patterson S, Butenko M, Kim J (2007) Ethylene responses in abscission and other processes of cell separation in Arabidopsis. In: Advances in Plant Ethylene Research. doi:https://doi.org/10.1007/978-1-4020-6014-4_60, p. 271–461

15. Basu MM, Gonzalez-Carranza ZH, Azam-Ali S et al (2013) The manipulation of auxin in the abscission zone cells of Arabidopsis flowers reveals that indoleacetic acid signaling is a prerequisite for organ shedding. Plant Physiol 162:96–106

16. Sexton R (1976) Some ultrastructural observations on the nature of foliar abscission in impatiens sultani. Planta 128:49–58

17. Stenvik GE, Butenko MA, Urbanowicz BR et al (2006) Overexpression of INFLORESCENCE DEFICIENT IN ABSCISSION activates cell separation in vestigial abscission zones in Arabidopsis. Plant Cell 18:1467–1476

18. Leslie ME, Lewis MW, Youn JY et al (2010) The EVERSHED receptor-like kinase modulates floral organ shedding in Arabidopsis. Development 137:467–476

Part VI

Systems Biology in Plant Senescence

Chapter 27

RNA-Seq Analysis of the Transcriptome of Leaf Senescence in Tobacco

Wei Li and Yongfeng Guo

Abstract

Leaf senescence is a complex developmental process which is under the control of a highly regulated genetic program and involves major changes in gene expression. During the past two decades, significant progress in molecular understanding of leaf senescence has been made through transcriptomic analysis, which, in comparison with traditional molecular biology, provides enormous amount of information on gene expression regulation at the whole genome level. In this chapter, we describe the protocol for RNA-seq, one of the most commonly used technologies in transcriptomic analysis, using tobacco leaf senescence as a model system.

Key words Leaf senescence, RNA-seq, Total RNA, Tobacco

1 Introduction

Leaf senescence is the final phase of leaf development. It is a complex developmental process under the control of a highly regulated genetic program which involves major changes in gene expression. Upon the onset of senescence, many genes expressed in green leaves, such as the genes encoding photosynthetic enzymes, are downregulated. Conversely, many *senescence-associated genes* (*SAGs*) are transcriptionally upregulated during leaf senescence [1, 2]. Identification and characterization of *SAGs* are critical in understanding the biochemical and physiological processes of leaf senescence.

With the availability of genome sequences and high-throughput tools since the early 2000s, thousands of *SAGs* have been identified [3–7]. Among the global gene expression profiling tools is RNA-seq (next- or second-generation sequencing), which allows unbiased quantification of expression levels of transcripts with a higher sensitivity and broader genome coverage [8, 9]. An analysis by Sîrbu et al. compared gene expression data of single-/dual-channel microarrays and RNA-seq for *Drosophila melanogaster* embryo

Yongfeng Guo (ed.), *Plant Senescence: Methods and Protocols*, Methods in Molecular Biology, vol. 1744,
https://doi.org/10.1007/978-1-4939-7672-0_27, © Springer Science+Business Media, LLC 2018

development time series. The results suggested that the RNA-seq technique is better in detecting transcripts with low abundance, and as a result, 2000 more differentially expressed genes were identified with RNA-seq [10].

In this chapter, we describe the method of RNA-seq, using transcriptomic analysis of tobacco (*Nicotiana tabacum*) leaf senescence as an example. We use the high-resolution power of RNA-seq coupled with the Illumina HiSeq™ 2000 sequencing in characterizing the transcriptome of tobacco leaves at different stalk positions and different developmental stages, aiming to screen a best group of *SAGs* to be validated by qRT-PCR.

2 Materials

2.1 Plant Materials

1. Plant materials: Common tobacco (*N. tabacum*) variety "Zhongyan 100" grown in the field.

2. Leaf samples: Collect leaf samples from different stalk positions (lower, middle, and upper leaves) at a series of developmental stages.

2.2 RNA Extraction

Prepare all solutions using diethyl pyrocarbonate (DEPC)-treated water. The DEPC-treated water is prepared by incubating with 0.1% DEPC and then autoclaved to remove the DEPC. DEPC should be handled in a fume hood. Prepare and store all reagents at room temperature (unless indicated otherwise).

1. 0.1% DEPC (v/v).

2. 1 M NaCitrate (pH 7.0): Add 5.882 g sodium citrate (dihydrate) to ~14 mL DEPC-treated water and stir to dissolve. Adjust pH to 7.0 with citric acid and dilute to a final volume of 20 mL with DEPC-treated water.

3. 20% Sarkosyl: N-lauroylsarcosine.

4. Extraction buffer stock solution: dissolve 47.28 g guanidinium thiocyanate in ~90 mL DEPC-treated water. Add 2.5 mL of 1 M NaCitrate (pH 7.0) and 2.5 mL of 20% Sarkosyl. Adjust the volume to 100 mL. The stock solution can be stored up to 3 months at room temperature.

5. Extraction buffer working solution: add 35 μL 2-mercaptoethanol (β-ME) to 5 mL of stock solution. The working solution should be freshly prepared with the final concentration of 4 M guanidinium thiocyanate, 25 mM NaCitrate (pH 7.0), 0.5% Sarkosyl, and 0.1 M β-ME.

6. 2 M NaAcetate (pH 4.0): add 16.406 g sodium acetate (anhydrous) to ~40 mL DEPC-treated water and 35 mL glacial acetic acid. Adjust pH to 4.0 with glacial acetic acid,

and adjust volume to 100 mL with DEPC-treated water. Store up to 1 year at room temperature.

7. 2 M NaAcetate (pH 5.0): weigh 16.406 g sodium acetate (anhydrous), and prepare a 100 mL solution as in previous step. Adjust pH to 5.0 with glacial acetic acid. Store up to 1 year at room temperature.

8. H_2O-saturated phenol.

9. Chloroform.

10. Isopropanol.

11. CES solution: 10 mM NaCitrate (pH 7.0), 1 mM EDTA (pH 8.0), 0.5% SDS. For 100 mL CES solution, mix 93.8 mL DEPC-treated water, 1 mL 1 M NaCitrate (pH 7.0), 200 μL 0.5 M EDTA, and 5 mL 10% SDS.

12. 75% ethanol.

2.3 Sample Preparation for Illumina Sequencing

1. Deoxyribonuclease I (Invitrogen).

2. Poly (T) oligo-attached magnetic beads.

3. RNA fragmentation kit.

4. cDNA synthesis kit.

5. Agilent 2100 bioanalyzer.

6. Illumina HiSeq™ 2000 sequencer.

3 Methods

3.1 Sample Harvest (See Note 1)

Upper leaves (UL, the 16th leaf position from the bottom) are harvested at 5, 15, 25, 35, 45, 55, 65, 75, and 85 days after topping (DAT). Middle leaves (ML, the 9–10th leaf positions) are harvested at 5, 15, 25, 35, 45, 55, 65, 75, and 85 DAT. Lower leaves (LL, the 4th leaf position) are harvested at 5, 15, 25, 35, and 45 DAT because after 45DAT most leaves at this position are desiccated as a result of senescence. Harvest the upper, middle, and lower leaves within the same replicate from the same plant. Twelve plants are harvested at each time point. All harvested leaves are wrapped in aluminum foil, immediately placed in liquid nitrogen, and stored at −80 °C until used.

3.2 RNA Extraction

All glassware, tubes, and pipette tips used in this section are dipped in 0.1% DEPC solution overnight and then autoclaved.

1. Grind 0.2–0.4 g tobacco leaf tissue in liquid nitrogen (*see* **Note 2**).

2. Add the ground tissue to a 15 mL Corning centrifuge tube containing 5 mL extraction buffer; vortex mix immediately to suspend the powder evenly.

3. Transfer the homogenate into a new 15 mL Corning centrifuge tube. Add 0.5 mL NaAcetate (pH 4.0) and mix thoroughly by inversion.

4. Add 5 mL H_2O-saturated phenol, mix thoroughly, and then add1 mL chloroform. Shake vigorously until the two phases form an emulsion. Incubate the suspension at room temperature for 10 min (*see* **Note 3**).

5. Centrifuge 10 min at $10,000 \times g$, 4 °C, and transfer the upper aqueous phase to a new 15 mL Corning centrifuge tube (*see* **Note 4**).

6. Add equal volume isopropanol to the new tube, screw the cap, and mix by inverting several times; incubate the sample at −20 °C for more than 1 h.

7. Centrifuge at $10,000 \times g$, 4 °C for 10 min, and discard the supernatant.

8. Dissolve the RNA pellet in 2 mL CES solution (*see* **Note 5**).

9. Add 2 mL (equal volume) chloroform, and mix thoroughly by inversion.

10. Centrifuge 10 min at $10,000 \times g$, 4 °C, and transfer the aqueous to a new tube (*see* **Note 6**).

11. Add 1/10 volume (~200 μL) of 2 M NaAcetate (pH 5.0) and equal volume of isopropanol, mix by inversion, and incubate the suspension at room temperature for 10 min.

12. Centrifuge 10 min at $10,000 \times g$, 4 °C, and discard the supernatant.

13. Wash the pellet with 1 mL 75% ethanol and centrifuge at $10,000 \times g$, 4 °C for 5 min, air-dry the pellet, and then dissolve the pellet with 50 μL DEPC-treated water (for RT-PCR, transcriptomic sequencing, or other analyses). The total RNA yield from senescent leaves is low, so a reduced final volume is suggested (*see* **Note 7**).

3.3 Sample Preparation for Illumina Sequencing (See Note 8)

1. RNA integrity is firstly assessed by electrophoresis in 1.0% native agarose gel.

2. The concentration and purity of the extracted total RNA is further analyzed on an Agilent 2100 bioanalyzer, which can also calculate the RNA integrity number (RIN value).

3. DNase treatment of total RNA is performed using deoxyribonuclease I.

4. Poly (A)-containing mRNA molecules are purified from total RNA using poly (T) oligo-attached magnetic beads.

5. mRNAs are fragmented into small pieces using an RNA fragmentation kit at elevated temperatures in Thermomixer,

followed by first- and second-strand cDNA synthesis using a cDNA synthesis kit.

6. Short cDNA libraries are constructed for Illumina HiSeq™ 2000 sequencing, according to the manufacturer's instruction, after the operation of ends reparation, A-Tailing, adaptor ligation, and product purification.

7. Samples are loaded to an Illumina HiSeq™ 2000 sequencer for sequencing.

3.4 De Novo Assembly of RNA-Seq Data (See Note 9)

1. The raw reads are filtered by removing adaptor sequences, reads with more than 5% unknown nucleotides, and low-quality reads containing more than 20% of bases with quality value ≤10.

2. The clean reads are assembled to the longest assembled sequences (called contigs) using Trinity [11].

3. The reads are mapped again back to contigs and get sequences that have the least Ns and cannot be extended on either end (called unigenes).

4. All the unigenes are assembled to form a single set of nonredundant unigenes using TGICL [12].

3.5 Function Annotation

1. The unigenes are analyzed by searching various databases such as nt, nr, KEGG, Swiss-Prot, and COG using BLAST (E-value < 1e-5) to predict their functions.

2. Unigenes are analyzed with the Blast2GO (v2.5.0) software to obtain GO annotations and classified according to GO functions using WEGO [13, 14].

3. For unigenes that do not align to any of the above databases, the ESTScan software is used to predict their coding regions and determine sequence direction [15].

3.6 Identification of Differentially Expressed Genes

1. SOAPaligner (Version 2.21) is used to map reads back to the unigenes at the parameters of "-m 0 -x 500 -s 40 -l 35 -v 5 -r 1," and the number of mapped reads for each unigene is recorded.

2. Expression profiles of the unigenes are analyzed with the FPKM method (fragments per kb per million reads). Differentially expressed genes (DEGs) between each of the samples are defined to have a threshold of false discovery rate (FDR) ≤ 0.001 and an absolute value of Log2 ratio ≥ 1.

3. The GO classifications are compared between upregulation and downregulation unigenes using WEGO, using the corrected p-value ≤ 0.05 as a threshold.

4. The identified differentially expressed genes can be further confirmed via qRT-PCR.

4   Notes

1. Samples from specific senescence stages and tissue types can be collected based on research purposes.

2. Samples are ground to fine powder in liquid nitrogen using a prechilled pestle and mortar (better grinding ensures a higher RNA yield). Do not let the frozen tissue thaw during grinding. Prechilling microspoon in liquid nitrogen will protect the ground frozen tissue from thawing during transfer. Add 0.2–0.4 g of frozen ground tissue to fresh extraction buffer. Excessive tissue will cause incomplete homogenization and RNA degradation.

3. Phenol and chloroform should be handled under a fume hood. The acidic pH is critical for the separation of RNA from DNA and proteins, so never use buffered phenol instead of water-saturated phenol, and mix thoroughly the organic phase with the acidic aqueous phase by shaking.

4. Carefully transfer the supernatant to a new tube. To ensure that high-quality RNA is obtained, avoid transferring the inter-phase or phenol-chloroform. The upper aqueous phase will be very viscous at this stage because of many polysaccharides and polyphenolic compounds. Re-extract the sample with an equal volume of chloroform from **step 9** of this section.

5. Dissolving of RNA can be assisted by passing the solution a few times through a pipette.

6. After centrifugation, a middle layer will be formed. Transfer aqueous phase as much as you can, but do not transfer anything from the middle layer. This step is very important and necessary for RNA isolation from senescent leaves. After two times extraction, most of polysaccharides and polyphenols are removed. A second chloroform extraction and RNA precipitation will facilitate removal of DNA and proteins from RNA. However, this step will slightly decrease the overall RNA yield.

7. Do not dry the pellet using a vacuum centrifuge. Never let the RNA pellet air-dry completely, as this can lose RNA solubility. After dissolving the pellet with DEPC-treated water, you could incubate RNA at 60 °C for 10–15 min to ensure complete solubilization.

8. This part can be done by a commercial sequencing company.

9. In the case of that a reference genome sequence is available, mapping of the RNA-seq data to the reference genome will have better results than de novo assembly.

Acknowledgments

This work was supported by the National Natural Science Foundation of China (31400267) (to WL), the Fundamental Research Funds for Central Non-profit Scientific Institution (2013ZL024), and the Agricultural Science and Technology Innovation Program (ASTIP-TRIC02).

References

1. Guo YF, Gan SS (2005) Leaf senescence: signals, execution, and regulation. Curr Top Dev Biol 71:83–112

2. Lim PO, Kim HJ, Nam HG (2007) Leaf senescence. Annu Rev Plant Biol 58:115–136

3. Zhang H, Zhou C (2013) Signal transduction in leaf senescence. Plant Mol Biol 82:539–545

4. Guo Y, Cai Z, Gan S (2004) Transcriptome of Arabidopsis leaf senescence. Plant Cell Environ 27:521–549

5. Zentgraf U, Jobst J, Kolb D et al (2004) Senescence-related gene expression profiles of rosette leaves of Arabidopsis thaliana: leaf age versus plant age. Plant Biol 6:178–183

6. Buchanan-Wollaston V, Page T, Harrison E et al (2005) Comparative transcriptome analysis reveals significant differences in gene expression and signalling pathways between developmental and dark/starvation-induced senescence in Arabidopsis. Plant J 42:567–585

7. Breeze E, Harrison E, Mchattie S et al (2011) High-resolution temporal profiling of transcripts during Arabidopsis leaf senescence reveals a distinct chronology of processes and regulation. Plant Cell 23:873–894

8. Mortazavi A, Williams BA, Mccue K et al (2008) Mapping and quantifying mammalian transcriptomes by RNA-Seq. Nat Meth 5:621–628

9. Hurd PJ, Nelson CJ (2009) Advantages of next-generation sequencing versus the microarray in epigenetic research. Brief Funct Genomic Proteomic 8:174–183

10. Sîrbu A, Kerr G, Crane M et al (2012) RNA-Seq vs dual- and single-channel microarray data: sensitivity analysis for differential expression and clustering. PLoS One 7:e50986

11. Grabherr MG, Haas BJ, Yassour M et al (2011) Trinity: reconstructing a full-length transcriptome without a genome from RNA-Seq data. Nat Biotechnol 29:644–652

12. Pertea G, Huang X, Liang F et al (2003) TIGR gene indices clustering tools (TGICL): a software system for fast clustering of large EST datasets. Bioinformatics 19:651–652

13. Conesa A, Gotz S, Garcia-Gomez JM et al (2005) Blast2GO: a universal tool for annotation, visualization and analysis in functional genomics research. Bioinformatics 21: 3674–3676

14. Ye J, Fang L, Zheng H et al (2006) WEGO: a web tool for plotting GO annotations. Nucleic Acids Res 34:W293–W297

15. Iseli C, Jongeneel C, Bucher P (1999) ESTScan: a program for detecting, evaluating, and reconstructing potential coding regions in EST sequences. Proc Int Conf Intell Syst Mol Biol 7:138–148

Chapter 28

Comprehensive Metabolomics Studies of Plant Developmental Senescence

Mutsumi Watanabe, Takayuki Tohge, Salma Balazadeh, Alexander Erban, Patrick Giavalisco, Joachim Kopka, Bernd Mueller-Roeber, Alisdair R. Fernie, and Rainer Hoefgen

Abstract

Leaf senescence is an essential developmental process that involves diverse metabolic changes associated with degradation of macromolecules allowing nutrient recycling and remobilization. In contrast to the significant progress in transcriptomic analysis of leaf senescence, metabolomics analyses have been relatively limited. A broad overview of metabolic changes during leaf senescence including the interactions between various metabolic pathways is required to gain a better understanding of the leaf senescence allowing to link transcriptomics with metabolomics and physiology. In this chapter, we describe how to obtain comprehensive metabolite profiles and how to dissect metabolic shifts during leaf senescence in the model plant *Arabidopsis thaliana*. Unlike nucleic acid analysis for transcriptomics, a comprehensive metabolite profile can only be achieved by combining a suite of analytic tools. Here, information is provided for measurements of the contents of chlorophyll, soluble proteins, and starch by spectrophotometric methods, ions by ion chromatography, thiols and amino acids by HPLC, primary metabolites by GC/TOF-MS, and secondary metabolites and lipophilic metabolites by LC/ESI-MS. These metabolite profiles provide a rich catalogue of metabolic changes during leaf senescence, which is a helpful database and blueprint to be correlated to future studies such as transcriptome and proteome analyses, forward and reverse genetic studies, or stress-induced senescence studies.

Key words Senescence, Metabolomics, *Arabidopsis*, GC/MS, LC/MS, HPLC, IC

1 Introduction

Senescence is an important developmental step in the plant life cycle, which involves a shift from nutrient assimilation to nutrient remobilization and recycling. Degradation of chlorophyll and macromolecules such as nucleic acids, proteins, and membrane lipids is one of the major events that occurs during leaf senescence. The metabolite changes accompanying degradation processes are important not only for nutrient recycling and remobilization but also as metabolic signals to progress the senescence processes since

Yongfeng Guo (ed.), *Plant Senescence: Methods and Protocols*, Methods in Molecular Biology, vol. 1744, https://doi.org/10.1007/978-1-4939-7672-0_28, © Springer Science+Business Media, LLC 2018

the concentration changes of some specific metabolites such as sugars and amino acids have been suggested to be key regulators during senescence [1–3]. Although targeted metabolite profiles focusing on specific groups of metabolites during plant senescence have been achieved (lipids, [4, 5]; sugars, amino acids, and nutrient ions, [6–14]), comprehensive metabolite profiles are rarely available. This is mainly due to the high degree of metabolite diversity. Measuring the complete set of metabolites in the complex process of plant senescence requires effective and specific extraction and analytical methods able to detect the respective metabolites and their derivatives. In this chapter we describe the details of extraction procedures and analytical methods used for the comprehensive metabolite profiles of leaf senescence in *Arabidopsis thaliana* published previously [15].

The visible symptom of leaf senescence, leaf yellowing, is caused by chlorophyll degradation, and loss of chlorophyll content has been used as a senescence marker. Although there are several methods for measuring chlorophyll content, we describe an easy spectrophotometric method to detect light absorbance by chlorophyll, and another method utilizing UPLC/ESI-MS [16, 17]. During leaf senescence, the loss of chlorophylls is generally accompanied by the massive reduction in leaf protein content due to the degradation of the chloroplast abundant protein, Rubisco (ribulose-1,5-bisphosphat-carboxylase/–oxygenase) [18]. The method for measuring the content of soluble proteins described in this chapter is based on the Bradford dye-binding procedure, which measures the color change of the Coomassie Brilliant Blue dye when it binds to protein [19]. The method for measuring starch, which is one of the major products of photosynthesis in plants, is an enzymatic method using a 96-well plate format [20, 21]. The method is based on the conversion of starch to glucose, followed by measurement of glucose using a hexokinase assay linked to NAD (β-nicotinamide adenine dinucleotide) reduction. The ion chromatography (IC) method is well established as a routine method for the analysis of ionic analytes in environmental samples. The IC method described in this chapter using a Dionex ICS-3000 system with anion- and cation-exchange columns and suppressed conductivity detection can easily be employed to measure the common nutrient anions such as nitrate, phosphate, and sulfate and cations such as potassium, sodium, ammonium, calcium, and magnesium in plant extracts. Thiols, especially glutathione, are key antioxidant metabolites and play a central part in plant stress responses such as scavenging of reactive oxygen species (ROS) which are also produced during leaf senescence [22, 23]. The method for measuring thiols is a monobromobimane derivatization method followed by high-performance liquid chromatography (HPLC) separation and fluorescence detection [24–26]. Monobromobimane reacts selectively with thiol groups yielding highly fluorescent and stable thioethers. Amino acid changes during leaf senescence are caused by various metabolic routes such as protein degradation, transport from source to sink tissues, interconversion to their transportable forms (e.g., conversions of glutamic acid and aspartic acid to glutamine and asparagine, respectively), and consumption of especially aromatic amino acids for the synthesis of a large number of secondary metabolites including flavonoids for protecting cells during senescence. The method for measuring amino acids is an *ortho*-phthalaldehyde derivatization method with HPLC and fluorescence detection [27, 28]. *Ortho*-phthalaldehyde reacts selectively with primary amines of amino acids yielding highly fluorescent products. Gas chromatography/time-of-flight-mass spectrometry (GC/TOF-MS) is widely used for metabolomics, allowing the measurement of complex mixtures of primary metabolites including organic acids, sugars, sugar alcohols, sugar acids, amino acids, and fatty acids in a single extract. The method for GC/TOF-MS is a two-step methoximation-silylation derivatization method to render the metabolites volatile [29–33]. GC/TOF-MS is a very sensitive and reproducible technique with high-rate scan, allowing deconvolution of closely overlapping peaks by using suitable mathematic algorithms (TagFinder software; [34, 35]). LC/MS can analyze a much wider range of metabolites than GC/MS since it facilitates the measurement of metabolites that are neither volatile nor derivatized. Liquid chromatography/electrospray ionization-mass spectrometry (LC/ESI-MS) is best suited for a discovery-based approach when researching unknown metabolites. A major advantage of the LC/ESI-MS analysis is that the molecular ion is produced by ESI, which helps metabolite identification. Furthermore, with ESI, it is possible to analyze both the positively and negatively charged molecules simultaneously in a single run which gives a broader coverage of the metabolites [36]. We describe here two methods using LC/ESI-MS, first, measurement of secondary metabolites such as flavonoids, phenylpropanoids, and glucosinolates by HPLC/ESI-MS analysis [37] with polar extract and, second, measurement of lipophilic metabolites such as chlorophylls, chloroplast-localized lipids, phospholipids, and storage lipids by UPLC/ESI-MS analysis with lipid extract [16, 17].

2 Materials

2.1 Plant Cultivation and Sample Preparation

1. *Arabidopsis* (*Arabidopsis thaliana*) accession Columbia-0.
2. Soil.
3. Nutrient solution: mix 10 mL of Previcur Energy (Bayer CropScience, Langenfeld, Germany) and 15 mL of 30 mM boric acid in 10 L of tap water.

4. Distilled, deionized (DDI) water.

5. Liquid nitrogen.

6. Eppendorf tubes: 1.5 and 2.0 mL round bottom.

7. Zirconia ball: 5 mm.

8. Mixer mill: MM 300 (Retsch).

2.2 Determination of Chlorophyll Contents

1. Absolute ethanol.

2. Microtest 96-well plate: Flat bottom.

3. Synergy HT Multi-Mode Microplate Reader (BioTek).

4. KC4 or Gen5 software (BioTek).

2.3 Determination of Soluble Protein Contents

1. Bradford reagent.

2. Bovine serum albumin (BSA).

3. Protein extraction buffer: 250 mM potassium phosphate buffer (KH_2PO_4/K_2HPO_4), pH 8.0, 0.5 mM ethylenediaminetetraacetic acid, 10 mM 2-mercaptoethanol.

4. Microtest 96-well plate: flat bottom.

5. ELX-800 or ELX-808 Microplate Reader (BioTek).

6. KC4 or Gen5 software (BioTek).

2.4 Determination of Starch Contents

1. 80% (v/v) ethanol.

2. 0.1 M NaOH.

3. 10× HCl/sodium-acetate solution: 0.5 M HCl, 0.1 M acetate/NaOH, pH 4.9.

4. HEPES buffer: 0.1 M HEPES/KOH, 3 mM $MgCl_2$, pH 7.0.

5. Starch degradation mix: centrifuge 500 µL of amyloglucosidase (Roche) and 5 µL of alpha-amylase (Roche) together at $20,000 \times g$ for 2 min at 4 °C. Remove the supernatant, and dissolve the enzyme pellet in 25 mL of 1× HCl/sodium-acetate solution.

6. Glucose determination mix (for one 96-well plate): centrifuge 80 µL of glucose-6-phosphate dehydrogenase (G6PDH) (Roche) for 2 min at $20,000 \times g$ and 4 °C to remove the supernatant, and dissolve the pellet in a mix containing 15.5 mL of 100 mM HEPES buffer, 480 µL of 100 mM ATP (adenosine-5′-triphosphate disodium salt trihydrate), and 480 µL of 45 mM NADP (β-nicotinamide adenine dinucleotide phosphate disodium salt).

7. Hexokinase (Roche).

8. Eppendorf tube 1.5 mL, round bottom.

9. Microtest 96-well plate: Flat bottom.

10. ELX-800 or ELX-808 Microplate Reader (BioTek).

11. KC4 or Gen5 software (BioTek).

2.5 Determination of Ion Contents

1. Water: Ultra liquid chromatography-mass spectrometry (ULC-MS) grade.

2. EGC III KOH (EluGen III Potassium Hydroxide Cartridge) (Dionex).

3. 0.1 M methanesulfonic acid: Concentrated solution.

4. HPLC elution buffer A for cation analysis: DDI water.

5. HPLC elution buffer B for cation analysis: 25 mM methanesulfonic acid.

6. Ultrafree MC 5000 MC NMWL Filter Unit.

7. Dionex ICS-3000 system.

8. Dionex Anion Self-Regenerating Suppressor ASRS 300 2 mm (Dionex).

9. IonPac AS11 Analytical Column 2 × 250 mm (Dionex).

10. IonPac AG11 Guard Column 2 × 50 mm (Dionex).

11. Dionex Cation Self-Regenerating Suppressor CSRS 300 2 mm (Dionex).

12. IonPac CS12A Analytical Column 2 × 250 mm (Dionex).

13. IonPac CG12A Guard Column 2 × 50 mm (Dionex).

14. Chromeleon software: Version 6.8 (Dionex).

2.6 Determination of Thiol Contents

1. Water: ULC-MS grade.

2. 0.1 M HCl.

3. 30 mM tris(2-carboxyethyl)phosphine.

4. 25 μM N-acetyl-cysteine.

5. 8.5 mM N-ethylmorpholine.

6. 30 mM monobromobimane: In acetonitrile.

7. HPLC elution buffer A: 0.25% acetic acid, pH 3.9 (NaOH).

8. HPLC elution buffer B: 100% methanol.

9. HPLC system: Dionex Summit.

10. Fluorescence detector: RF-2000 (Dionex).

11. C18 column: 250 × 4 mm.

12. Chromeleon software.

2.7 Determination of Amino Acid Contents

1. *Ortho*-phthalaldehyde solution.

2. 1 M borate buffer: pH 10.7.

3. 0.1 M HCl.

4. 80% (v/v) aqueous ethanol: Buffered with 2.5 mM HEPES/KOH, pH 6.2.

5. 50% (v/v) aqueous ethanol: Buffered with 2.5 mM HEPES/KOH, pH 6.2.

6. 80% (v/v) ethanol.

7. HPLC elution buffer A: 2% (v/v) tetrahydrofolate, 8.5 mM sodium phosphate buffer, pH 6.8.

8. HPLC elution buffer B: 32.5% (v/v) methanol, 20.5% (v/v) acetonitrile, 18.5 mM sodium phosphate buffer, pH 6.8.

9. Hyperclone C18 (ODS) column: 150 × 4.6 mm.

10. HPLC system: Dionex Summit.

11. Fluorescence detector: RF-2000.

12. Chromeleon software.

2.8 Measurement of Primary Metabolites by GC/TOF-MS Analysis

1. Water: ULC-MS grade.

2. Extraction buffer: 300 μL methanol, 30 μL methyl nonadecanoate (2 mg/mL in $CHCl_3$) for quantitative internal standardization for the lipid phase, 30 μL internal standard solution made from, e.g., $^{13}C_6$-sorbitol (0.2 mg/mL in methanol) or any stable isotope labeled internal standard for quantitative internal standardization for the polar phase.

3. Chloroform.

4. Methoxyamination reagent: dissolve 4-(dimethylamino)pyridine at a final concentration of 5 mg/mL in pure pyridine, and then add methoxyamine hydrochloride at final concentration of 40 mg/mL.

5. BSTFA: N,O-bis(trimethylsilyl)trifluoroacetamide.

6. Retention index (RI) standard mixture: dissolve n-alkanes in pyridine at a final concentration of 0.22 mg/mL each. The following substances are combined: n-decane (RI 1000), n-dodecane (RI 1200), n-pentadecane (RI 1500), n-octadecane (RI 1800), n-nonadecane (RI 1900), n-docosane (RI 2200), n-octacosane (RI 2800), n-dotriacontane (RI 3200), and n-hexatriacontane (RI 3600).

7. Speed vacuum concentrator.

8. Crimp cap vial.

9. Pegasus III TOF mass spectrometer (LECO Instrumente GmbH, Moenchengladbach, Germany).

10. Agilent 6890 N gas chromatograph, split/splitless injector with electronic pressure control up to 150 psi (Agilent, Boeblingen, Germany).

11. NIST mass spectral search and comparison software (Version 2.0, NIST).

12. TagFinder software [34, 35].

2.9 Measurement of Secondary Metabolites by LC/ESI-MS Analysis

1. Extraction buffer: dissolve isovitexin in 80% methanol at a final concentration of 5 μg/mL.

2. HPLC elution buffer A: formic acid in water at a final concentration of 0.1% formic acid (v/v).

3. HPLC elution buffer B: formic acid in acetonitrile at a final concentration of 0.1% formic acid (v/v).

4. Linear ion trap (IT) ESI-MS system Finnigan Ltq (Thermo Finnigan).

5. Xcalibur (Version 2.10, Thermo Fisher, Bremen, Germany).

2.10 Measurement of Lipophilic Metabolites by UPLC/ ESI-MS

1. Corticosterone solution: 1 mg/mL in methanol.

2. 1,2-ditetradecanoyl-sn-glycero-3-phosphocholine solution: 1 mg/mL in chloroform.

3. Extraction buffer: 25 mL methanol with 45 μL corticosterone solution for quantitative internal standardization for the polar phase, 75 mL methyl-*tert*-butyl-ether with 45 μL 1,2-ditetradecanoyl-*sn*-glycero-3-phosphocholine solution.

4. Acetonitrile/isopropanol mixture (7:3).

5. UPLC elution buffer A: 1% 1 M ammonium acetate, 0.1% acetic acid in water.

6. UPLC elution buffer B: 1% 1 M ammonium acetate, 0.1% acetic acid in acetonitrile/isopropanol (7:3) mixture.

7. Speed vacuum concentrator.

8. Ultra-sonication bath.

9. Waters Acquity UPLC system (Waters, Milford, MA, USA).

10. Xcalibur (Version 2.10, Thermo Fisher).

11. ToxID (Version 2.1.1, Thermo Fisher).

12. Refiner MS software (Version 6.0, Genedata, Basel, Switzerland).

3 Methods

3.1 Plant Cultivation and Sample Preparation

1. Suspend *Arabidopsis* seeds in 1 mL of DDI water in 1.5 mL round bottom tubes, and stratify at 4 °C in the dark for 2–4 days.

2. Sow seeds on soil supplemented with nutrient solution.

3. Grow plants under long-day (16 h/8 h day/night) or short-day (8 h/16 h day/night) conditions in a growth chamber or green house, with 140–160 μmol m^{-2} s^{-1}, 20 °C for the day and 16 °C, 60% humidity for the night (*see* **Note 1**).

4. Water plants every day using tap water.

5. Harvest plant samples, and freeze the samples immediately in liquid nitrogen (*see* **Note 2**).

6. Grind the samples. During the grinding, keep the samples frozen (e.g., 30 s at 25–30 Hz with one zirconia ball in 2.0 mL round bottom tube using the Mixer Mill).

7. Store the samples at −80 °C until ready for analysis.

3.2 Determination of Chlorophyll Contents

3.2.1 Determination of Chlorophyll Contents from Frozen Ground Materials

1. Aliquot frozen ground material (10 mg) in a 2.0 mL round bottom tube with one zirconia ball.

2. Add 1 mL of absolute ethanol to the frozen ground material and vortex for 5 s.

3. Homogenize the sample with the Mixer Mill for 1 min at 25 Hz.

4. Centrifuge the sample for 5 min at $20,000 \times g$ and 4 °C.

5. Transfer 200 μL of the supernatant into a 96-well plate.

6. Measure absorbance at $649, 665$, and 750 nm by a plate reader.

7. Record and process the data with KC4 or Gen5 software.

8. Calculate chlorophyll content using these equations:

   Chlorophyll *a* (g/mL) = $13.70 \times (A665 − A750) − 5.76 \times (A649 − A750)$;

   Chlorophyll *b* (g/mL) = $25.80 \times (A649 − A750) − 7.60 \times (A665 − A750)$;

   Chlorophyll *a* + *b* (g/mL) = $6.10 \times (A665 − A750) + 20.04 \times (A649 − A750)$. A, Absorbance.

3.2.2 Determination of Chlorophyll Contents from GC/TOF-MS Extracts

1. Dry the lipid phase (100 μL) from GC/TOF-MS extraction (**step 8**, Subheading 3.8.1) in a 1.5 mL round bottom tube in the speed vacuum concentrator for 3 h to overnight without heating.

2. Resuspend the sample in 500 μL of absolute ethanol.

3. Centrifuge the sample for 5 min at $20,000 \times g$ and 4 °C.

4. Continue from **step 5** in Subheading 3.2.1.

3.2.3 Determination of Chlorophyll Contents from UPLC/TOF-MS Extracts

1. Dry the lipid phase (100 μL) from UPLC/TOF-MS extraction (**step 9**, Subheading 3.10.1) in a 1.5 mL round bottom tube in the speed vacuum concentrator for 3 h to overnight without heating.

2. Resuspend the sample in 250 μL of absolute ethanol.

3. Centrifuge the sample for 5 min at $20,000 \times g$ and 4 °C.

4. Continue from **step 5** in Subheading 3.2.1.

3.3 Determination of Soluble Protein Contents

1. Aliquot frozen ground material (10 mg) in a 2.0 mL round bottom tube with one zirconia ball.

2. Add 100 µL of protein extraction buffer to the frozen ground material and vortex for 5 s.

3. Homogenize the sample with the Mixer Mill for 1 min at 25 Hz.

4. Centrifuge the sample for 5 min at 20,000 × g and 4 °C.

5. Prepare Bio-Rad Bradford reagents by diluting one part of the reagent concentrate with four parts DDI water.

6. Mix 997–995 µL of the diluted reagent with 3–5 µL of the supernatant of protein extracts.

7. Incubate at room temperature for 10 min.

8. Transfer 200 µL of the supernatant into a 96-well plate.

9. Measure absorbance at 595 nm by a plate reader.

10. Record and process the data with KC4 or Gen5 software.

11. Calculate protein contents using a protein standard curve (e.g., BSA).

3.4 Determination of Starch Contents (see Note 3)

1. Aliquot frozen ground material (20 mg) in a 1.5 mL round bottom tube.

2. Add 1 mL of 80% (v/v) ethanol to the frozen ground material and heat at 80 °C for 1 h.

3. Centrifuge the sample for 5 min at 20,000 × g and 4 °C to discard the supernatant.

4. Repeat **steps 2** and **3**.

5. Solubilize the pellet with 400 µL of 0.1 M NaOH by heating at 95 °C for 30 min.

6. Acidify the samples with 80 µL of HCl/sodium-acetate solution.

7. Digest part of the suspension (40 µL) with 110 µL of starch degradation mix in a 96-well plate overnight (10–16 h) at 37 °C.

8. Centrifuge the sample in the 96-well plate for 2 min at 4,000 × g and 4 °C.

9. Determine the glucose content using part of the supernatant (50 µL) with 150 µL of glucose determination mix in a 96-well plate.

10. Read the plate at 340 nm (1 min intervals and 1 h kinetic) by a plate reader until optical density is stabilized, and then add 1 µL hexokinase.

11. Record and process the data with KC4 or Gen5 software.

12. Calculate glucose contents using a glucose standard curve.

3.5 Determination of Ion Contents

3.5.1 Determination of Ion Contents from Frozen Ground Materials

1. Aliquot frozen ground material (20 mg) in a 2.0 mL round bottom tube with one zirconia ball.

2. Add 200 μL of water to the frozen ground material and vortex for 5 s.

3. Homogenize the sample with the Mixer Mill for 1 min at 25 Hz.

4. Centrifuge the sample for 10 min at $20,000 \times g$ and 4 °C.

5. Transfer the supernatant to an Ultrafree MC 5000 MC NMWL Filter Unit and centrifuge for 90 min at $20,000 \times g$ and 4 °C.

6. Dilute the sample ten times with water.

7. Analyze ions with 25 μL injection by the Dionex ICS-3000 system with a KOH gradient for anions (e.g., chloride, sulfate, nitrate, and phosphate) (parameters listed in Table 1) and with a methanesulfonic acid gradient for cations (e.g., potassium, ammonium, sodium, calcium, and magnesium) (parameters listed in Table 2).

8. Record and process the chromatograms with the Chromeleon software.

9. Calculate ion contents using a standard curve of each ion.

3.5.2 Determination of Ion Contents from GC/TOF-MS Extracts

1. Dry the polar phase (100 μL) from GC/TOF-MS extractions (**step 8**, Subheading 3.8.1) in a 1.5 mL round bottom tube in the speed vacuum concentrator for 3 h to overnight without heating.

2. Resuspend the sample in 500 μL of water (ULC-MS grade).

3. Centrifuge sample for 15 min at $20,000 \times g$ and 4 °C.

4. Continue from **step 7** in Subheading 3.5.1.

Table 1
LC parameters and gradient for anion analysis

LC parameters		
Column oven	35 °C	
Compartment oven	20 °C	
Sample tray	4 °C	
LC gradient		
Time (min)	Flow rate (μL/min)	KOH eluent *(mM)
0.0	250	6
10.0	250	45
12.0	250	55
17.0	250	6

*Dionex™ Eluent Generator Cartridges

Table 2
LC parameters and gradient for cation analysis

LC parameters			
Column oven	35 °C		
Compartment oven	20 °C		
Sample tray	4 °C		
LC gradient			
Time (min)	*Flow rate (μL/min)*	*Buffer A (%)*	*Buffer B (%)*
0.0	300	40	60
7.0	300	0	100
15.0	300	0	100
15.2	300	40	60
20.0	300	40	60

3.6 Determination of Thiol Contents

1. Aliquot frozen ground material (20 mg) in a 2.0 mL round bottom tube with one zirconia ball.

2. Add 60 μL of 0.1 M HCl to the frozen ground material and vortex for 5 s.

3. Homogenize the sample with the Mixer Mill for 1 min at 25 Hz.

4. Centrifuge the sample for 10 min at 20,000 × g and 4 °C.

5. For measuring total thiols (oxidized and reduced thiols), in 20 μL extracts, add 3 μL of 30 mM tris(2-carboxyethyl)phosphine, 40 μL of 25 μM *N*-acetyl-cysteine as the internal standard, and 10 μL of 8.5 mM *N*-ethylmorpholine. Incubate at 37 °C for 20 min in 1.5 mL round bottom tubes. For measuring reduced thiols, add 3 μL of DDI water instead of tris(2-carboxyethyl)phosphine in this step.

6. Label thiols by adding 3 μL of 30 mM monobromobimane in acetonitrile and incubating at 37 °C in dark for 20 min.

7. Terminate the labeling reaction by adding 10 μL acetic acid.

8. Measure the thiols (e.g., cysteine, *gamma*-glutamyl cysteine, and glutathione) by reverse-phase HPLC on a C18 column connected to the Dionex Summit HPLC system and the fluorescence detector RF-2000 with 5 μL injection following the parameters listed in Table 3.

9. Record and process the chromatograms with the Chromeleon software.

Table 3
LC/detector parameters and LC gradient for thiol analysis

LC parameters			
Column oven	25 °C		
Sample tray	4 °C		
LC gradient			
Time (min)	Flow rate (µL/min)	Buffer A (%)	Buffer B (%)
0.0	1000	85	15
4.0	1000	85	15
17.0	1000	77	23
20.0	1000	67	33
24.0	1000	0	100
27.0	1000	0	100
27.1	1000	85	15
30.0	1000	85	15
Detector parameters			
Excitation wavelength	380 nm		
Emission wavelength	480 nm		

10. Calculate thiol contents using a standard curve of each thiol. Normalize the data by the internal standard.

11. Calculate the contents of total thiols and reduced thiols and then the contents of oxidized thiols by subtracting the contents of reduced thiols from the contents of total thiols.

3.7 Determination of Amino Acid Contents

3.7.1 Determination of Amino Acid Contents from Frozen Ground Materials

1. Aliquot frozen ground material (20 mg) in a 2.0 mL round bottom tube with one zirconia ball.

2. Add 200 µL of 80% (v/v) aqueous ethanol to the frozen ground material and vortex for 5 s.

3. Homogenize the sample with the Mixer Mill for 1 min at 25 Hz.

4. Centrifuge the sample for 10 min at 20,000 × *g* and 4 °C. Take the supernatant to a new 1.5 mL round bottom tube.

5. Add 200 µL of 50% (v/v) aqueous ethanol to the pellet and vortex for 5 s. Repeat **steps 3** and **4**. Combine the supernatant from this step into the tube from **step 4**.

6. Add 100 μL of 80% (v/v) ethanol to the pellet and vortex for 5 s. Repeat **steps 3** and **4**. Combine the supernatant from this step into the tube from **step 4**.

7. Measure amino acids (e.g., standard amino acids) by reverse-phase HPLC on a Hyperclone C18 (ODS) column connected to the Dionex Summit HPLC system and fluorescence detector RF-2000 with pre-column online derivatization with 5 μL injection. Perform the derivatization by mixing 30 μL of the extract, 30 μL of *ortho*-phthalaldehyde solution diluted tenfold with water, and 25 μL of 1 M borate buffer, pH 10.7. HPLC parameters are listed in Table 4.

8. Record and process the chromatograms with the Chromeleon software.

9. Calculate amino acid contents using a standard curve of each amino acid. Normalize the data by the internal standard.

Table 4
LC/detector parameters and LC gradient for amino acid analysis

LC parameters			
Column oven	25 °C		
Sample tray	4 °C		
LC gradient			
Time (min)	Flow rate (μL/min)	Buffer A (%)	Buffer B (%)
0.0	800	100	0
2.0	800	100	0
21.0	800	87	13
28.0	800	85	15
37.0	800	50	50
48.0	800	40	60
54.0	800	0	100
59.0	800	0	100
60.0	800	100	0
65.0	800	100	0
Detector parameters			
Excitation wavelength	330 nm		
Emission wavelength	450 nm		

3.7.2 Determination of Amino Acid Contents from GC/TOF-MS Extracts

1. Dry the polar phase (100 µL) from GC/TOF-MS extraction (**step 8**, Subheading 3.8.1) in a 1.5 mL round bottom tube in the speed vacuum concentrator for 3 h without heating.

2. Resuspend the sample in 40 µL of 0.1 M HCl.

3. Centrifuge the sample for 15 min at 20,000 × *g* and 4 °C.

4. Continue from the **step 7** in Subheading 3.7.1.

3.8 Measurement of Primary Metabolites by GC/TOF-MS Analysis

3.8.1 Extraction

1. Aliquot frozen ground material (50 mg) in a 2.0 mL round bottom tube.

2. Add 360 µL of extraction buffer to the frozen ground material and vortex for 5 s.

3. Homogenize the sample at 70 °C for 15 min in a block shaker (500 rpm).

4. Cool the samples for 1 min at room temperature.

5. Add 200 µL of chloroform and heat at 37 °C for 5 min in a block shaker (500 rpm).

6. Add 400 µL of water and vortex for 15 s for liquid partitioning.

7. Centrifuge the sample for 10 min at 20,000 × *g* and room temperature.

8. Dry the polar phase (100 µL) in a 1.5 mL round bottom tube in the speed vacuum concentrator for 3 h to overnight without heating.

3.8.2 Chemical Derivatization

1. Resuspend the sample in 40 µL of methoxyamination reagent and agitate for 90 min at 30 °C.

2. Add 70 µL of BSTFA and 10 µL of a retention index (RI) standard mixture, and then agitate the sample for 45 min at 37 °C.

3. Transfer 80 µL of the sample into a crimp cap vial.

4. Inject 1 µL of the derivatized sample for GC/TOF-MS analysis.

3.8.3 GC/TOF-MS Analysis

1. Perform GC/TOF-MS analysis with 1 µL sample injection at 230 °C in splitless mode (parameters listed in Table 5). The flow rate is kept constant with electronic pressure control enabled (*see* **Note 4**).

2. Evaluate the chromatograms and mass spectra using the TagFinder software and NIST software.

3. Normalize peak heights of the mass fragments on the basis of the fresh weight of the sample and the internal standard.

4. Metabolite identification is manually supervised using the mass spectral and retention index collection of the Golm Metabolome Database (GMD) (http://gmd.mpimp-golm.mpg.de/).

Table 5
GC/TOF-MS setting for primary metabolite profiling

Injection parameters	
Helium carrier gas flow	0.6 mL/min
Purge time	1 min at 20 mL/min flow
GC parameters	
Capillary column	30 m Vf-5 ms 0.25 mm ID 0.25 μm film thickness + 10 m EZ-guard
Temperature program	isothermal mode (1 min at 70 °C)
Isothermal step	9 °C/min ramp to 350 °C
After final temperature	kept constant for 5 min
Cooling	as fast as instrument specifications allow
Transfer line temperature	250 °C and matches ion source conditions
MS parameters	
Mass range	70–600 *m/z*
Mass scan rate	20 scans/s
Manual mass defect	0
Filament bias current	−70 V
Detector voltage	1700–1850 V (depending on detector age)
Maximum instrument specifications	250 °C
Mass spectrometric solvent	6.6–7.5 min

3.9 Measurement of Secondary Metabolites by LC/ESI-MS

3.9.1 Extraction

1. Aliquot frozen ground material (50 mg) in a 2.0 mL round bottom tube with one zirconia ball.

2. Add 250 μL of extraction buffer to the frozen ground material and vortex for 5 s.

3. Homogenize the sample with the Mixer Mill for 2 min at 25 Hz.

4. Centrifuge the sample for 10 min at 20,000 × g and 4 °C.

3.9.2 Determination of Secondary Metabolites by LC/ESI-MS

1. Perform LC/ESI-MS analysis with 5 μL sample injection using the parameters listed in Table 6.

2. Record and process the chromatograms with Xcalibur.

3. Process detected peak area using Quan Browser of Xcalibur.

4. Normalize peak area of the mass fragments on the basis of the fresh weight of the sample and the internal standard.

Table 6
LC/ESI-MS setting for secondary metabolite profiling

LC parameters			
Column oven	35 °C		
Sample tray	10 °C		
LC gradient			
Time (min)	Flow rate (μL/min)	Buffer A (%)	Buffer B (%)
0.0	200	100	0
2.0	200	100	0
4.0	200	85	15
14.0	200	68	32
19.0	200	50	50
19.0	200	0	100
21.0	200	0	100
21.0	200	100	0
23.0	200	100	0
MS parameters			
Scan mode			positive/negative
Mass range			200–1500 m/z
Capillary temperature			275 °C
Spray voltage			4.00 kV

3.9.3 Determination of Secondary Metabolites from GC/TOF-MS Extract

1. Dry the polar phase (100 μL) from GC/TOF-MS extraction (**step 8**, Subheading 3.8.1) in a 1.5 mL round bottom tube in the speed vacuum concentrator for 3 h without heating.

2. Resuspend the sample in 100 μL of extraction buffer for LC/ESI-MS analysis.

3. Continue from the **step 4** in Subheading 3.9.1.

3.10 Measurement of Lipophilic Metabolites by UPLC/ ESI-MS

1. Aliquot frozen ground material (50 mg) in a 2.0 mL round bottom tube with one zirconia ball.

2. Add 250 μL of extraction buffer to the frozen ground material and vortex for 5 s.

3.10.1 Extraction

3. Homogenize the sample with the Mixer Mill for 2 min at 25 Hz.

4. Add 750 μL of extraction buffer to the sample.

5. Incubate the sample for 10 min at 4 °C in a block shaker (500 rpm).

6. Incubate the sample for 10 min in an ultra-sonication bath at room temperature.

7. Add 500 µL of water and vortex 5 s.

8. Centrifuge the sample for 3 min at 20,000 × g and 4 °C. Take the supernatant as much as possible into a new 1.5 mL round bottom tube.

9. Dry the lipid phase in the speed vacuum concentrator for 3 h to overnight without heating.

3.10.2   UPLC/ESI-MS
Analysis for Lipid Profiling

1. Resuspend the sample in 500 µL of acetonitrile/isopropanol mixture.

2. Perform UPLC/ESI-MS analysis with 2 µL sample injection with the parameters listed in Table 7.

Table 7
UPLC/ESI-MS setting for lipid profiling

LC gradient			
Time (min)	Flow rate (µL/min)	Buffer A (%)	Buffer B (%)
0.0	400	45	55
1.0	400	45	55
4.0	400	65	35
12.0	400	89	11
15.0	400	99	1
18.0	400	99	1
18.0	400	45	55
20.0	400	45	55
MS parameters			
Scan mode			positive
Mass range			200–1500 m/z
Mass resolution			25,000
Scan rate			10 scans/s
Scan loading time			100 ms
Capillary voltage			3 kV
Sheath gas flow			60
Auxiliary gas flow			35
Capillary temperature			150 °C
Electrospray source			350 °C
Skimmer voltage			25 V
Tube lens			130 V

3. Record and process the chromatograms with Xcalibur, ToxID, or automatically with the Refiner MS software.

4. Normalize peak heights of the mass fragments on the basis of the fresh weight of the sample and the internal standard.

4 Notes

1. Short-day conditions have an advantage to eliminate effects that might be caused by the transition to flowering.

2. Harvest samples using measurable indexes (e.g., days after germination, days after flowering, degree of leaf yellowing). It is highly recommended to check the expression of suitable senescence marker genes (e.g., *SAG12*, *CAB1*, *RBCS1A*) by qRT-PCR for evaluating the developmental stage and senescence status before taking or analyzing samples. This allows making comparisons with other metabolite and transcript data obtained in different laboratories or experiments. Harvest samples at the same time (e.g., in the middle of the day) since many metabolites do show strong diurnal variations. Harvesting samples should be as quick as possible and ideally should be finished for all samples within 1–2 h. The selection of the tissue for harvest depends on the purpose of the experiment. In the case of leaves, there are the following options: (a) When the analysis is focused on senescence processes occurring within a single leaf, divide leaves into distinct parts (e.g., tip/middle/base or proximal half/distal half/petiole). (b) When the analysis is focused on the senescence of complete leaves, either harvest consecutive individual leaves of the rosette at a defined time point or sample single leaves (defined by leaf number, emergence after germination) from plants going through the senescence stages of the plant over several consecutive days. (c) When senescence of the whole rosette is to be analyzed, combine all leaves of the rosette for analysis. Such analysis can, e.g., be performed using whole rosettes before bolting or after bolting.

3. Glucose obtained by the hydrolysis of starch by amyloglucosidase is phosphorylated to G6P by hexokinase in the presence of ATP. G6P is then oxidized by G6PDH to 6-phosphogluconate in the presence of NAD. During this oxidation, an equimolar amount of NAD is reduced to NADH. The increase in absorbance of the product NADH at 340 nm is directly proportional to the glucose concentration.

4. Optional and recommended in cases of high metabolite concentrations, injection is either performed in split or splitless mode.

Acknowledgments

B. M.-R. and S.B. thank the Deutsche Forschungsgemeinschaft (DFG) for funding (FOR 948; MU 1199/14-1 and 14-2, and BA4769/1-2). We thank the Max Planck Society (MPG) for funding, the Max Planck Institute of Molecular Plant Physiology (MPI-MP) for providing metabolomics and bioinformatics services, and the greenteam of the institute for growing plants.

References

1. Noodén L (1980) Senescence in whole plants. In: Thimann KV (ed) Senescence in plants. CRC Press, Boca Raton, FL, pp 219–258

2. Yoshida S (2003) Molecular regulation of leaf senescence. Curr Opin Plant Biol 6:79–84

3. van Doorn WG (2004) Is petal senescence due to sugar starvation? Plant Physiol 134:35–42

4. Yang ZL, Ohlrogge JB (2009) Turnover of fatty acids during natural senescence of Arabidopsis, brachypodium, and switchgrass and in Arabidopsis beta-oxidation mutants. Plant Physiol 150:1981–1989

5. Seltmann MA, Stingl NE, Lautenschlaeger JK et al (2010) Differential impact of lipoxygenase 2 and jasmonates on natural and stress-induced senescence in Arabidopsis. Plant Physiol 152:1940–1950

6. Masclaux C, Valadier MH, Brugiere N et al (2000) Characterization of the sink/source transition in tobacco (*Nicotiana tabacum* L.) shoots in relation to nitrogen management and leaf senescence. Planta 211:510–518

7. Quirino BF, Reiter WD, Amasino RD (2001) One of two tandem Arabidopsis genes homologous to monosaccharide transporters is senescence-associated. Plant Mol Biol 46:447–457

8. Stessman D, Miller A, Spalding M et al (2002) Regulation of photosynthesis during Arabidopsis leaf development in continuous light. Photosynth Res 72:27–37

9. Diaz C, Purdy S, Christ A et al (2005) Characterization of markers to determine the extent and variability of leaf senescence in Arabidopsis. A metabolic profiling approach. Plant Physiol 138:898–908

10. Masclaux-Daubresse C, Carrayol E, Valadier MH (2005) The two nitrogen mobilisation- and senescence-associated *GS1* and *GDH* genes are controlled by C and N metabolites. Planta 221:580–588

11. Masclaux-Daubresse C, Purdy S, Lemaitre T et al (2007) Genetic variation suggests interaction between cold acclimation and metabolic regulation of leaf senescence. Plant Physiol 143:434–446

12. Pourtau N, Jennings R, Pelzer E et al (2006) Effect of sugar-induced senescence on gene expression and implications for the regulation of senescence in Arabidopsis. Planta 224:556–568

13. Wingler A, Purdy S, MacLean JA et al (2006) The role of sugars in integrating environmental signals during the regulation of leaf senescence. J Exp Bot 57:391–399

14. Wingler A, Delatte TL, O'Hara LE et al (2012) Trehalose 6-phosphate is required for the onset of leaf senescence associated with high carbon availability. Plant Physiol 158:1241–1251

15. Watanabe M, Balazadeh S, Tohge T et al (2013) Comprehensive dissection of spatiotemporal metabolic shifts in primary, secondary, and lipid metabolism during developmental senescence in Arabidopsis. Plant Physiol 162:1290–1310

16. Giavalisco P, Li Y, Matthes A et al (2011) Elemental formula annotation of polar and lipophilic metabolites using C-13, N-15 and S-34 isotope labelling, in combination with high- resolution mass spectrometry. Plant J 68:364–376

17. Hummel J, Segu S, Li Y et al (2011) Ultra performance liquid chromatography and high resolution mass spectrometry for the analysis of plant lipids. Front Plant Sci 2:54

18. Peoples MB, Dalling MJ (1978) Degradation of ribulose 1,5-bisphosphate carboxylase by proteolytic enzymes from crude extracts of wheat leaves. Planta 138:153–160

19. Bradford MM (1976) rapid and sensitive method for the quantitation of microgram quantities of protein utilizing the principle of protein-dye binding. Anal Biochem 72: 248–254

20. Stitt M, Lilley RMC, Gerhardt R et al (1989) Determination of metabolite levels in specific cells and subcellular compartments of plant leaves. Methods Enzymol 174:518–552

21. Gibon Y, Blaesing OE, Hannemann J et al (2004) A robot-based platform to measure multiple enzyme activities in Arabidopsis using a set of cycling assays: Comparison of changes of enzyme activities and transcript levels during diurnal cycles and in prolonged darkness. Plant Cell 16:3304–3325

22. Noctor G, Foyer CH (1998) Ascorbate and glutathione: keeping active oxygen under control. Annu Rev Plant Physiol Plant Mol Biol 49:249–279

23. Zagorchev L, Seal CE, Kranner I et al (2013) A central role for thiols in plant tolerance to abiotic stress. Int J Mol Sci 14:7405–7432

24. Anderson ME (1985) Determination of glutathione and glutathione disulfide in biological samples. Methods Enzymol 113:548–555

25. Fahey RC, Newton GL (1987) Determination of low-molecular-weight thiols using mono-bromobimane fluorescent labeling and high-performance liquid-chromatography. Methods Enzymol 143:85–96

26. Saito K, Kurosawa M, Tatsuguchi K et al (1994) Modulation of cysteine biosynthesis in chloroplasts of transgenic tobacco overexpressing cysteine synthase [O-acetylserine(thiol)-lyase]. Plant Physiol 106:887–895

27. Lindroth P, Mopper K (1979) High-performance liquid-chromatographic determination of subpicomole amounts of amino-acids by precolumn fluorescence derivatization with *ortho*-phthaldialdehyde. Anal Chem 51:1667–1674

28. Kim H, Awazuhara M, Hayashi H et al (1997) Analysis of O-acetyl-L-serine in *in vitro* cultured soybean cotyledons. In: Cram WJ, De Kok LJ, Stulen I et al (eds) Sulphur metabolism in higher plants: molecular, ecophysiological and nutrition aspects. Backhuys Publishers, Leiden, pp 307–309

29. Lisec J, Schauer N, Kopka J et al (2006) Gas chromatography mass spectrometry-based metabolite profiling in plants. Nat Protoc 1:387–396

30. Erban A, Schauer N, Fernie AR et al (2007) Nonsupervised construction and application of mass spectral and retention time index libraries from time-of-flight gas chromatography-mass spectrometry metabolite profiles. In: Weckwerth W (ed) Methods in molecular biology. Humana Press, Totowa, NJ, pp 19–38

31. Strehmel N, Hummel J, Erban A et al (2008) Retention index thresholds for compound matching in GC-MS metabolite profiling. J Chromatogr B Analyt Technol Biomed Life Sci 871:182–190

32. Allwood JW, Erban A, de Koning S et al (2009) Inter-laboratory reproducibility of fast gas chromatography-electron impact-time of flight mass spectrometry (GC-EI-TOF/MS) based plant metabolomics. Metabolomics 5:479–496

33. Sanchez DH, Szymanski J, Erban A et al (2010) Mining for robust transcriptional and metabolic responses to long-term salt stress: a case study on the model legume *Lotus japonicus*. Plant Cell Environ 33:468–480

34. Luedemann A, Strassburg K, Erban A et al (2008) TagFinder for the quantitative analysis of gas chromatography - mass spectrometry (GC-MS)-based metabolite profiling experiments. Bioinformatics 24:732–737

35. Luedemann A, von Malotky L, Erban A et al (2012) TagFinder: preprocessing software for the fingerprinting and the profiling of gas chromatography-mass spectrometry based metabolome analyses. Methods Mol Biol 860:255–286

36. Dettmer K, Aronov PA, Hammock BD (2007) Mass spectrometry-based metabolomics. Mass Spectrom Rev 26:51–78

37. Tohge T, Fernie AR (2010) Combining genetic diversity, informatics and metabolomics to facilitate annotation of plant gene function. Nat Protoc 5:1210–1227

INDEX

Yongfeng Guo (ed.), *Plant Senescence: Methods and Protocols*, Methods in Molecular Biology, vol. 1744,
https://doi.org/10.1007/978-1-4939-7672-0, © Springer Science+Business Media, LLC 2018

Printed in the United States
By Bookmasters